OF BRITAIN AND NORTHERN EUROPE

Text by

Richard Fitter · **Alastair Fitter**

Illustrated by

Marjorie Blamey

HarperCollins*Publishers*

HarperCollins*Publishers* 1993

First Published in 1974 by William Collins Sons & Co Ltd
London · Glasgow · Sydney · Auckland
Toronto · Johannesburg

The Authors assert the moral right
to be identified as the authors of this work

First edition 1974
Second edition 1974
Third edition 1978
Fourth edition 1985
Reprinted 1986
Reprinted 1987
Reprinted 1989
Reprinted 1993

Paperback ISBN: 0 00 219715 4

Printed and bound in Great Britain by HarperCollins*Publishers*

Contents

Introduction 7

Main Key 13

Key to Trees and Tall Shrubs 23

Conifer Families 24
Willow Family: Willows, Sallows, Poplars 26
Other catkin-bearing Families 30
Other tree Families 34

Birthwort Family to **Nettle** Family 38
Dock Family: Bistorts, Docks, Sorrels etc 40-44
Purslane Family 42
Goosefoot Family: Goosefoots, Oraches etc 46

Key to Chickweeds and Allies 50

Pink Family: Chickweed-like plants 52
Campions, Pinks etc 58

Water-lily Family 66
Buttercup Family: Hellebores etc 66
Buttercups 68
Water Crowfoots 70
Meadow-rues, Clematises 72
Columbine, Pheasant's-eyes etc 74
Anemones 76
Fumitory Family: Fumitories, Corydalises 78
Poppy Family 80

Key to Crucifers 82

Cabbage Family: Yellow Crucifers 84
White and Mauve Crucifers 90

Mignonette Family 100
Sundew Family 100
Stonecrop Family 100, 102
Saxifrage Family 100, 104, 106
Rose Family: Meadowsweet to Lady's Mantles 108
Roses, Brambles 110
Strawberries, Cinquefoils 112
Cherries, Pears etc 116, 118

Key to Peaflowers 120

Pea Family: Milk-vetches 122
    Shrubs 124
    Vetches 126
    Peas, Vetchlings 128
    Clovers 132

Wood-sorrel Family 136
Flax Family 136
Geranium Family: Cranesbills 136
Spurge Family 140
Balsam Family 142
Milkwort Family 142
Daphne Family 142
Currant and other shrub Families 144
Mallow Family 146
St John's Wort Family 148
Violet Family: Violets, Pansies 150
Rock-rose Family 152
Willowherb Family: Willowherbs, Evening Primroses 152
Dogwood Family 158

Key to Umbellifers 156

Carrot Family: Sanicle, Sea Holly 158
    White Umbellifers 160
    Yellow Umbellifers 170

Wintergreen Family 172
Heath Family 174
Primrose Family 178
Sea-lavender Family 182
Gentian Family: Centauries, Gentians 184
Bindweed Family: Bindweeds, Dodders 188
Bedstraw Family: Bedstraws, Woodruffs 188
Borage Family: Comfreys, Gromwells, Buglosses 192
    Forgetmenots 194

Key to Two-lipped Flowers (Labiate and Figwort Families) 198

Labiate Family: Skullcaps, Germanders 200
    Self-heals, Catmints 202
    Dead-nettles 204
    Claries, Woundworts 206
    Mints, Thymes 208
Nightshade Family 210

Figwort Family: Mulleins 212
                    Figworts, Toadflaxes 214
                    Foxgloves, Bartsias 216
                    Speedwells 218
                    Louseworts, Cow-wheats 220

Broomrape Family 222
Plantain Family 224
Valerian Family 226
Honeysuckle Family 226
Teasel Family: Teasels, Scabiouses 228
Bellflower Family: Bellflowers, Campions 230

Key to Composites (Daisy Family) 232

Daisy Family: Hemp Agrimony, Golden-rod etc 234
              Daisy-like Flowers, white and mauve 236
              Cudweeds 238
              Daisy-like Flowers, yellow 240
              Butterbur, Yarrow, Mugwort 242
              Ox-eye Daisy, Marigolds, Coltsfoot 244
              Ragworts, Groundsel 246
              Thistles, Knapweeds 248
              Dandelion-like Flowers 254

Water-plantain Family 262
Lily Family: Lilies, Asphodels 264
            Fritillary, Gageas 266
            Leeks, Garlics 268
            Squills, Grape Hyacinths 270
            Solomon's Seals, Stars of Bethlehem 272
Daffodil Family: Daffodils, Snowflakes 274
Iris Family: Flags, Irises, Crocuses 276
Orchid Family 278

Key to Waterweeds 292

Waterweed Families 294

Appendix 1 Introduced shrubs 304
Appendix 2   Aggregates 305
Plant Ecology 306
Glossary 312
Plant Photography 316
Societies to Join 316
Further Reading 316
Index of English Names 317
Index of Scientific Names 326

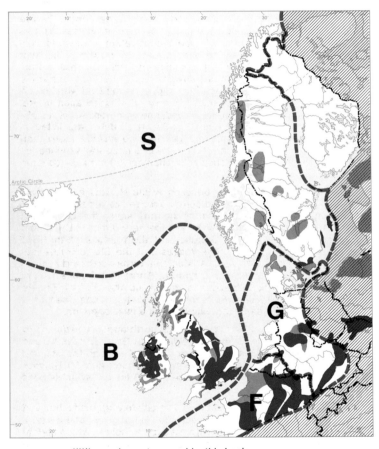

region not covered by this book

— — boundary between distribution areas

■ major outcrops of chalk and limestone

■ scattered outcrops of chalk and limestone

land over 1000 m. (3250 ft.)

The map does not show limestone in the Alps.

Introduction

The Text

All the native flowering plants of the north-western quadrant of Europe are described in the text, with the exception only of grasses, sedges and rushes. A small number of plants penetrating westwards only as far as Finland, and of alpine species descending down the river shingle to the neighbourhood of Munich are also omitted. Ferns, horsetails and their allies are not flowering plants and are therefore omitted, although seven species which might be confused with flowering plants are briefly mentioned. All widely established alien herbaceous plants and trees are also described, and some less widely established alien shrubs most of them familiar in gardens, are listed in Appendix 1, p.304. It has proved impossible to include more than a few of the very large number of alien trees, shrubs and herbaceous plants that are locally established in a small part of the area covered by the book.

The **area chosen** lies roughly between Arctic Norway in the north and the river Loire in western and central France and the river Danube in Germany in the south. Its boundary runs along the Loire from its mouth on the Atlantic coast of France eastwards to Dijon and Basle and thence along the Danube and the foothills of the Alps to Munich, so as to include the Vosges and the Black Forest, but excluding the Jura and the Alps. At Munich it turns northwards across Germany through Regensburg, Bayreuth, Erfurt and Braunschweig, to reach the Baltic at Lübeck, and thence up the Gulf of Finland and the eastern frontier of Finland to the Arctic Ocean. Iceland is included. These boundaries are shown on the map opposite.

To indicate distribution in the text, this north-western segment of Europe has been divided into four regions. (1) Great Britain, Ireland and the Isle of Man; (2) France, Belgium, Luxemburg and the Channel Isles; (3) Germany, the Netherlands, and Denmark; and (4) Norway, Sweden, the Faeroes, and Iceland. See page 9 for the symbols used to indicate these regions.

Description. Since the illustration is immediately opposite the text relating to each plant, a detailed description is not given, and descriptive details are largely confined to points such as height, which cannot easily be brought out in the illustration, to factual information such as habitat and distribution, which, of its nature, cannot be illustrated, and to stressing certain points which are crucial for identification.

Each text description includes certain standardised information, such as the height of the plant, flowering period and the habitat.The major diagnostic features distinguishing it from similar plants with which it might be confused, are shown in italics. A number of assumptions are made in the descriptions: unless otherwise stated plants are erect and herbaceous, leaves are stalked, and flowers are open.

The descriptions are compiled as follows:

Group descriptions of families and genera (see p. 10) are given wherever possible, so as to avoid constant repetition of common characters, for example the shape of a peaflower or the nature of a composite flower. They should therefore always be read carefully and in conjunction with the species descriptions. The group descriptions refer only to species mentioned in the text.

English names follow, in most cases and where possible, the list approved by the Botanical Society of the British Isles. Where no English names exist, we have not invented any. Nor have we always included English names which are mere translations of the Latin ones.

Latin scientific names in general follow *Flora Europaea*, with minor later adjustments (see also Further Reading, p. 316).

British status: plants native to, or commonly naturalised in the British Isles are preceded by an asterisk.

Number: each plant has a number corresponding with its number on the plate opposite, and elsewhere may be referred to, where necessary, by this number preceded by the page number, so that the Primrose becomes 178/1. Where several plants are very similar, only one is illustrated and the others are described in the same paragraph. They are given the number of the illustrated species and a letter. The descriptions of all such **subsidiary species**, however many are listed, refer back to the main species unless otherwise stated, and only the principal characters that differ are mentioned. On a few pages the descriptions necessary to identify the plants overrun the page. Rather than cut these to a point which would endanger identification, the descriptions have instead been placed on previous or following pages of the text.

Hairiness or hairlessness is indicated in most cases, often an important clue to identification

Height is indicated as follows:

Tall
over 60 cm (2 ft)

Medium
30-60 cm
(1-2 ft)

Short
10-30 cm
(4-12 in)

Low
0-10 cm (0-4 in)

Plant sizes can vary greatly according to altitude, climate and soil. Plants are more likely to be found smaller than is indicated here than larger.

Status: annual, biennial or perennial status is shown in each case. Perennials are usually stouter than annuals, and of course are much more likely to be seen above ground in winter.

Leaf-shapes: see Glossary (p.312) for definition of terms such as *pinnate*.

Flower shapes and arrangements: see Glossary (p.312). Flower sizes are diameters unless otherwise stated. Flower colour refers to petals or to sepals when there are no petals. It is more variable in some species than others. Most pink, mauve, purple and blue flowers produce white forms from time to time, and many white flowers can be tinged pink. Some, such as milkworts (p.142), are exceptionally variable. But a normally coloured flower is usually to be found nearby.

Flowering-time refers to the central part of the area, and may be earlier in the south, near the sea and in forward seasons, and later in the north, on mountains and in backward seasons.

Fruits are usually only described where important for identification.

Habitat: for definition of terms such as fen and bog, here used strictly, see Glossary (p.312). The major mountain regions and areas with lime (chalk or limestone) in the soil are shown in the map on p.6.

Distribution: the following symbols show whether the plant occurs, either commonly or uncommonly, within each region. If it has only a very few localities, "rare" is added, and if it is mainly found, e.g. in the south of the region, the word "southern" is also added.

T – Throughout the area covered by the book (see p.7).
B – Great Britain, Ireland, Isle of Man.
F – France, Belgium, Luxemburg, Channel Isles.
G – Germany, the Netherlands, Denmark.
S – Norway, Sweden, Iceland, the Faeroes.

If parentheses enclose the symbol – for example, (G) – the plant is introduced, not native. The great majority of introduced plants occur on waste or disturbed ground, roadsides, or other habitats much affected by human activity. Only rarely, as with New Zealand Willowherb (p.154), do they succeed in invading natural or semi-natural habitats, such as heaths, moors and calcareous grassland.

The Illustrations

Over 1000 species are illustrated in colour. Many illustrations are life-size, but, where they are not, a single flower of the plant in question is also shown life-size. Only the *Iris* plate (page 277) is painted to a smaller scale. Sometimes the distinctive character of a subsidiary species is also illustrated. The symbol □ is used after the scientific name of all subsidiary species with an illustration either on the plate opposite or on the text page.

Since the illustrations are arranged in the scientific order, it will often be necessary to identify a plant via the keys. A plant should never be identified simply from the picture, especially since several similar species may be described under the one illustrated. Always remember, too, that a flower may sometimes have a different colour from the one illustrated, and that some important identification features, such as smell or hairiness, may not be of the type that can be shown in an illustration.

Classification and Scientific Nomenclature

The seed plants or Spermatophyta are in two main groups, a small number of gymnosperms or conifers and a much larger number of angiosperms or flowering plants. The European conifers (p. 24) are all trees or shrubs, evergreen (except the larches) with narrow leaves ("needles"), male and female flowers always separate but usually on the same plant, and the seeds, unlike those of the flowering plants, not enclosed in an ovary. The flowering plants are themselves divided into two groups, dicotyledons and monocotyledons. Most of the plants in the book are dicotyledons (pp. 26-260), which have two seed-leaves (cotyledons), mature leaves which are usually broad, often stalked, and nearly always net-veined, and flower-parts in fours or fives. The monocotyledons (pp. 262-290) are represented mainly by lilies, orchids, and allied plants. They have only one seed-leaf, their later leaves are usually narrow and unstalked, often parallel-sided and nearly always parallel-veined, their flower-parts are almost always in multiples of three. Except for Butcher's Broom (p. 272), none of them are woody.

Within these major divisions flowering plants, like all other plants, are arranged in families, genera, and species, groups of increasingly close affinity, in what, according to current scientific opinion, corresponds to the way in which they have evolved. Family names usually end in ... aceae. In the book the plants are mainly arranged in the currently accepted order of families. The chief exceptions to this rule are the grouping together, for the reader's convenience, of trees and shrubs, mainly at the beginning, and of waterweeds at the end.

Individual scientific plant names are binomial; the generic name comes first, always with a capital initial, and the specific name second, always with a small initial, and since it is an adjective, agreeing in gender with the generic name. This system, devised by the great 18th-century Swedish naturalist, Linnaeus, represents the international scientific nomenclature, and using Latin or Latinised Greek, enables naturalists in any country to know exactly what plant is being referred to.

How to Identify Wild Flowers

A great deal of pleasure may be had in identifying wild flowers without any equipment at all, and as far as possible in this book we have relied on characters that can clearly be seen by the naked eye. Nevertheless a small pocket lens, magnifying 5 or 10 times, is an invaluable aid, and is no trouble to carry, unlike the binoculars used by birdwatchers – though these can be useful to the botanist in helping to identify inaccessible plants. With the aid of a lens such details as hairiness or the number of styles and stamens, sometimes critical for distinguishing otherwise similar plants, can more readily be determined.

Wild flowers, being static, have many advantages over animals for those wishing to identify them. It is much easier to bring a knowledgeable friend back to see an unusual plant, than to hope a rare bird or butterfly will still be there next day. We would urge our readers to do this wherever possible, and *not to pick specimens.* Be careful not to wait too long, as some plants wither after a short time in flower. If a discovery is really rare, the plant's survival may be endangered by picking a piece to show somebody for identification, so the slogan is:

"Take the book to the plant, not the plant to the book."

If you are not likely to be able to return to the plant, perhaps because you are on a day trip into the country, make a sketch on the spot and take detailed notes on the following points:

● Height and shape of the plant; whether annual or perennial.

● Shape of leaves, arrangement on stem, length of stalk.

● Shape and colour of flowers; number of petals, stamens and styles.

● Arrangement of flowers, e.g. solitary or in clusters or spikes.

● Shape and colour of fruits.

● Degree of hairiness or downiness on all parts of the plant.

● Habitat and abundance.

For this reason always carry a small field note-book with you. A good colour photograph may also be valuable (see p.316).

Wherever possible we have confined ourselves to easily ascertainable characters, as above, which do not require the plant to be picked to see if the stem is hollow or its rootstock tuberous, for example. In a few instances, however, where these are the best known or most reliable characters, we have given such details.

Abbreviations used in the Text

Agg.	Aggregate	Lf, Lvs	Leaf, leaves
Fl, fls	Flower, flowers	Sp.	Species (singular)
Flg.	Flowering	Spp.	Species (plural)
Flhds	Flowerheads	Ssp.	Subspecies
Fr	Fruit, fruits, fruiting	Var.	Variety
×	Indicates a hybrid, e.g. 1 × 2 is a hybrid between species 1 and species 2.		
*	Native or commonly naturalised in Britain.		
□	This symbol after the name of a subsidiary species (see p.8) indicates that it is illustrated.		
B, F, G, S, and T	Found in Britain, France, Germany, Scandinavia, and Throughout respectively. For precise definition of these areas see introduction (pp.7,9).		
( )	Indicates the plant is introduced, not native.		

The Keys

The Main Key (opposite) is based on flower-shape and colour. It includes all plants in the book except: trees and tall shrubs with catkins or small flowers: water weeds with inconspicuous flowers; and the introduced shrubs in Appendix 1, p.304. Subsidiary species are covered by the reference to the main species.

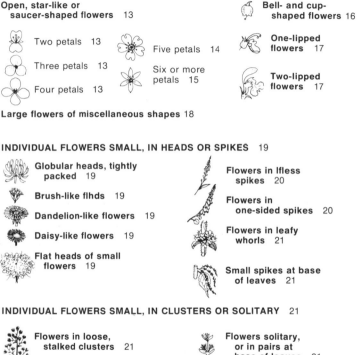

INDIVIDUAL FLOWERS LARGE OR CONSPICUOUS 13

Open, star-like or saucer-shaped flowers 13

Two petals 13

Three petals 13

Four petals 13

Five petals 14

Six or more petals 15

Bell- and cup-shaped flowers 16

One-lipped flowers 17

Two-lipped flowers 17

Large flowers of miscellaneous shapes 18

INDIVIDUAL FLOWERS SMALL, IN HEADS OR SPIKES 19

Globular heads, tightly packed 19

Brush-like flhds 19

Dandelion-like flowers 19

Daisy-like flowers 19

Flat heads of small flowers 19

Flowers in lfless spikes 20

Flowers in one-sided spikes 20

Flowers in leafy whorls 21

Small spikes at base of leaves 21

INDIVIDUAL FLOWERS SMALL, IN CLUSTERS OR SOLITARY 21

Flowers in loose, stalked clusters 21

Flowers in small clusters at base of leaves 21

Flowers solitary, or in pairs at base of leaves 21

In addition, eight special keys are scattered through the text, appearing with the families or groups concerned. They are cross-referenced in the main key.

Trees and tall shrubs 23

Chickweeds and allies 50-1

Crucifers (Cabbage Family) 82-3

Peaflowers 120-1

Umbellifers (Carrot Family) 156-7

Two-lipped flowers of the Mint and Figwort Families 198-9

Composites (Daisy Family) 232-3

Waterweeds with small flowers 292

The Main Key

Individual flowers large or conspicuous   see below
Individual flowers small   see page 19
(n.b. Composites, actually tight heads of small flowers, also have a separate key, pp.232-3)

INDIVIDUAL FLOWERS LARGE OR CONSPICUOUS

Open, star-like or Saucer-shaped flowers

Two petals

 Enchanter's Nightshades **152**

 Goldilocks Buttercup **70**

Three petals

 Mossy Stonecrop **102**, Water Plantains, Star-fruit, Arrow head, Frogbit, Water Soldier **262**

 Ranunculus hyperboreus **68**, Goldilocks Buttercup **70**, Rannoch Rush **224**

 Water Plantains **262**

 Water Plantains **262**

Four petals

 Bastard toadflaxes **38**, Mossy Sandwort **52**, Water Tillaea **102**, Dwarf cornel **158**, Squinancywort, Woodruff **188**, Bedstraws, Cleavers **190**

 Cranberry **176**, Squinancywort **188**, Slender Marsh Bedstraw **190**

 Wall Bedstraw **190**

Crosswort, Lady's Bedstraw **190**

Spurge Laurel **142**, False Cleavers, Wall Bedstraw **190**, Herb Paris **272**

 Blue Woodruff **188**

 Mezereon **142**, Dwarf Cornel **158**, Field Madder **188**

Pearlworts **56**, Crucifers **82**, Allseed **136**, Water Chestnut **262**, May Lily **264**

 Poppies, Red-horned Poppy **80**, *Capsella rubella* **96**, Garden Cress **98**

 Red-horned Poppy **80**, Wallflower **86**

Poppies, Greater Celandine **80**, Crucifers **82**, Tormentil **114**, Evening Primroses **152**, Yellow Centauries **184**

 Mistletoe **38**, Pearlworts **56**, Narrow-leaved Pepperwort **98**, Golden Saxifrages **100**

Opium Poppy **80**, Crucifers **82**

 Meadow-rues, Baneberry, *Clematis recta* **72**, Shoreweed **294**

 Lesser Meadow-rue **72**, Roseroot **100**

 Traveller's Joy **72**

 Meadow-rues, Alpine Clematis **72**

 Crucifers **82**, Willowherbs **154**

 Willowherbs **154**

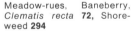

Willowherbs **154**

 Speedwells **218**

 Pink Water Speedwell **218**, *Veronica urticaefolia* **218**

 Speedwells **218**

Heath Speedwell **218**

Five petals

 Spring Beauty **42**, Chickweeds and allies **50**, Catchflies **58** and **62**, White Campion **60**, Tunic Flower **62**, Roses **110**, Rock Cinquefoil **112**, Holyhock **146**

 Pink Purslane **42**, Spurreys **56**, Catchflies **58-62**, Red Campion **60**, Cow Basil, Proliferous Pink, Tunic Flower **62**, Roses **110**, Pink Barren Strawberry **112**, Cranesbills **138**, Mallows **146**, Birdseye Primrose **178**

 Sticky catchfly **60**, Small-flowered Catchfly **62**

 Cowslip **178**

Portulaca oleracea **42**, Cinquefoils **114**, Primrose, Oxlip, Cowslip **178**

Mallows **146**, Primroses **178**

 Chickweeds and allies **50**, Bladder Campion **58**, Catchflies **58-62**

 Catchflies **58-62**, Ragged Robin **60**

 Spanish Catchfly **58**

Berry Catchfly **62**

Northern Catchfly **58**

 Wild Pink **64**, White Campion **60**

 Red Campion **60**, Pinks **64**, Cranesbills, Herb Robert, **138**, Mallows **146**

Pinks **64**

Bastard Toadflax **38**, Orpine, Stonecrops **102**, Purging Flax **136**, Burning Bush **142**, Chickweed Wintergreen **180**, Bogbean, Vincetoxicum **182**, Nightshades **210**

 Orpine, Pink Stonecrop **102**, Burning Bush **142**, Bogbean **182**

 Bastard Toadflax **38**, Mousetail **74**, Orpine, Stonecrops **102**, Vincetoxicum **182**, Wild Madder **190**, Tomato **210**

 Hop **38**, Saltwort **48**, Lesser Chickweed **54**, Lady's Mantles **108**, Gooseberry **144**, Ivy **158**, False Cleavers, Wall Bedstraw **190**

Lomatogonium **186**

 Fastigiate Gypsophila **62**, Marsh Cinquefoil **112**, Marsh Felwort **186**, Bittersweet **210**

Knotgrass **40**, Gypsophila **62**, Buttercups, Water Crowfoots **70**, Sundews, Grass of Parnassus **100**, Saxifrages **106**, Meadowsweet Goatsbeard Spiraea **108**, Strawberries **112**, Brambles **110**, *Linum tenuifolium*, White Flax **136**, White Rock-rose **152**, Common Wintergreen **172**, Labrador Tea **176**, Brookweed **180**, Heliotrope **192**, White Mullein **212**

Knotgrass **40**, Soapwort **60**, Annual Gypsophila **62**, Bramble **110**, Cotoneasters **118**, Pink Oxalis, *Linum viscosum*, *L. tenuifolium*, **136**, Cranesbills **138**, Sea-heath **152**, Umbellate Wintergreen **172**, Sea Milkwort **180**, Nonea **192**, Lungworts **194**, Viper's Bugloss, Oyster Plant **196**

Hawkweed Saxifrage **106**, Houndstongues **192**

 Nonea **192**

 Amsinckia **192**, Orange Mullein **212**

 Buttercups, Spearworts **68-70**, Marsh Saxifrage **106**, Agrimonies **108**, Cinquefoils, Silverweed **114**, Yellow Oxalis, Bermuda Buttercup, *Tribulus terrestris*, Yellow Flax **136**, Rockroses, Common Fumana **152**, Yellow Wintergreen, Diapensia **172**, Fringed Water-lily **182**, Amsinckia, Yellow Alkanet **192**, Mulleins **212**

 Hemp **38**, Polycnemum **46**, Hairy Seablite **48**, Tufted Saxifrage, Hawkweed Saxifrage **106**, White Bryony **152**

 Flaxes **136**, Meadow Cranesbill **138**, Pimpernels **180**, Jacob's Ladder **188**, Lungworts, Blue-eyed Mary, Forgetmenots, Madwort **194**, Buglosses, Alkanets, Borage, Oyster Plant **196**, Large Speedwell **218**

 Corncockle **60**, Purple Saxifrage **104**, Dusky Cranesbill **138**, Sea-heath **152**, Water Violet **180**, Jacob's Ladder **188**, Houndstongues, Heliotrope **192**, Purple Viper's Bugloss, Abraham Isaac and Jacob, Alkanet, Vervain **196**, Purple Mullein **212**

 Chickweeds and allies **50**, Snowdrop Windflower **76**, Saxifrages **104, 106**, Brambles, Raspberry, Cloudberry **110**, Storksbills **136**, Oneflowered Wintergreen **172**, Androsaces **180**, Gromwells **192**

 Sea Spurrey **56**, Storksbills **136**, Wild Azalea **176**, Northern Androsace **180**, Centauries **184**

Arctic Bramble **110**, Scarlet Pimpernel **180**

Cyphel **52**, Pearlworts **56**, Marsh Marigold **68**, Yellow Saxifrage **106**, Herb Bennet **112**, Sibbaldia **114**, St John's Worts, Tutsan, Rose of Sharon **148**, Loosestrifes, Yellow Pimpernel **178**, Changing Forgetmenot **194**

 Sea Sandwort **52**, Knawels **56**, Green Hellebore **66**, Sea Storksbill **136**

 Love-in-a-mist **66**, Periwinkles **182**, Gentians **186**, Purple Gromwell **192**, Forgetmenots **194**

 Cranesbills **138**, Purple Gromwell **192**, Venus's Looking Glasses **230**

 Violets **150**

Yellow Wood Violet, Pansies **150**

Violets **150**

Violets, Pansies **150**

Six or more petals

White Water-lily **66**, Wood Anemone, Pale Pasque Flower **76**, Mountain Avens **112**, Chickweed Wintergreen **180**, Snowdon Lily, St. Bernard's Lily, False Helleborine **264**, Drooping and Common Stars of Bethlehem **272**

Winter Aconite **66**, Lesser Celandine **68**, Yellow Anemone **76**, Alpine Avens **112**, Great Yellow Gentian **184**

False Helleborine, Asphodels **264**

Blue Anemone **76**, Blue-eyed Grass **276**

Pasque Flowers, Hepatica **76**, Gladiolus **276**

Corn Mignonette **100**, Kerry Lily **264**, Spiked Star of Bethlehem, Butcher's Broom **272**

 Montbretia **276**

Mousetail **74**, Wild Mignonette **100**, Black Bryony **276**, Spiked Star of Bethlehem, Butcher's Broom **272**

 Purple Loosestrife **154**, Kerry Lily **264**

 Meadowsweet **108**, Mountain Avens **112**

 Grass Poly **154**, Crowberry **176**, Flowering Rush **262**

 Pheasant's-eyes **74**

 Yellow Water-lily **66**, Pheasant's eyes **74**

 St. Bernard's Lily, Kerry Lily **264**, Ramsons **268**, Spiked Star of Bethlehem **272**

 Reflexed Stonecrop, Hen-and-chickens Houseleek **102**, Yellow-wort **184**, Bog Asphodel **266**

 Gageas **266**

 Squills **270**, Blue-eyed Grass **276**

 Pasque Flowers **76**, Kerry Lily **264**

Bell- and cup-shaped flowers

 Wood Sorrel **136**, Cassiope **176**, Alpine Snowbell **178**, Snowdon Lily **264**

 Pink Oxalis **136**, Bog Pimpernel **180**, Twinflower **226**

 Yellow Oxalis, Bermuda Buttercup **136**, Creeping Jenny **178**

 Stinking Hellebore **66**

Alpine Snowbell **178**, *Polemonium acutiflorum* **186**, Harebell, Bellflowers **230**

Heather **174**

Round-leaved and intermediate Wintergreens **172**

 Spurges **140**, Yellow Wintergreen **172**

 Navelwort **100**, Common and Toothed Wintergreens **172**, Cowberry, Black Bearberry, Cassiope **176**, Comfreys **188**, Lily of the Valley **264**, Wild Leek **268**, Solomon's Seals, Wild Asparagus **272**, Snowflakes **274**

 Fringe Cups **104**, Common Wintergreen **172**, Bog Rosemary, Heaths, Dodders **188**

 Pick-a-back Plant **104**

 Hen-and-chickens Houseleek **102**, Yellow Birdsnest **172**, Wild Asparagus **272**

 Navelwort **100**, Fringe Cups **104**, Bilberry **176**, Deadly Nightshade **210**, Solomon's Seals **272**

 Comfreys **192**, Harebell **230**, Bluebells, Grape Hyacinths **270**

 Asarabacca **38**, Arctic Rhododendron, Heather, Heaths **174**, Comfreys **192**, Deadly Nightshade **210**, Wild Leek **268**

Alpine Snowbell **178**, Spring Crocus **276**

 Foxglove **216**, Meadow Saffron **270**

Spotted Gentian **186**

Foxgloves **216**

 Deadly Nightshade **210**

Alpine Snowbell **178**, Gentians **186**, Viper's Bugloss **196**, Harebell, Bellflowers **230**

 Asarabacca **38**, Gentians **186**, Bellflowers **230**, Meadow Saffron **270**, Crocuses **276**

Wild and 3-cornered Leeks, Welsh Onion **268**, Drooping Star of Bethlehem, Wild Asparagus **272**, Snowflakes, Snowdrop **274**

 Small Pasque-flower **76**, Water Avens **112**, Fritillary **266**, German Garlic **268**

Creeping Jenny **178**, Wild Tulip **264**, Welsh Onion **268**, Wild Asparagus **272**

 Water Avens **112**, Fritillary **266**, Wild Leek, German Garlic **268**

 Bindweeds **188**, Thorn-apple **210**

Dodders, Bindweeds **188**

Flax Dodder **188**, Henbane, Small Tobacco Plant, **210**

Gentians **186**, Apple of Peru **210**, Harebell, Bellflowers **230**

Apple of Peru **210**, Bellflowers, Venus's Looking Glasses **230**

Lipped flowers
One-lipped flowers

Germanders **200**

Mountain Germander, Ground Pine, Wood Sage **200**

 Bugles **200**

 Birthwort **38**

Two-lipped flowers (see also Peaflowers **120**)

 Labiates and Figworts **198**, Globularia **222**

 Labiates and Figworts **198**, Toothwort **222**, Black-berried Honeysuckle **226**

Figworts **212**, Broomrapes **222**, *Lonicera alpigena* **226**

Labiates and Figworts **198**, Yellow Figwort **212**, Broomrapes **222**, Honeysuckles **226**, Bladderworts **294**

Labiates and Figworts **198**, Globularia **222**

 Vervain **196**, Labiates and Figworts **198**, Broomrapes **222**

 Corydalises, Fumitories **78**

 Bulbous Corydalis, Fumitories **78**

 Yellow Corydalis **78**, Skullcaps **200**, Cow-wheats **220**

Skullcaps **200**

  Milkworts **142**

 Alpine Butterwort **222**

 Himalayan Balsam **142**, Pale Butterwort **222**

 Orange Balsam **142**

 Balsams **142**

 Butterworts **222**

Toadflaxes **214**

Toadflaxes **214**

 Dense-flowered Orchid **284**, Butterfly Orchids, Creeping Lady's Tresses **286**, Violet Helleborine **288**, Coralroot **290**

 Bee and Spider Orchids **278**, Lady, Loose-flowered, Early Purple and Green-winged Orchids **280**, Marsh and Spotted Orchids **282**, Black Vanilla Orchid, Calypso **290**

 Dark Red and Broad-leaved Helleborines **288**

17

Lady's Slipper Orchid **278,** Pale-flowered Orchid **280,** Early and Broad-leaved Marsh Orchids **282,** Bog Orchids **290**

 Dense-flowered Orchid **284,** Bog Orchids **290**

 Lady's Slipper, Spider, Fly and Black Vanilla Orchids **278,** Violet Bird's-nest Orchid **290**

 Military, Burnt-tip, Monkey and Bug Orchids **280,** Pyramidal and Fragrant Orchids **282**

 Bug Orchid **280,** Lesser Twayblade **282**

 Man and Lizard Orchids **284,** Bird's-nest Orchid **290**

Man and Lizard Orchids, Twayblades **284**

Ghost Orchid **290**

 Dense-flowered Orchid **284,** White and Narrow-leaved Helleborines, Creeping Lady's Tresses **286**

False Musk Orchid **284,** Red Helleborine **286**

Frog Orchid **284**

False Musk, Musk and Frog Orchids **284**

 White Frog Orchid **286,** Coralroot Orchid **290**

 Coralroot, Fen, Bog and One-leaved Bog Orchids **290**

 Coralroot, Fen, Bog and One-leaved Bog Orchids **290**

 Heath Lobelia **230**

Water Lobelia **294**

Large flowers of miscellaneous shapes

Sowbread **180**

Cyclamen **180**

Columbine **74**

Daffodils **274**

Foxgloves **216**

Bellflowers **230**

Martagon Lily **266**

Pyrenean Lily **266**

Calypso **278**

Lady's Slipper Orchid **278**

Globe Flower **68**

Aconitum variegatum **74**

Wolfsbane **74**

Monkshood, Northern Wolfsbane, Larkspurs **74**

Gladiolus **276**

Irises **276**

Lords and Ladies **274**

INDIVIDUAL FLOWERS SMALL, IN HEADS OR SPIKES

Globular Heads tightly packed

 Sticky Mouse-ear **54**, Common Cudweed **238**, Leeks **268**

 Proliferous Pink **62**, Thrift **74**, Garlics, Chives **268**

 Common Cudweed **238**

 Cockleburs **234**, Crow Garlic, Onion **268**

 Globularia **222**, Cornsalads **226**, Grape Hyacinths **270**

 Chives, Garlics, Leeks **268**, Small Grape Hyacinth **270**

 Pirri-pirri Bur **108**, Clovers **134**, Small Teasel **228**

 Clovers **134**

 Great Burnet **108**, Crimson Clover **134**

 Branched Plantain **224**

 Dwarf Willow **28**, Medicks, Trefoils **132**, Mountain Clover **134**, Pineapple Mayweed **240**

 Hop **38**, Salad Burnet **108**

 Scabiouses, Rampions **228**

 Red Clover **134**, Devilsbit Scabious **228**

 Sanicle, Astrantia, Field Eryngo **158**

 Sanicle, Astrantia **158**

 Tufted Loosestrife **178**

 Sanicle, Field Eryngo **158**

 Sea Holly **158**, Globularia **222**, Scabiouses, Rampions **228**

 Devilsbit Scabious **228**

 Marsh Pennywort **158**

 Marsh Pennywort **158**, Moschatel **224**

 Dodders **188**

Brush-like heads

 Astrantia, Sanicle **158**, Composites **232**, Pipewort **264**

 Pipewort **264**

 Astrantia, Sanicle **158**, Composites **232**

 Cudweeds **238**, *Artemisia austriaca* **242**

 Safflower **252**

 Slender Hare's-ear **168**, Yellow Scabious **228**, Composites **232**

 Ragweed, Cockleburs **234**

 Composites **232**

 Cornsalads, Scabiouses, Teasels **228**, Composites **228**

Dandelion-like flowers

 Composites **232**, **256-60**

Daisy-like flowers

 Composites **232**

Flat heads of small flowers

 Hoary Cress **98**, Orpine, White Stonecrop **102**, Umbellifers **156**, Dwarf Elder **226**, Yarrow, Sneezewort **242**

 Orpine **102**, Umbellifers **156**, Dwarf Elder **226**, Hemp Agrimony **234**, Mountain Everlasting **238**

Roseroot **100**, Reflexed Stonecrop, Orpine **102**, Spurges **140**, Umbellifers **156**, Goldilocks, Ploughman's Spikenard **234**,

 Petty Spurge **140**, Umbellifers **156**

 Valerianella coronata **226**

Cornsalads **226**, Ploughman's Spikenard **234**

Flowers in leafless spikes

 Redshank, Waterpepper, Buckwheat **40**, Copse Bindweed, Knotweeds **42**

Redshank, Waterpepper, Buckwheat **40**, Black Bindweed, Himalayan Knotweed **42**

Pale Persicaria **40**, Docks, Sorrels **44**

Willows **28**, Bog Myrtle, Dwarf Birch **30-32**, Rannoch Rush **224**, Mugworts, Sea Wormwood **242**

Bulrushes **302**

Willows **26-28**, Pale Persicaria, Waterpepper **40**, Black Bindweed **42**, Docks, Sorrels **44**, Goosefoots,

Oraches, Sea Beet **46**, Sea Purslane **48**, Dog's Mercury **140**, Currants **144**, Arrowgrasses **224**, Sweet Flag **308**

 Sea-lavenders **182**

 Alpine Bistort **40**, Goatsbeard Spiraea **108**, Haresfoot Clover **134**, Hoary Plantain **224**, Spiked Rampion **228**, White Frog Orchid **286**

 Bistorts **40**, Haresfoot and Crimson Clovers **134**

 Green Amaranth **48**, Crimson Clover **134**

 Ribwort and Buckshorn Plantains **224**

Mousetail **74**, Greater and Buckshorn Plantains **224**, Asphodels **264**, Musk Orchid **284**, Bog Orchids **290**

 Hop, Hemp, Nettles **38**, Goosefoots, Oraches **46**, Glassworts, Amaranths **48**, Sea Plantain, Arrow-grasses **224**

Sea-lavenders **182**, Spearmint **208**, Black Vanilla Orchid **278**

 Fumitories **78**, Mignonettes **100**, White Melilot **130**, Thyme-leaved Speedwell **218**, May Lily **264**

Fumitories **78**, Pink Water Speedwell **218**

Spanish Catchfly **58**, Alpine Meadow-rue **72**, Barberry **78**, Weld, Wild Mignonette **100**, Melilots **128**, Goldenrods **234**

Hop, Hemp **38**

Speedwells **218**, Grape Hyacinth **270**

Alpine Meadow-rue **72**, Purple Fumitory **78**, Heath Speedwell **218**, Small Grape Hyacinth **270**, Black Vanilla Orchid **278**

Flowers in one-sided spikes

 Nottingham Catchfly **58**, Navelwort **100**, Vetches **126**, Toothed Wintergreen **172**, Three-cornered Leek **268**, Lily of the Valley, Drooping Star of Bethlehem **272**, Lady's Tresses **286**

Vetches **126**, Heaths **174**, Toothwort **222**

Montbretia **276**

Yellow Birds-nest **172**, Canadian Goldenrod **234**

Navelwort **100**

Scutellaria columnae **200**, *Campanula bononiensis* **230**, Bluebell **270**

Vetches **126**, Heaths **174**, Sea-lavenders **182**

Flowers in leafy whorls

 Bedstraws **190**, Labiates **198**

Labiates **198**

Docks **44**

Crosswort **190**

Docks **44**

Labiates **198**

 Milk-vetches **124**, Water Violet **180**, Labiates **198**

Flowers in small spikes at base of leaves

 Hemp, Nettles **38**, Goosefoots, Oraches **46**, Amaranths **48**

 Amaranths **48**

INDIVIDUAL FLOWERS SMALL, IN CLUSTERS OR SOLITARY

Flowers in loose stalked clusters

 Wall Bedstraw **190**, Red Valerian **230**

Meadow-rues **72**, Dropwort, Meadowsweet **108**, Bedstraws **190**, Red Valerian **226**, Canadian Fleabane **234**

Lesser Meadow-rue **72**, Lady's Bedstraw, Wild Madder **190**, Black Bryony **270**

Slender Marsh Bedstraw **190**, Valerians **226**, Hemp Agrimony **234**

Maple-leaved Goosefoot **46**, 4-lvd Allseed **56**, Lady's Mantles **108**, Spurges **140**, False Cleavers, Wall Bedstraw **190**

Lesser Meadow-rue **72**

Flowers in small clusters at base of leaves

Strapwort, Coral Necklace **56**, Swinecresses **98**, Mossy Stonecrop **102**, Cudweeds **238**

Cudweeds **238**

Heath Cudweed **238**

Joint Pine **24**, Narrow Hare's-ear **170**, Cudweeds **238**

Pellitory **38**, Iceland Purslane **42**, Knawels, Ruptureworts **56**, Golden Saxifrages **100**, Parsley Piert **108**, Annual Thymelaea **142**

Field Madder **190**

Flowers solitary or in pairs at base of leaves

Bastard Toadflax **38**, Knotgrass **40**, Blinks **42**, Strapwort **56**, Mossy Stonecrop **102**, Guernsey Centaury **184**, Cleavers **190**, Mudworts **294**

Mind Your Own Business **38**, Knotgrass **40**, Black Bindweed **42**, Grass Poly **154**, Crowberry **176**, Chaffweed, Sea Milkwort **180**, Guernsey Centaury **184**, Cornish Moneywort **220**, Water Purslane, Mudworts **294**

Juniper, Joint Pine **24**, Guernsey Centaury **184**, Cornish Moneywort **220**

Juniper **24**, *Polygonum patulum* **40**, Iceland Purslane **42**, Polycnemum **46**, Glassworts, Seablites, Saltworts **48**, Annual Mercury **140**, Annual Thymelaea **142**, Gooseberry **144**, Hampshire Purslane **294**

Key to Trees and Tall Shrubs

(Pages 26-37)

This is a key, based on leaf-shape, to all trees and tall shrubs (shrubs more than 1 m high) in the book.

A tree is a tall woody plant with a single, usually stout, stem. A shrub is a woody plant with several fairly stout stems from a single rootstock, usually not so tall as a small tree. When trees are coppiced, however, as beech and hornbeam sometimes are, they may have several stems like a shrub. A pollarded tree, on the other hand, is one which has been beheaded 6-7 ft from the ground, so as to produce a crop of thin branches. An undershrub is a low woody plant, such as heather or rock-rose.

Leaves oval or pointed oval

Bay Willow, Goat Willow, Sallows 26-28; Hornbeam, Beech, Evergreen Oak 32; Elms, Holly 34-36; *Apples, Pears, Whitebeams* 116; *Blackthorn, Cherries* 118; Spindle-tree, Buckthorns 144; *Dogwood,* Cornelian Cherry 158; *Strawberry Tree* 176; *Wayfaring Tree* 226; Shrubby Orache, Escallonia, Pittosporum, Spiraea, *Quince Pyracantha, Aronia,* Fuchsia, Lilac, *Shallon,* Prickly Heath 304.

N.B. Those in italics have white flowers.

Leaves ace-of-spades

Black Poplar, Birches 30; Limes 36; Flowering Nutmeg 305

Leaves elliptical

Sweet Chestnut 32; Privets 182; Medlar, Rhododendrons, Butterfly Bush 304

Leaves rounded

Aspen, Dwarf Birch, Alders, Hazel 30-32; Barberry 78; Amelanchiers 116; Cotoneasters 118 and 304; Box 144; Physocarpus 304; Snowberry 305

Leaves jagged

Oaks 32; Holly 36; Wild Service Tree, Hawthorns 116

Leaves palmately lobed

Grey Poplar 30; London Plane, Sycamore, Maples 34; Currants 144 and 304; Guelder Rose 226; Fig, Vine 304

Leaves long and narrow

Yew 24; Willows 26-28; Bog Myrtle 30; Sea Buckthorn 144

Leaves needles

Firs, Spruces, Larches, Pines, Cypress, Western Red Cedar, Juniper, Yew 24; Gorses 124; Tamarisk 144

Leaves pinnate or trefoil

Walnut 30; Ash, Tree of Heaven 36; Oregon Grape 78; Roses, Brambles 110 and 304; Rowan 116; Broom, Laburnum, Bladder Senna, False Acacia 124; Bladder-nut 144; Elders 226; Sorbaria, Box Elder 304

Leaves palmate

Horse Chestnut 36; Tree Lupin 124; Virginia Creeper 304

CONIFERS _Coniferae_ (see p. 23).

Pine Family Pinaceae

Fr. a woody cone. Widely planted.

1* SILVER FIR _Abies alba_. Pyramidal tree to 50 m, branches regularly whorled; bark scaly, greyish. Lvs single, silvery beneath in two strips, grooved, leaving an _oval scar_. Fl Apr-May. Cones erect, 10-20 cm, with bracts between the scales. Mainly in mountains. (B),F,G,(S). _A. procera_ and _A. grandis_ from N America are widely planted. **1a* Douglas Fir** _Pseudotsuga menziesii_ □ is taller, to 100 m, with branches irregularly whorled, bark ridged and red-brown, lvs fragrant when bruised and shorter, hanging cones. (T) from N America.

2* NORWAY SPRUCE _Picea abies_. Pyramidal tree to 60 m, branches regularly whorled; bark scaly, red-brown. Lvs single, in two rows, 4-sided, grass-green, falling to leave a _peg-like projection_. Fl May-June. Cones hanging, 10-18 cm, with no bracts protruding. Forests, especially in mountains. T, but (B). **2a* Sitka Spruce** _P. sitchensis_ has sharply pointed blue-green lvs, and shorter cones with toothed scales. (T) from N America. **2b* Western Hemlock** _Tsuga heterophylla_ □ has branches irregularly whorled, leading shoots drooping, flat lvs on cushion-like projections, and much shorter oval cones. (T) from N America.

3* EUROPEAN LARCH _Larix decidua_. Pyramidal tree to 35 m; bark rough, grey-brown. Lvs single and tufted, pale green, _deciduous_. Male fls yellow, female pink; Mar-Apr. Cones short, egg-shaped. Mountain forests. (B),F,G,(S). **3a* Japanese Larch** _L. kaempferi_ has twigs often red-brown, lvs bluer green and female fls green. (T) from Japan. **3 × 3a*** is also widely planted.

4* SCOTS PINE _Pinus sylvestris_. Dome-shaped tree to 40 m, but pyramidal when young; bark reddish, flaking. Lvs in _pairs,_ greyish, twisted, 30-70 mm; buds sticky, resinous. Male fls orange-yellow, female pinkish-green; May-June. Cones conical, hanging, 30-60 mm. Forests, moors, heaths.T. **4a* Austrian Pine** _P. nigra_ is more pyramidal, with grey-black bark, and longer (to 190 mm) straighter greener lvs.(T) from S Europe. **4b* Maritime Pine** _P. pinaster_ □ has more ridged bark, lvs much longer (to 250 mm), curved and greener, buds not sticky and larger, broader cones. (B,F) from S Europe. _P. contorta, P. radiata,_ and _P. strobus_ from N America, and _P. mugo_ from the Alps are also widely planted.

Cypress Family Cupressaceae

5* LAWSON'S CYPRESS _Chamaecyparis lawsoniana_. Evergreen columnar tree to 65 m; bark red-brown. Lvs _scale-like,_ opposite, closely pressed to flattened stems, often greyish, parsley-scented, leading shoots drooping. Male fls pink, female yellow-brown; Apr. Cones globular, 8 mm. Widely planted, sometimes naturalised. (T) from N America. **5a* Western Red Cedar** _Thuja plicata_ has lvs resin-scented, leading shoots erect and larger cones. Less often naturalised.

6* JUNIPER _Juniperus communis_. Evergreen shrub or small tree to 6 m, bushy columnar or (ssp. _nana_) prostrate. Lvs greyish, in whorls of 3, _spine-tipped._ Fls yellow, male and female on separate plants; May-June. Fr berry-like, green at first, blue-black in its second year. Coniferous woods, moors, heaths, scrub; ssp. _nana_ on mountains and by sea.T.

Yew Family Taxaceae

7* YEW _Taxus baccata_. Evergreen tree or shrub to 20 m; bark red-brown, flaking. Lvs dark green, in two rows. Fls green; male, with many yellow stamens, and female on separate trees; Feb-Apr. Fr in a fleshy _reddish-pink_ cup. Woods, scrub, screes, often on lime; widely planted.T.

Joint Pine Family Ephedraceae

8 JOINT PINE _Ephedra distachya_. Well-branched undershrub with green twigs, appearing lfless because lvs so small. Fls small, yellow-green, often unstalked; male and female fls on different plants; May-June. Fr fleshy, red. Sandy shores, rocks, walls. F, western.

Willow Family   Salicaceae

Deciduous trees or shrubs. Lvs usually alternate and finely toothed. Fls small, petalless but with a scale, in catkins, often appearing before the lvs; male and female on different plants, male usually yellow, female usually green. Fr woolly with long hairs. Willows and Sallows *Salix* have buds with only one outer scale, catkins usually on lfy shoots, scales untoothed and usually 2 stamens. Hybrids are very frequent.

1* BAY WILLOW *Salix pentandra.* Shrub or small tree to 7 m, *hairless;* bark grey, rugged; twigs shiny. Lvs usually *broad* elliptical, dark glossy green, sticky and fragrant when young. Catkins slender, stamens 5 or more, stigmas purple; May-June, with lvs. By fresh water. T.

2* CRACK WILLOW *Salix fragilis.* Spreading tree to 25 m, often pollarded; bark grey, rugged; twigs hairless, easily breaking. Lvs *narrow* elliptical, hairless, paler beneath; stipules soon falling. Catkins slender; Apr-May, with lvs. By fresh water. T. **2a* White Willow** *S. alba* □ is less spreading and looks silvery-grey from silky hairs on young lvs and beneath mature lvs; twigs not fragile; May. Distinct subspecies, often planted, have blue-green lvs (ssp. *coerulea,* the Cricket-bat Willow) and bright orange-yellow twigs (ssp. *vitellina* □, the Golden Osier). **2 × 2a*** is common. **2b* Weeping Willow** *S. babylonica* with drooping branches and narrower lvs is also widely planted. **2c* Almond Willow** *S. triandra* is a shrub or small tree to 10 m, with smooth flaking cinnamon-brown bark, darker green lvs with persistent stipules and fatter catkins with 3 stamens.

3* GOAT WILLOW or GREAT SALLOW *Salix caprea* agg. Small tree or shrub to 10 m; twigs downy at first, later (except in the N) hairless. Lvs *oval with pointed tip,* usually rounded at base, hairless above (except in the N), softly grey downy beneath. Catkins stout, unstalked, the scales black-brown with white hairs, which show silvery white in bud, then known as 'pussy willow' or 'palm'; Mar-Apr, before lvs. Woods, scrub, by fresh water. T. **3a* Grey Willow** or **Grey Sallow** *S. cinerea* agg. □ is a shrub to 6 m, with twigs ridged under the bark and often continuing downy, and lvs narrower, tapering to base and either greyish above or with only a few rust-coloured hairs beneath. Usually in damp places. **3b* Eared Willow** or **Eared Sallow** *S. aurita* □ is a much smaller shrub, to 2 m, with twigs widely angled and ridged under the bark, more rounded, markedly wrinkled lvs, and green catkin scales. **3c** *S. appendiculata* has faint ridges under bark, and narrower wrinkled lvs. G, southern.

4* PURPLE WILLOW *Salix purpurea.* Shrub to 5 m, with straight slender twigs, often purplish at first. Lvs *bluish-green,* lanceolate, often opposite, hairless, blackening when dried. Male catkins with dark, purple-tipped scales and reddish or purplish anthers, female often red-purple; Mar-Apr, before lvs. Fens, by fresh water. T, but (S). **4a* Violet Willow** *S. daphnoides* □ is taller, to 10 m, and may be a tree, with bluish waxy bloom on twigs, narrow, more sharply toothed lvs, not blackening, and yellow anthers. (B, F, G), S.

5* OSIER *Salix viminalis.* Shrub or small tree to 5 m; twigs long, straight, flexible, used in basket-making as withies. Lvs very *long and narrow,* to 25 cm, silky white beneath, untoothed, the margins down-turned. Catkins longer than 3, scales brown with white hairs; Mar-Apr, before lvs. By fresh water. T, but (S). Many hybrids are cultivated and become naturalised. **5a** *S. elaeagnos* has twigs reddish or yellowish and lvs finely toothed. G, southern.

Golden Osier

Willow Family *(contd.)*

1* DARK-LEAVED WILLOW *Salix nigricans* agg. Shrub or small tree to 4 m; twigs usually downy. Lvs *pointed oval,* mostly greyish beneath, almost hairless, usually blackening when dried. Catkins short, scales green and black-brown; Apr-May, usually before lvs. By fresh water; in south of range only in the hills. T. **1a* Tea-leaved Willow** *S. phylicifolia* agg. is always a shrub, with twigs almost hairless and lvs shiny above, all greyish beneath and not blackening. B,G,S. **1b** *S. starkeana* agg. may have rounder lvs, often reddish when young, sometimes downy, not blackening. G,S.

2* DOWNY WILLOW *Salix lapponum.* Shrub to 4 m, with *grey down.* Lvs pointed oval, to broad lanceolate, *untoothed,* grey on both sides. Catkin scales brown with white hairs, anthers reddish at first; May-June, with lvs. Tundra, mountains. B,S, northern. **2a Bluish Willow** *S. glauca* agg. is shorter, with longer sparser hairs, narrower blunter lvs bluish beneath, and yellow catkins. S.

3* WOOLLY WILLOW *Salix lanata* agg. Shrub to 3 m, twigs *thickly downy.* Lvs broad oval, *untoothed,* yellow-hairy at first, then grey-hairy, finally almost hairless above. Catkins with long yellow hairs; May-July, with lvs. Tundra, mountains. B,S, northern.

4* MOUNTAIN WILLOW *Salix arbuscula.* Shrub to 2 m; twigs hairless, ridged under the bark. Lvs *pointed oval,* shiny above, greyish and often hairless beneath. Catkin scales with rusty hairs, anthers reddish at first, stigmas purple-brown; May-June, with lvs. Mountains. B,S. **4a** *S. hastata* □ has mature lvs larger, sometimes roundish, not shiny or hairy, and stipules much larger and more persistent. S. **4b* Whortle-leaved Willow** *S. myrsinites* □ is prostrate with twigs downy at first, not ridged, lvs shiny on both sides, hairy only on margins and sometimes untoothed, and anthers purple.

5* CREEPING WILLOW *Salix repens* agg. Shrub with *creeping* stems up to 2 m high; twigs silky hairy at first. Lvs narrow to broad oval, white usually with *silky hairs* beneath and sometimes above, almost untoothed. Catkins short, scales dark green tipped purple-brown, female green to purple-brown; Apr-May, before lvs. Swamps, bogs, fens, dune slacks. T. **5a** *S. myrtilloides* is more prostrate, with twigs brown-hairy at first, and lvs often rounder and not silky hairy, the margins down-turned. G,S.

6* DWARF WILLOW *Salix herbacea.* Prostrate *patch-forming* undershrub; twigs 2-3 cm long. Lvs *rounded,* toothed, hairless, shiny above. Catkins short, scales yellowish, sometimes tinged red; June, after lvs. Mountains, tundra. B,G,S. **6a** *S. polaris* has narrower, longer-stalked, often untoothed lvs, purplish scales and purple male catkins. S. **6b* Net-leaved Willow** *S. reticulata* □ has larger, darker green, wrinkled, longer-stalked, untoothed lvs, whitish and conspicuously net-veined beneath, and brownish scales with grey hairs; anthers and stigmas purple.

Bog-Myrtle Family Myricaceae

7* BOG MYRTLE or SWEET GALE *Myrica gale.* Shrub to 2.5 m, *fragrant* from yellow dots on red-brown twigs and narrow oval lvs, downy beneath. Orange male and shorter reddish female catkins on different plants; Apr-May, before lvs. Wet heaths, bogs. T, western. **7a* Bayberry** *M. carolinensis* has downy twigs and lvs, catkins with the lvs, and waxy white berry-like fr. (B, G) from N America.

All species on this and the next page (32) are deciduous (except Evergreen Oak, p.32) and have alternate lvs and separate male and female fls – Poplars on separate plants, the others on the same plant.

Willow Family *(contd.)*

POPLARS *Populus* have buds with several unequal outer scales, and catkins, appearing before the *broad* lvs, with toothed scales, the male reddish and with 4 or more stamens, the female greenish-yellow.

1* GREY POPLAR *Populus canescens.* Spreading tree to 30 m, with black gashes on smooth grey bark, suckering freely; young twigs and buds *white downy.* Spring lvs pointed oval, broadest above the middle, with large teeth, almost hairless in summer; summer and sucker lvs more sharply lobed and maple-like, always downy. Catkin scales deeply cut; March. Damp woods; widely planted. B, F, G. **1a* White Poplar** *P. alba* ☐ has young twigs and buds more thickly downy, spring lvs rounder, broadest below middle, always whitely downy beneath, summer and sucker lvs more deeply palmately lobed ☐, and catkin scales less deeply cut. (B, F), G.

2* ASPEN *Populus tremula.* Spreading tree to 20 m, suckering freely; bark smooth, grey-brown, twigs and slightly sticky buds almost hairless. Lvs *rounded,* bluntly toothed, soon hairless, swaying in every breeze on thin flattened stalks; sucker lvs very like 1. Catkin scales deeply cut; March. Damp woods and heaths, fens, moors, mountains. T.

3* BLACK POPLAR *Populus nigra.* Spreading tree to 30 m, with rugged blackish bark and large bosses on trunk, rarely suckering; twigs and sticky buds hairless. Lvs *ace-of-spades,* shorter and broader on short shoots. Fl March-April, before lvs. By fresh water. B, F, G. The tall slender Lombardy Poplar var. *italica* is widely planted. **3a* Italian Poplar** *P.* × *canadensis* has a more fan-shaped crown, no bosses on trunk, more frequent suckers, and larger lvs with one or two pimples at base; female trees rare. Very widely planted. **3b* Balsam Poplar** *P. gileadensis* ☐ is like 3a with buds and young lvs balsam-fragrant and lvs without pimples, paler beneath, with hairs on veins. Less widely planted.

4* WALNUT *Juglans regia* (Walnut Family, *Juglandaceae*). Spreading tree to 30 m; bark grey, smooth at first. Lvs pinnate, aromatic. Fls yellow-green, male in hanging catkins, female in short erect spikes; Apr-May. Fr a brown nut in a green case. Hedgerows, widely planted, often bird-sown. (B, F, G) from S E Europe.

Birch Family   Betulaceae

5* SILVER BIRCH *Betula pendula.* Erect tree to 30 m, with papery peeling *white* bark and dark bosses on trunk; young twigs hairless, with resinous dots. Lvs ace-of-spades, teeth of different depths. Catkins yellowish, male hanging, female erect, shorter; Apr-May, with lvs. Woods, scrub, heaths, moors. T. **5a* Downy Birch** *B. pubescens* ☐ is shorter and may be a shrub, with bark grey or brown, no bosses, young twigs downy and without dots, and lvs uniformly toothed. Usually wetter places, not on lime. Northern. 5 × 5a* is frequent.

6* DWARF BIRCH *Betula nana.* Semi-prostrate shrub to 1 m; twigs hairless. Lvs *rounded,* deeply toothed, downy when young. Catkins as 5 but smaller; June-July, after lvs. Moors, bogs, tundra. B, G, S. 6 × 5a occurs. **6a** *B. humilis* is taller, with twigs and narrower lvs often downy. Bogs, Fens. G.

7* ALDER *Alnus glutinosa.* Spreading tree or shrub to 20 m; bark dark brown, rugged, twigs hairless. Lvs *roundish,* toothed, blunt-tipped, almost hairless. Male catkins yellowish, female egg-shaped, purplish; Feb-March, before lvs. Fr cone-like. Fens, by fresh water. T. **7a* Grey Alder** *A. incana* ☐ has smooth bark, and young twigs and narrower pointed lvs downy at least when young. Drier places. T, but (B). 7 × 7a is found. **7b Green Alder** *A. viridis* is always a shrub, with narrower lvs sticky when young, and fls with lvs. Mountains. F, G.

See note on p. 30.

Hazel Family   Corylaceae

1* HORNBEAM *Carpinus betulus.* Spreading tree to 25 m; bark smooth, grey, twigs slightly downy. Lvs pointed oval, toothed, scarcely downy. Catkins greenish; Apr-May with lvs. Fr a nut within *lfy bracts.* Woods; widely planted. T, south-eastern.

2* HAZEL *Corylus avellana.* Shrub to 6 m; bark red-brown, peeling. Lvs oval to rounded, with a small point, toothed, downy. Male fls in hanging catkins, pale yellow; female erect, bud-like, with bright red styles; *Jan-Mar,* before lvs. Fr a nut, enclosed in a lfy husk. Woods, scrub, hedges. T.

Beech Family   Fagaceae

3* BEECH *Fagus sylvatica.* Spreading tree to 30 m; bark smooth, grey, buds red-brown. Lvs pointed oval, shallowly toothed, veins prominent at edge, silky white beneath at first. Fls greenish, male in a roundish *tassel,* female erect, in pairs; Apr-May, with lvs. Fr a warm brown nut ("mast") enclosed in a bristly husk. Woods, usually in pure stand; widely planted. T, southern.

4* SWEET CHESTNUT *Castanea sativa.* Spreading tree to 30 m; bark rugged, grey-brown. Lvs *elliptical,* toothed. Catkins yellowish, with a sickly fragrance, male fls at the top, female below; July. Fr a brown nut encased in a softly spiny cup. Widely planted, sometimes naturalised. (T) from S Europe.

5* EVERGREEN OAK or HOLM OAK *Quercus ilex.* Evergreen tree to 25 m; bark greyish, scaly, twigs grey downy. Lvs pointed oval, untoothed or with holly-like spines, grey downy beneath. Catkins greenish-yellow, male long, female short; May-June. Fr a nut (acorn) in a cup. (B), F.

6* PEDUNCULATE OAK *Quercus robur.* Spreading tree to 45 m; bark rugged, grey-brown, twigs and lvs almost hairless. Lvs oblong, *lobed,* the basal lobes usually overlapping the very short stalk. Catkins greenish-yellow, male longer than female; Apr-May, with lvs. Fr a nut (acorn) in a long-stalked scaly cup. Woods, usually in pure stand; widely planted. T. **6a* Sessile Oak** or **Durmast Oak** *Q. petraea* □ has lvs tapering to the unlobed base and acorns scarcely stalked. Less often planted. Southern. **6b* Turkey Oak** *Q. cerris* □ has downy young twigs, larger, more jaggedly lobed lvs and short-stalked acorns in bristly cups. (B), F, G. **6c White Oak** *Q. pubescens* is shorter or a shrub with thickly downy twigs, lvs downy beneath when young, and acorn cups downy. F, G. **6d* Red Oak** *Q. rubra* □ with dark red twigs and much larger lvs turning rich red in autumn is the most frequent of many planted N American oaks.

1* Hornbeam   **3*** Beech   **4*** Sweet Chestnut   **6*** Pedunculate Oak

Elm Family   Ulmaceae

1* WYCH ELM *Ulmus glabra.* Spreading tree to 40 m, with rugged bark and no suckers; twigs roughly hairy when young. Lvs broad oval to elliptical, alternate, roughly hairy. Fls small, unstalked, with a *tuft of reddish stamens;* Mar-Apr, before lvs. Fr a pale green notched disc, the seed in the centre. Woods, hedges. T. **1a* English Elm** *U. procera* □ suckers freely and has twigs always hairy, lvs smaller and rounder, sometimes almost hairless above, and fr (rare in B) with seed above middle. Rarely in woods. B, F. **1b* Small-leaved Elm** *U. minor* □ suckers, and has hairless twigs, narrower lvs hairless beneath, and fr with seed above middle. Rarely in woods. B, F, G. **1c Fluttering Elm** *U. laevis* has twigs and undersides of lvs hairless or softly downy, fls and fr long-stalked and fr with hairy margins. F, G, S, southern.

Plane Family   Platanaceae

2* LONDON PLANE *Platanus hybrida.* Spreading tree to 30 m, with smooth grey bark, constantly flaking off. Lvs alternate, sharply palmately lobed. Fls and fr in *roundish heads,* fls green. Widely planted, occasionally self-sown. (B, F, G), of unknown origin.

Maple Family Aceraceae

Trees or shrubs, with long-stalked, palmately-lobed, ivy-like opposite lvs. Fls small, greenish-yellow, 5-petalled, male and female separate on same tree. Fr a pair of keys (samara), each with a single long wing.

3* SYCAMORE *Acer pseudoplatanus.* Spreading tree to 30 m, with smooth grey bark, flaking when old. Lvs 5-lobed to about half-way, well toothed. Fls in *hanging* clusters; May. Fr with wings diverging at right angles. Woods, hedges, widely planted. (B), F, G, (S). **3a* Norway Maple** *A. platanoides* □ has lvs with more sharply pointed lobes and fewer, sharper teeth, brighter yellow fls in erect clusters, appearing in April, before lvs, and fr as 4. T, but (B). **3b Italian Maple** *A. opalus* □ has rather leathery, much less toothed lvs and fr with wings diverging more narrowly. F, G.

4* FIELD MAPLE *Acer campestre.* Small tree or shrub to 20 m, with roughish grey bark; twigs downy, later corky. Lvs bluntly 5-lobed, untoothed, much smaller than 3, red when young. Fls in *erect* clusters; May. Fr with wings diverging horizontally, usually downy. Woods, hedges. T, southern. **4a Montpelier Maple** *A. monspessulanus* □ has lvs 3-lobed, rather leathery, shiny above, fls sometimes before lvs, and fr with wings almost parallel. F, G, eastern.

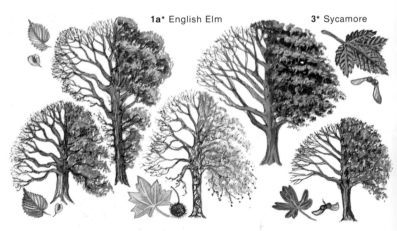

1a* English Elm   3* Sycamore

1* Wych Elm   **2* London Plane**

4* Field Maple

Horse-Chestnut Family   Hippocastanaceae

1* HORSE CHESTNUT *Aesculus hippocastanum*. Tree to 25 m, with grey bark, smooth at first, and sticky buds. Lvs opposite, palmate with 5-7 *lflets*. Fls white, with a yellow to pink spot, 5-petalled, in conspicuous erect stalked clusters; May. Fr a large warm-brown nut, in a fleshy green, stoutly but softly spiny case. Widely planted, often self-sown. (B, F, G) from SE Europe.

Holly Family   Aquifoliaceae

2* HOLLY *Ilex aquifolium*. Evergreen tree or shrub to 10 m, with grey bark, smooth at first. Lvs oval, usually *spiny,* leathery, dark green. Fls 4-petalled, white, often purplish, in clusters, male and female on different trees; May-Aug. Fr a red berry. Woods, scrub, hedges. T, southern.

Lime Family   Tiliaceae

3* SMALL-LEAVED LIME *Tilia cordata*. Spreading tree to 30 m, with smooth dark brown bark and large bosses on trunk; young twigs more or less hairless. Lvs heart-shaped, greyish beneath, with faint side-veins and small tufts of reddish hairs. Fls 5-petalled, yellowish, *fragrant,* in small clusters, on stalks half-joined to an oblong bract; July. Fr a globular nut. Woods. T. **3a* Large-leaved Lime** *T. platyphyllos* □ has no trunk bosses, young twigs usually downy, larger lvs with all veins prominent and uniformly grey downy beneath, and fewer larger fls to the cluster. Southern. **3b* Common Lime** *T.* × *vulgaris* is 3 × 3a, more like 3, but with larger lvs, more prominent veins and no red hairs. Widely planted, occasionally a natural hybrid.

Olive Family   Oleaceae

4* ASH *Fraxinus excelsior*. Spreading tree to 30 m, with grey bark, becoming rugged; buds *black.* Lvs opposite, *pinnate,* the lflets unstalked. Fls petalless, with tufts of purple-black stamens, becoming greenish; Apr-May, before lvs. Fr a key (samara) with a single wing. Woods, sometimes in pure stand, hedges. T. **4a Flowering Ash** *F. ornus* is much smaller, to 10 m, with lflets usually shortly stalked and fragrant whitish fls appearing with lvs. F, G.

Quassia Family   Simaroubaceae

5* TREE OF HEAVEN *Ailanthus altissima*. Tree to 20 m, with smooth grey bark; *suckering.* Lvs alternate, pinnate. Fls 5-petalled, greenish, in stalked clusters; *July.* Fr a key, reddish at first. Widely planted, frequently naturalised. (B, F, G) from China.

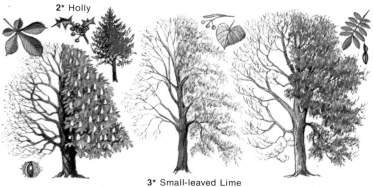

2* Holly

3* Small-leaved Lime

1* Horse Chestnut

4* Ash

37

Birthwort Family Aristolochiaceae

1* BIRTHWORT *Aristolochia clematitis.* Medium/tall hairless stinking perennial. Lvs bluntly heart-shaped, alternate, untoothed. Fls dull yellow, irregularly *tubular* with a flattened lip, swollen at the base, in small clusters at base of upper lvs; June-Sept. Waste places, often by water. (B),F,G, eastern.

2* ASARABACCA *Asarum europaeum.* Low creeping patch-forming downy perennial. Lvs kidney-shaped, shiny, long-stalked. Fls dull purple, solitary, *bell-shaped,* 3-lobed; Mar-Aug. Woods on lime-rich soils. (B),F,G, southern, scattered.

Mistletoe Family Loranthaceae

3* MISTLETOE *Viscum album.* Woody, regularly branched parasite on tree boughs (mainly apples and poplars, rarely conifers). Lvs elliptical, in pairs, yellowish, *leathery.* Fls inconspicuous, green, 4-petalled, stamens and styles on separate plants; Feb-Apr. Fr a sticky *white berry*; Nov-Feb. T, southern.

Sandalwood Family Santalaceae

4* BASTARD TOADFLAX *Thesium humifusum.* Short, often prostrate, parasitic perennial. Lvs linear, 1-veined, alternate, conspicuously yellow in late summer. Fls dull white inside, yellowish-green outside, in stalked spikes, with three bracts but no sepals; petal tube 5-lobed, persistent and much shorter than fr; June-Aug. Dry grassland on lime. B,F,G, western. **4a** *T. ebracteatum* has only one bract. G. **4b** *T. alpinum* has purer white fls, the petal tube 4-lobed and 2-3 times as long as fr. F,G,S, mountains. **4c** *T. rostratum* is like 4b but has only one bract. G. **4d** *T. pyrenaicum* has petal-tube equalling fr. Dry places. F,G. **4e** *T. linophyllon* has runners and usually 3-veined lvs. Sandy places on acid soils. F,G, eastern. **4f** *T. bavarum* has dark green lanceolate 3-5-veined lvs. Mountains. S.

Hemp Family Cannabaceae

5* HOP *Humulus lupulus.* Tall, roughly hairy, square-stemmed climbing perennial, twining *clockwise.* Lvs deeply *palmately* lobed, toothed. Fls green, the branched clusters of males and the cone-like females on separate plants; July-Sept. Often in hedges, also cultivated. T,lowland.

6* HEMP *Cannabis sativa.* Tall, *strong smelling* annual. Lvs deeply 3-9-lobed, almost to the base. Fls green, male in a branching cluster, female, on separate plants, in stalked spikes; July-Sept. Casual on waste ground, also illegally sown as a source of marijuana. T, lowland.

Nettle Family Urticaceae

7* NETTLE *Urtica dioica.* Medium/tall perennial, armed with *stinging* hairs. Lvs heart-shaped, toothed, opposite, longer than their stalks. Fls green with yellow stamens, in catkins, stamens and styles on separate plants; June-Sept. Woods, waste ground, often near houses. T. **7a* Annual Nettle** *U. urens*□ is annual and smaller, with sting less powerful, lower lvs shorter than their stalks, and stamens and styles in separate fls on same plant. Cultivated ground.

8* PELLITORY OF THE WALL *Parietaria judaica.* Short/medium, much-branched, softly hairy perennial; stems reddish. Lvs untoothed, alternate, to 70 mm, not more than three times as long as their stalks. Fls green with yellow stamens, in small forked clusters, stamens and styles in separate fls; June-Oct. Walls, banks, rocks. B,F,G,western. **8a** *P. officinalis* is larger and scarcely branched, with lvs to 120 mm, at least four times as long as their stalks. Open places. F,G, eastern. **8b* Mind Your Own Business** *Soleirolia soleirolii* is much smaller, with tiny roundish lvs and solitary pink fls; May-Aug.

Dock Family   Polygonaceae

pp. 40-44. Frequent weeds of waste places and arable land, also wet places. Lvs alternate. Characteristically there is a sheath at the leaf base forming a tube around the stem (the *ochrea*). In bistorts *Polygonum* the usually five sepal-like petals of the small sepal-less fls persist and enclose the fr.

1* BISTORT *Polygonum bistorta.* Medium patch-forming, unbranched, almost hairless perennial. Lvs narrow *triangular,* their stalks winged in the upper part. Fls pink, in a dense spike, up to 15mm broad; June-Oct. Meadows and woods, usually away from lime, often near water. B,F,G,(S).

2* AMPHIBIOUS BISTORT *Polygonum amphibium.* Short/medium creeping perennial with two forms. In water, hairless and floating, rooting at the lf-junctions, with lvs oblong, not tapering at the base. The terrestrial form☐ is smaller, slightly hairy, the longer, thinner lvs with a rounded base. Both have pink fls in dense spikes; June-Sept. Near or in still or slow-flowing water or on waste ground. T, lowland.

3* REDSHANK *Polygonum persicaria.* Medium, sprawling, hairless, branched annual. Lvs lanceolate, usually *dark-spotted,* tapering to the base. Fls pink or whitish, in densely crowded spikes; June-Oct. Bare ground, often near water. T. **3a* Pale Persicaria** *P. lapathifolium*☐ is larger and slightly hairy, with pale dots on lvs and fl-stalks, and stouter spikes of greenish-white or brick-red fls.

4* ALPINE BISTORT *Polygonum viviparum.* Short unbranched hairless perennial. Lvs narrow lanceolate, tapering at base. Fl spikes with pale pink or white fls at top and red *bulbils* at base; June-Aug. Mountain grassland and lowland in the north. B,G,S.

5* WATER-PEPPER *Polygonum hydropiper.* Medium hairless annual with a *burning taste*; ochreae not fringed. Lvs narrow lanceolate. Fls white, tinged pink or green, in a narrow, yellow-dotted nodding spike; July-Sept. Damp meadows, wet mud, shallow water. T, lowland. **5a* Tasteless Water-pepper** *P. mite* is tasteless with fringed ochreae and an erecter undotted spike of pinker fls. Southern. **5b* Small Water-pepper** *P. minus* agg. is like 5a. but has much narrower lvs, less than 10 mm broad, and a slenderer erect fl-spike.

6* KNOTGRASS *Polygonum aviculare.* Low hairless annual, with silvery, ragged few-veined ochreae. Lvs lanceolate, larger on main stem than on branches. Fls pink or white, 1-6 together at base of upper lvs; petals persistent, *enclosing fr*; June-Nov. Bare ground, including sea-shores. T. **6a* Sea Knotgrass** *P. maritimum*☐ is a perennial with many-veined ochreae, down-rolled leaf-margins, and dead petals shorter than fr; July-Sept. Atlantic coasts. B(rare),F. **6b* Ray's Knotgrass** *P. oxyspermum* also has fr longer than dead petals. Coastal sand and shingle. Not east coasts. **6c*** *P. boreale* has broader lvs, longer lf-stalks emerging from the ochreae, and pink-edged petals. Coasts. B,S, northern. **6d*** *P. rurivagum* has long (10 mm) red-brown ochreae and much narrower lvs. Dry places on lime. Southern. **6e*** *P. arenastrum* is mat-forming, with lvs all similar, shortly oval. Flg later, July-Sept. Open ground. Lowland. **6f** *P. patulum* has greenish fls and no lvs at top of fl-spike. F,G.

7* BUCKWHEAT *Fagopyrum esculentum.* Medium hairless annual, reddening. Lvs arrow-shaped. Fls white or pink in branched clusters; June-Sept. Cultivated and naturalised in open places. T, southern. **7a** *F. tataricum* is usually taller and less red with smaller fls in a looser head. A local weed in cultures of 7. F,G,S.

Dock Family *(contd.)*

1* BLACK BINDWEED *Bilderdykia convolvulus.* Tall, climbing annual, with broader lvs and totally different fls from the true bindweeds (p. 184); twines *clockwise.* Lvs heart-shaped, powdery below. Fls greenish-pink, in loose spikes; July-Oct. Fr stalks 2 mm, jointed above the middle; fr dull-black, narrowly winged white. Cultivated and disturbed ground. T, lowland. **1a* Copse Bindweed** *B. dumetorum* □ is taller and slenderer with whiter fls, fr stalks up to 8 mm, jointed below the middle, and fr shiny black, broadly winged white. Often in hedges, in warm sunny places. T, southern.

2* JAPANESE KNOTWEED *Reynoutria japonica.* Tall, vigorous perennial, often forming dense thickets; stems stout, somewhat zigzag. Lvs broad triangular. Fls whiter in branched spikes; Aug-Oct. Waste ground. (T), from Japan. **2a* Giant Knotweed** *R. sachalinensis* is taller, reaching 3 m; lvs very long, to 300 mm, pointed oval with bluntly arrow-shaped base; fls in dense clusters, a shy flowerer. Garden escape. (B, F, G), local, from Sakhalin, N E Asia.

3* HIMALAYAN KNOTWEED *Polygonum polystachyum.* Tall stout patch-forming perennial. Lvs lanceolate, *tapering* at the base, often slightly hairy beneath. Fls pink or white with red stalks, in long branched spikes; July-Oct, a shy flowerer. Waste places, often by rivers. (B, F, G), from Assam.

4* ICELAND PURSLANE *Koenigia islandica.* Insignificant prostrate annual. Resembles 6, but has minute *broad* oval unstalked lvs and often reddish stems, whole plant reddening in late summer. Fls 3-petalled, green, short-stalked, in small clusters; July-Sept. Damp mud and bare ground. B (rare), S, northern.

5* MOUNTAIN SORREL *Oxyria digyna.* Short hairless perennial; stem almost lfless. Lvs fleshy, kidney-shaped. Fl June-Aug. Fr with *two* valves, not swollen. Damp rock ledges and by streams on mountains; to sea level in the north. T. Dock Family contd. on p. 44.

Mesembryanthemum Family   Aizoaceae

6* HOTTENTOT FIG *Carpobrotus edulis.* Prostrate perennial with long creeping woody stems, and stout *fleshy* lvs, *triangular* in section. Fls usually magenta with a yellow centre, sometimes all yellow; May-Aug. Cliffs and dunes. (B, F), western, from S Africa.

Purslane Family   Portulacaceae

7* BLINKS *Montia fontana.* Very variable, but always insignificant straggling annual/perennial; stems often reddish. Lvs *narrow* oval, tapering into the stalk. Fls tiny, white, 5-petalled, long-stalked, in loose clusters; April-Oct. Bare, usually wet places, avoiding lime. T.

8* SPRING BEAUTY *Montia perfoliata.* Short hairless annual. Lvs pointed oval, long-stalked at base, one pair completely *encircling* the stem. Fls white, the five petals *not* deeply notched; April-July. Acid sandy soils, often in quantity on disturbed ground. (B, F, G), northern, from N America. **8a*** *Portulaca oleracea* has spoon-shaped lvs crowded below yellow fls. Cultivated and as a weed. (B), F, G.

9* PINK PURSLANE *Montia sibirica.* Short annual, hairless and rather fleshy. Stem lvs unstalked, but *not* encircling the stem. Fls pink, the five petals deeply *notched*; April-July. Damp woods. (T), scattered, from N America and N E Asia.

On p. 45

7* MARSH DOCK *Rumex palustris* (see p. 45). Medium annual/perennial, much branched and turning *yellow-brown* in fr; stems rather wavy. Root lvs tapering, wavy-edged, c.6 times as long as wide. Fls in lfy whorls, well separated, but more crowded at the tips; June-Sept. Fr-stalks thick, stiff, shorter than valves; valves toothed, all swollen. Bare ground near water. T, southern. **7a* Golden Dock** *R. maritimus* turns golden-yellow in fr; fr-stalks slender and longer than the longer-toothed valves. Ponds, sometimes sea-shores. T.

Dock Family *(contd.)*

DOCKS *Rumex* are mostly hairy perennials with alternate lvs and ochreae (see p.40). Fls small, green, in whorls (the lower accompanied by a small lf) in branched spikes, often turning red. Petals and sepals alike, six, the three inner larger, enlarging and becoming the 'valves' around the 3-sided nut. Fr characters are important for identification. Hybrids are frequent.

1* COMMON SORREL *Rumex acetosa.* Short/tall acid-tasting perennial. Lvs arrow-shaped, 2-6 times as long as wide, the lobes at the base pointing backwards, the upper clasping the stem; ochreae with fringed margin. Fl-head little branched, loose; May-Aug. Grassland and woods. T. **1a** *R. thyrsiflorus* has lvs 4-14 times as long as wide; fl-head much branched, dense; fls up to six weeks later. Dry places. F, G, S, eastern. **1b* French Sorrel** *R. scutatus* is shorter, with lvs often greyish, about as long as broad. Mountain screes in the south, also on walls and in waste places. T, but (B). **1c** *R. arifolius* has lvs about twice as long as wide, the lobes very short; ochreae not fringed. Mountain meadows. F, G, S.

2* SHEEP'S SORREL *Rumex acetosella* agg. Slender low/short perennial. Lvs *arrow-shaped,* sometimes very narrow (**2a** □), the basal lobes spreading or pointing *forwards,* the upper stalked. Fl May-Aug. Dry, rather bare places, not on lime. T.

3* BROAD-LEAVED DOCK *Rumex obtusifolius.* Tall perennial with broad oblong lvs up to 25 cm long, hairy on veins beneath, the lower *heart-shaped* at the base. Fl-head lax; June-Oct. Fr valves triangular, toothed, one swollen. Very common on bare and disturbed ground. T. **3a* Curled Dock** *R. crispus* □ has narrower lvs with strongly wavy margins, not distinctly heart-shaped at the base; fl-head dense; valves more oval, not toothed, usually all three swollen. **3b* Northern Dock** *R. longifolius* has broad lvs, with strongly wavy margins, very dense fl-spikes, and large valves kidney-shaped, not toothed or swollen; June-July. Often by streams. T, northern. **3, 3a,** and **3b** hybridise. **3c* Monk's Rhubarb** *R. alpinus* is less tall with much broader roundish lvs; valves oval, not toothed or swollen. Usually near houses, upland. (B), F, G. **3d* Willow-leaved Dock** *R. triangulivalvis* has narrower pointed lvs, widely branching fl-stems and triangular untoothed valves, all swollen; July-Aug. (T) from N. America. **3e* Argentine Dock** *R. frutescens* is short and creeping, with lvs wavy-edged and leathery, and valves triangular, untoothed and swollen; July-Aug. (B, G) from S America. **3f** *R. pseudonatronatus* is like **3b**, but has much narrower lvs. S(rare), northern.

4* CLUSTERED DOCK *Rumex conglomeratus.* Medium/tall perennial with *zig-zag* stem, and branches spreading. Lower lvs oblong with *rounded* base, long-stalked, the upper narrower, rarely red-fringed. Fls in whorls, Lfy nearly to the top, well spread out; June-Aug. Valves untoothed, oblong, all swollen. Woods, grassy and waste places. T, southern. **4a* Wood Dock** *R. sanguineus* has lvs rarely red-veined and only one valve swollen. Damp, usually shady places. 4a and 4b have stems straight, branches erect and only the lowest whorls lfy. **4b* Shore Dock** *R. rupestris* is stouter and shorter with greyish lvs. Shores, dune-slacks. B(rare). F. western.

5* FIDDLE DOCK *Rumex pulcher.* Short/medium, *spreading* perennial, with lvs often *waisted* like a fiddle. Fls as in 4; June-Aug. Fl-stems drooping, stiff and tangled in fr; valves toothed, all swollen. Dry, often waste places. T, southern, scattered.

6* WATER DOCK *Rumex hydrolapathum.* Very tall perennial reaching 2 m; much branched. Lvs narrow, sometimes *over 1 m* long, 4-5 times as long as wide. Fl July-Sept. Valves triangular, not toothed, all swollen. Rivers and wet places. T. **6a* Scottish Dock** *R. aquaticus* is slenderer with broader, triangular lvs, less than three times as long as wide, and valves unswollen. T, scattered, rare in B. **6b* Patience Dock** *R. patientia* has more rounded valves, often only one swollen; May-June. (B) from SE Europe.

7* MARSH DOCK *Rumex palustris*: see p. 42.

45

Goosefoot Family Chenopodiaceae

Goosefoots *Chenopodium* and Oraches *Atriplex* are annuals (except 1), often mealy, with alternate toothed lvs and spikes of inconspicuous green petalless fls, with joined sepals (except 3) and yellow anthers. There are few good diagnostic characters for the naked eye. In Goosefoots leaf-shape is important. Oraches are told from Goosefoots by their separate male and female fls, and by their sepals being enclosed by 2 triangular bracts. Bare ground, often by the sea.

1* GOOD KING HENRY *Chenopodium bonus-henricus.* Medium *perennial,* often reddish. Lvs large to 10 cm, mealy when young, *triangular,* almost untoothed. Fls in almost lfless spikes; May-Aug. Cf 6b, which has smoother lvs and stamens and styles in separate fls. Waste places. T, southern.

2* FAT HEN *Chenopodium album.* Medium/tall and deep green under a thick *white mealy* covering; stem often reddish. Lvs variable, lanceolate to diamond-shaped. Fl June-Oct. Tall robust plants with bright purple spots at stem junctions have been named *C.* × *variabile* (? *C. album* × *C. berlandieri*). B, G. *C. berlandieri* is a rare casual. Abundant on disturbed ground. T. **2a* Fig-leaved Goosefoot** *C. ficifolium* has large, deeply 3-lobed lower lvs, the middle lobe much the longest. B, F, G. **2b* Grey Goosefoot** *C. opulifolium* has lvs broader than long though sometimes with small lateral lobes, as in 2b. Scattered. **2c* Green Goosefoot** *C. suecicum* is brighter green than 2b, the broader lvs with sharp, ascending teeth. B, S. 2a-2d have stems always green and are less common.

3* RED GOOSEFOOT *Chenopodium rubrum.* Variable, medium/tall, hairless, and often red-tinged. Lvs diamond-shaped, irregularly toothed. Sepals *not* joined together; Aug-Oct. Disturbed ground, manure heaps, common near the sea. T, southern. **3a* Small Red Goosefoot** *C. botryodes* is shorter and spreading, with less toothed lvs and fls persisting, covering fr. Drier parts of salt-marshes. B, F, G, scattered. **3b* Oak-leaved Goosefoot** *C. glaucum* has narrower, more evenly lobed, somewhat oak-like lvs, darker and bluish-green above and conspicuously white-mealy beneath.

4* MAPLE-LEAVED GOOSEFOOT *Chenopodium hybridum.* Tall and hairless, with lvs usually triangular, heart-shaped at base, conspicuously though sparsely toothed. Fls in scattered lfless clusters; July-Oct. Disturbed ground. T, scattered. **4a* Nettle-leaved Goosefoot** *C. murale* is stouter with diamond-shaped lvs; fls more crowded. **4b* Upright Goosefoot** *C. urbicum* is almost hairless with smaller, less deeply toothed lvs, tapering gradually at the base. T, southern, rare in B.

5* MANY-SEEDED GOOSEFOOT *Chenopodium polyspermum.* Variable, short, spreading; stems square, often reddish. Lvs oval, *untoothed,* turning purple in autumn. Fl July-Oct. Disturbed ground. T. **5a* Stinking Goosefoot** *C. vulvaria* is smaller and more prostrate and smells strongly of rotting fish. Also in salt-marshes. Southern, local.

6* COMMON ORACHE *Atriplex patula.* Very variable, medium/tall, much branched, usually mealy. Lvs lanceolate to triangular, the basal lobes pointing forwards, tapering into the stalk. Bracts toothed or not; July-Oct. Disturbed ground, sea-shores. T. **6a* Grass-leaved Orache** *A. littoralis* □ has almost linear lvs, the upper unstalked. Salt-marshes. **6b* Spear-leaved Orache** *A. hastata* □ agg. often reddens and has ascending branches, triangular lvs, the basal lobes at right angles to the stalk, and bracts always toothed. **6c** *A. calotheca* is like 6b, but with bracts deeply toothed. G, S. **6d** *A. oblongifolia* has narrower lvs and bracts always untoothed. F(rare), G, southern.

7* FROSTED ORACHE *Atriplex laciniata.* Short/medium, sprawling, *silvery*; stems reddish, much branched, rather weak. Lvs diamond-shaped. Bracts 3-lobed; July-Sept. Dunes and beaches near high-tide mark. T, southern. **7a*** *A. rosea* is erecter with pointed bracts. Also inland. (B), F, G.

8* SEA BEET *Beta vulgaris* ssp. *maritima.* Sprawling, hairless, often red-tinged perennial with tall, upright shoots. Lvs dark green, *leathery,* untoothed. Fls as in Goosefoots (above); June-Sept. Wild ssp. of beets, mangolds, etc. Coasts. T, southern.

1

2

3

4

5

6a

6b

6

7

8

2a

Goosefoot Family *(contd.)*

1* **SEA PURSLANE** *Halimione portulacoides.* Short/medium silvery *undershrub*; stems brownish. Lvs elliptical, untoothed, the lower opposite. Fls green, in short branched spikes, stamens and styles in separate fls; July-Oct. Fr unstalked. Banks of channels in salt-marshes. B, F, G, southern. **1a* Stalked Orache** *H. pedunculata* is a short annual, not woody, with all lvs alternate, and fr stalked. Drier parts of salt-marshes. T, southern, but extinct in B.

2* **GLASSWORT** *Salicornia europaea* agg. Low/medium fleshy annual, sometimes unbranched, often reddening or yellowing in fr. Lvs scale-like, opposite, *succulent,* translucent, fused to envelop the stem. Fls obscure, green with yellow anthers, usually in groups of three; Aug-Sept. Muddy salt-marshes, sometimes on sandy soils inland. T. **2a* Perennial Glasswort** *Arthrocnemum perenne* is a creeping woody perennial, forming tussocks up to 1 m across. Salt-marshes, gravelly shores. B, F, southern.

3* **ANNUAL SEABLITE** *Suaeda maritima.* Variable, short/medium, sometimes prostrate branched annual. Lvs cylindrical, alternate, fleshy with *pointed* tip and tapering base. Fls small, green, 1-3 in a cluster at base of upper lvs; stigmas 2; Aug-Oct. Coasts, especially salt-marshes. T, southern.

4* **SHRUBBY SEABLITE** *Suaeda vera.* Small greyish hairless *shrub* to 120 cm. Lvs cylindrical, rounded at base and *tip*. Fls as in 3; stigmas 3; June-Oct. Sand and shingle or among rocks above high-tide mark. B, F.

5* **PRICKLY SALTWORT** *Salsola kali.* Variable, short/medium, prickly, usually semi-prostrate annual, hairless or not; stem often striped. Lvs linear, unstalked, succulent, usually *sharp-tipped*. Fls green, solitary with lf-like bracts; fl persisting and covering the fr; July-Oct. Sandy coasts, in the south sometimes inland. T.

6 HAIRY SEABLITE *Bassia hirsuta.* Short/medium, often prostrate, slightly hairy annual. Lvs linear, fleshy. Fls resembling 3, but ripe fr *spiny.* Meadows near the sea. G. **6a** *Kochia laniflora* has thread-like lvs and spineless winged fr. Dry sandy places. G. **6b** *K. scoparia* is like 6a, but with broader 3-veined lvs. (F, G), from Asia. **6c** *Corispermum leptopterum* is like 6b with 1-petalled fls and flat seeds. (F, G), from S Europe. **6d** *C. marschallii* is like 6c but softly-hairy and petal-less. G.

7 POLYCNEMUM *Polycnemum majus.* Short, often sprawling hairless annual. Lvs sharp-tipped, triangular in section. Fls greenish, petalless, solitary, at base of upper lvs; July-Oct. Dry places, bare ground. F, G, eastern. **9a** *P. arvense* has warty, spirally twisted stems, softer spines and longer bracts than *P. majus.* G.

Amaranth Family   Amaranthaceae

8* **GREEN AMARANTH** *Amaranthus hybridus.* Medium/tall yellow-green hairless annual; much branched. Lvs pointed oval, alternate, untoothed. Fls tiny, 5-petalled, green or red, in very dense erect spikes, often branched, mixed with pointed bracts; July-Oct. Waste places. (B, F, G), from tropical America. **8a* Common Amaranth** *A. retroflexus* is shortly hairy, seeming greyish; fl-spike largely lfless, crowded and always green. From N America. **8b* White Amaranth** *A. albus* is hairless with whitish stems, sharp-tipped lvs and a lfy fl-spike. From N America. **8c* Love Lies Bleeding** *A. caudatus*, with brilliant red-purple drooping tail-like fl-spikes, is a frequent garden escape. **8d** *A. blitoides, A. bouchonii, A. cruentus, A. graecizans, A. lividus* and other species also occur as casuals in the south of the area. They are a difficult group to identify.

49

Key to Chickweeds and Allies

(Pages 52-57)

(Fls white, 5-petalled).
A group of often confusingly similar plants of the Pink Family, Caryophyllaceae (pp. 52-64), all with white or whitish flowers and separate sepals. The shape of the leaves and whether the petals are notched or not are two important clues to identification. Sometimes, as in the stitchworts, the petals are so deeply notched as to appear to be ten instead of five.

Plant

greyish: Thyme-leaved Sandwort 52; Marsh Stitchwort, Umbellate Chickweed, Upright Chickweed 54; Strapwort 56

stickily hairy: Sticky Mouse-ear 54; Corn Spurrey 56

hairs in lines on stem: Common Chickweed, Starwort Mouse-ear 54

hairless: Arctic Sandwort, Fine-leaved Sandwort 52; Sea Spurreys, Procumbent Pearlwort, Knawels, Strapwort, Smooth Rupturewort, Four-leaved Allseed 56

square-stemmed: Greater Stitchwort, Bog Stitchwort 54

stems woody: Curved Sandwort 52

Leaves

all alternate: Strapwort 56

all unstalked: Thyme-leaved Sandwort 52; Greater Stitchwort, Bog Stitchwort 54

linear: Spring Sandwort, Fine-leaved Sandwort 52; Corn Spurrey, Sea Spurreys, Pearlworts, Strapwort 56

curved: Curved Sandwort 52

whorled: Corn Spurrey 56

with stipules: Spurreys 56

tufted: Spring Sandwort 52; Knotted Pearlwort 56

fleshy: Arctic and Sea Sandworts 52; Stitchworts 54; Sea Spurreys, Sea Pearlwort 56

Flowers

tinged green or yellow: Sea Sandwort 52; Lesser Chickweed 54; Procumbent Pearlwort, Ruptureworts 56

pink: Spurreys 56

more than 10 mm: Greater Stitchwort, Water Chickweed, Field Mouse-ear 54

with notched petals: Stitchworts, Chickweeds, Mouse-ears 54

with no petals: Lesser Chickweed 54; Pearlworts, Knawels 56

petals shorter than sepals: Sandworts 52; Bog Stitchwort, Mouse-ears 54; Sea Spurreys, Pearlworts 56

Habitat

mountains: Sandworts 52; Mouse-ears 54; Pearlworts 56

walls: Sandworts 52; Umbellate Chickweed 54

lead-mines: Spring Sandwort 52

salt-marshes: Sea Sandwort 52; Stitchworts 54; Sea Spurreys 56

Pink Family   Caryophyllaceae

pp. 52-64. Lvs in opposite pairs, undivided, usually untoothed. Flg shoots repeatedly forked; petals 4-5, sepals may be separate (pp. 52-56) or fused into a tube (pp. 58-64). Fr dry. Sandworts have fls white (except 7), five *unnotched* petals and three styles.

1* THYME-LEAVED SANDWORT *Arenaria serpyllifolia.* Variable low/short, prostrate to bushy, hairy greyish annual. Lvs pointed oval, unstalked. Fls 5-8 mm, petals shorter than sepals; anthers yellow; April-Nov. Dry, bare places, walls. T.

2* ARCTIC SANDWORT *Arenaria norvegica.* Low, *tufted* perennial, almost hairless. Lvs oval, rather fleshy, not clearly veined. Fls 8-12 mm, petals longer than sepals, anthers yellowish; June-Aug. Bare places, usually on lime. B, S, scattered, northern. **2a* Fringed Sandwort** *A. ciliata* is not tufted, and slightly hairy, with lvs not fleshy but clearly 1-veined. Mountains. T, rare in B (Ireland). **2b* Balearic Sandwort** *A. balearica* is mat-forming and has rounded lvs. Walls (B) from Mediterranean. **2c** *A. gothica* has narrower lvs. S (Gotland). **2d** *A. humifusa* has runners and much smaller, solitary fls with lilac anthers. S, northern.

3* THREE-VEINED SANDWORT *Moehringia trinervia.* Low/short, downy, rather straggling annual. Lvs pointed oval, with three or five *prominent veins.* Fls 6 mm, petals shorter than sepals; May-June. Woods on rich soils. T. **3a Mossy Sandwort** *M. muscosa* is hairless with linear, 1 to 3-veined lvs, and larger fls with four petals longer than sepals. Mountains, F, G.

4* SPRING SANDWORT *Minuartia verna.* Low mat-forming perennial; stems usually downy. Lvs stiff, linear, 3-veined in tufts. Fls 8-9 mm, petals longer than sepals, anthers pink; May-Sept. Dry rocky places, often on lime; old mine-tips. B, F, G, scattered. **4a* Mountain Sandwort** *M. rubella* is much smaller and tufted with almost lfless fl-stems; fls 5-8 mm, solitary, petals shorter than sepals; July-Aug. Mountain rock-ledges. B(rare), S. **4b** *M. biflora* is like 4a but with blunt 1-veined lvs and fls usually in pairs; petals longer than sepals; July. S. **4c** *M. setacea* is more tufted and hairless above. F, G, southern.

5* FINE-LEAVED SANDWORT *Minuartia hybrida.* Low, very slender, erect annual, usually much branched and hairless. Lvs linear, mostly near the base. Fls 6 mm, petals much shorter than narrow white-edged sepals, anthers pink; May-Sept. Dry stony and sandy places. B, F, G. **5a* Teesdale Sandwort** *M. stricta* is a tufted perennial, unbranched, but with several almost lfless fl-stems; fls 8 mm, usually three to a stem, petals equalling sepals; June-July. Wet places in mountains. B(rare), S, scattered. **5b** *M. viscosa* has stems stickily hairy and usually branched from above the middle. F, G, S, south-eastern. **5c** *M. rubra* has fls in a dense terminal cluster. G. **5d** *M. mediterranea* has fls on short crowded stalks. F, western, rare.

6* CURVED SANDWORT *Minuartia recurva.* Short, tufted perennial, slightly downy with woody stems and dense, somewhat *curved* lvs. Petals slightly longer than sepals; June-Oct. Mountains. Ireland (rare).

7* CYPHEL *Minuartia sedoides.* Low, cushion-forming, hairless perennial. Lvs linear, overlapping. Fls almost flush with the cushion, *yellow-green* often without petals, distinguishing it from Moss Campion (p. 58); June-Aug. Bare, rocky places, usually in mountains. B, F, G.

8* SEA SANDWORT *Honkenya peploides.* Creeping, rather yellowish perennial; stems and oval lvs *fleshy.* Fls 6-10 mm, greenish-white, solitary among the lvs along the stems, petals not longer than sepals; May-Aug. Fruits like small green peas. Sand and shingle by sea. T.

On p. 55

7* STICKY MOUSE-EAR *Cerastium glomeratum* (see p. 55). Low, yellowish, stickily hairy annual. Fls in *tight* heads, rarely opening fully, sepals very hairy; Apr-Oct. Bare ground. T.

8* UPRIGHT CHICKWEED *Moenchia erecta* (see p. 55). Low greyish hairless annual. Fls white with petals not notched, four styles and sepals with broad *white* margins. Apr-June. Bare sandy places. B, F, G.

Pink Family (contd.)

STITCHWORTS and CHICKWEEDS *Stellaria*. Fls white, petals five, notched or cleft, styles three.

1* GREATER STITCHWORT *Stellaria holostea*. Short, straggly perennial; stems square. Lvs narrow lanceolate, *unstalked*. Fls 20-30 mm, petals split to halfway; April-June. Woods, hedges, on heavier soils. T. **1a* Marsh Stitchwort** *S. palustris* is half the size and greyish; fls 12-18 mm, petals split to base; May-July. Wet, grassy places in lowlands. **1b* Lesser Stitchwort** *S. graminea* □ has fls 5-12 mm, petals equalling sepals; May-Aug. Rarely on lime. **1c** *S. longifolia* is like 1b with rough stems; fls 2-4 mm; June-July. Damp woods. G, S. **1d** *S. crassifolia* has slightly fleshy lvs, fls 5-8 mm, petals longer than sepals; July. Wet places. G, S, mainly coastal. **1e** *S. humifusa* is like 1d, but has smaller lvs and larger fls. S. northern.

2* COMMON CHICKWEED *Stellaria media*. Low/short, very variable, often prostrate annual, with a line of hairs down the weak round stem. Lvs *oval*, the lower long-stalked. Fls 8-10 mm, petals cleft to base, equalling sepals, or absent; all year. Bare and cultivated ground. T. **2a* Wood Stitchwort** *S. nemorum* □ is a short perennial with stem hairy all round and petals twice as long as sepals; May-July. Damp woods. **2b* Greater Chickweed** *S. neglecta* □ is short and sometimes perennial, with fls 10 mm, petals slightly longer than sepals; Apr-July. Cf. 3. Damp shady places. Local. **2c* Lesser Chickweed** *S. pallida* □ has all lvs short-stalked, and fls 4-8 mm, always without petals; Mar-May. Sandy places. Southern. **2d* Bog Stitchwort** *S. alsine* □ has a square stem, unstalked lvs, fls 4-6 mm, and petals shorter than sepals; May-Aug. Wet places. **2e** *S. crassipes* is perennial, with thick unstalked lvs, fls usually solitary and sepal-teeth recurved in fr; June-July. S, northern. **2f** *S. calycantha* is perennial, with square stems, narrower yellowish lvs, and smaller fls; June-July. S, northern.

3* WATER CHICKWEED *Myosoton aquaticum*. Straggling medium perennial, downy above. Lvs pointed oval, the lower stalked. Fls 15 mm, petals deeply split, much longer than sepals; differs from 2b in having five styles; June-Oct. Wet places. T, southern.

4* UMBELLATE CHICKWEED *Holosteum umbellatum*. Low greyish annual. Lvs lanceolate. Fls in umbel-like heads, white; petals ragged, longer than sepals; Mar-May. Dry sandy places, walls. F, G, S, southern; extinct in B.

MOUSE-EARS *Cerastium*. Fls white, five notched petals, five styles (except 5d and 5e).

5* FIELD MOUSE-EAR *Cerastium arvense*. Low perennial, slightly hairy. Lvs narrow lanceolate, not tapering at base. Fls *12-20 mm*, petals twice as long as sepals; April-Aug. Dry open places on lime. T. **5a* Dusty Miller** *C. tomentosum* is densely white-hairy. Waysides. (T), from S E Europe. **5b* Alpine Mouse-ear** *C. alpinum* is shorter with long white hairs, lvs narrowed at base, bracts white-edged, and fls 18-25 mm; June-Aug. Mountain rocks. B, G, S. **5c* Arctic Mouse-ear** *C. arcticum*, as 5b but white hairs much shorter and bracts all green, lflike. **5d* Starwort Mouse-ear** *C. cerastoides* □ is hairless except for line of hairs down stem; fls 9-12 mm, styles 3; July-Aug. Mountains. B, S. **5e** *C. dubium* is an annual resembling 5d but larger and with slightly hairy stems. Damp places. F, G, lowland.

6* COMMON MOUSE-EAR *Cerastium fontanum*. Very variable, low/short perennial, slightly hairy, with lfy non-flowering shoots. Lvs lanceolate. Petals deeply notched, equalling sepals; lower bracts lflike, upper sometimes white-edged; Apr-Nov. Grassland, bare ground. T. **6a* Sea Mouse-ear** *C. diffusum* low with 3-7 styles and all bracts lflike; Mar-July. Dry places, especially near sea. **6b* Grey Mouse-ear** *C. brachypetalum* has deeply-notched petals shorter than sepals, and all bracts lflike. Dry, open places on lime. T but (B), south-eastern. **6c* Little Mouse-ear** *C. semidecandrum* semi-prostrate, with slightly notched petals shorter than sepals and all bracts white-edged; Mar-May. Sandy ground. **6d* Dwarf Mouse-ear** *C. pumilum* like 6c but erect and often reddish, with lower bracts lflike; Apr-June. On lime. Southern. 6a-6d are annual.

7* STICKY MOUSE-EAR *Cerastium glomeratum*: see p. 52.

8* UPRIGHT CHICKWEED *Moenchia erecta*: see p. 52.

Pink Family *(contd.)*

1* **CORN SPURREY** *Spergula arvensis.* Short/medium, stickily hairy annual. Lvs linear, blunt-tipped, in whorls, channelled on underside, with small stipule at base. Fls 4-8 mm, white, petals 5, not notched, styles 5; May-Sept. Disturbed ground, not on lime. T, lowland. **1a* Pearlwort Spurrey** *S. morisonii* is more erect, with unchannelled lvs and blunt petals; Apr-June. Bare sandy ground. T. but (B, rare). **1b** *S. pentandra* resembles 1a but has pointed, narrower petals and longer. less densely clustered lvs. F. G.

2* **LESSER SEA SPURREY** *Spergularia marina.* Low/short, rather fleshy annual. Lvs slightly pointed, with small stipule at base. Fls 5-8 mm, usually pink, with 5 pointed petals shorter than sepals; styles 5; May-Sept. Dry parts of salt-marshes. T, rare inland. **2a* Rock Sea Spurrey** *S. rupicola* is a stickily hairy perennial with shorter lvs and fls 8-10 mm, petals as long as sepals. Cliffs and rocks by sea. B, F, western. **2b* Greater Sea Spurrey** *S. media* □ is an almost hairless perennial; fls 9-12 mm, petals a paler, bluer pink, longer than sepals. Salt-marshes. T, inland in G. **2c* Sand Spurrey** *S. rubra* is a low/short, stickily hairy annual/biennial; lvs not fleshy, ending in a short, stiff hair-point; stipules silvery; fls 3-6 mm, pale pink. Sandy places, not on lime or salty soils. T. **2d** *S. segetalis* is low, with minute white fls, the petals shorter than the pointed sepals; May-July. Arable fields. F, G. **2e Greek Sea Spurrey** *S. bocconii* has stipules triangular and not silvery and 2 mm white or pink and white fls. Coastal sand and rocks. F. (G). **2f** *S. echinosperma* is like 2c, with black seeds. F, G.

3* **KNOTTED PEARLWORT** *Sagina nodosa.* Short, tufted perennial. Lvs linear, largest at base, forming stiff *clusters* ("knots") up stem. Fls white, 5-10 mm, petals 5, not notched, much larger than sepals; styles 5; July-Sept. Damp, sandy places, mainly on lime. T.

4* **PROCUMBENT PEARLWORT** *Sagina procumbens.* Low hairless tufted perennial, spreading from a non-flowering rosette. Lvs linear, ending in a minute bristle. Fls tiny, greenish-white, long-stalked, petals usually 4, much smaller than sepals, often absent; styles usually 4; May-Sept. Sepals spreading when fr ripe. Damp, bare places. T. **4a* Snow Pearlwort** *S. intermedia* agg. is compacter with petals 4-5, white, almost equalling sepals, styles 4-5, and sepals erect when fr ripe. B (rare), S, mountains. **4b* Alpine Pearlwort** *S. saginoides* resembles a large 4a with the habit of 4; fls on central rosette; July-Oct. **4 × 4b*** occurs. B, S, mountains. **4c* Annual Pearlwort** *S. apetala* is a variable low/short lax annual, not forming a rosette; petals usually absent, sepals sometimes erect in fr; Apr-Aug. Dry, sandy places. Southern. **4d* Sea Pearlwort** *S. maritima* is a dark green fleshy annual, like 4c but with lvs blunt, stubby, and sepals not spreading in fr. Open ground. T, coasts. **4e* Heath Pearlwort** *S. subulata* □ is slightly hairy with fls white, petals 5, equalling slightly sticky sepals; styles 5. Dry, sandy places. Scattered.

5* **ANNUAL KNAWEL** *Scleranthus annuus.* Low, rather spiky-looking annual. Lvs linear, *meeting round stem.* Fls in clusters, no petals, sepals 5, usually pointed, green, with narrow white margin; May-Oct. Dry, sandy places. T. **5a* Perennial Knawel** *S. perennis* is a stouter perennial; branches woody at base; sepals very blunt, with broad white margin. T. rare in B.

6* **STRAPWORT** *Corrigiola litoralis.* Low greyish annual with stems often reddish. Lvs linear, all *alternate.* Fls white, minute and crowded, petals 5, as long as often partly reddish sepals; June-Oct. Damp, sandy and gravelly places. T, southern. scattered, rare in B.

7* **SMOOTH RUPTUREWORT** *Herniaria glabra.* Variable prostrate bright green, more or less hairless annual/perennial. Fls *green,* in clusters at base of oval lvs, petals green, blunt; May-Oct. Fr pointed, longer than sepals. Dry, bare places, often on lime. T, southern, rare in B. **7a* Fringed Rupturewort** *H. ciliolata* is a slightly hairy perennial; fr blunt, shorter than sepals; June-Aug. Maritime sand and rocks. B (rare), F, south-western. **7b* Hairy Rupturewort** *H. hirsuta* is a slightly hairy perennial; fr blunt, shorter than sepals; June-Aug. Maritime sand and rocks. B (rare), F, south-western.

8* **CORAL NECKLACE** *Illecebrum verticillatum*: see p. 58.

9* **FOUR-LEAVED ALLSEED** *Polycarpon tetraphyllum*: see p. 58.

57

Pink Family (contd.)

CAMPIONS and CATCHFLIES *Silene* and *Lychnis* (pp. 58-60) mostly have conspicuous fls with five petals, and sepals joined in a tube.

1* BLADDER CAMPION *Silene vulgaris*. Medium greyish perennial, usually hairless; all shoots flowering, woody at base. Lvs pointed oval, often wavy-edged. Fls white, petals deeply cleft, sepal-tube inflated to form a bladder; many fls in each cluster; styles 3; May-Sept. Arable, waste, and grassy places, especially on lime. T. **1a* Sea Campion** *S. maritima* □ is short, with non-flowering shoots and broader petals; clusters few-fld. Cliffs and shingle by sea, also on mountains.

2* NOTTINGHAM CATCHFLY *Silene nutans*. Variable medium unbranched perennial, downy below and sticky above. Lvs broader at tip. Fls white, drooping, opening and fragrant at night, all usually pointing one way; petals *rolled back*, very narrow and deeply cleft; May-Aug. Dry, undisturbed places. T. **2a* Italian Catchfly** *S. italica* is branched, with erect fls not all pointing one way; petals less recurved. (B, rare), F, local.

3 WHITE STICKY CATCHFLY *Silene viscosa*. Medium biennial/perennial, densely covered with *sticky hairs*. Lvs narrow oval, the lower with wavy margins. Fls white, 20 mm, in a long spike, petals deeply cleft, sepals hairy; June-July. Dry, grassy places. G, S, eastern. **3a** *S. tatarica* is almost hairless, with non-flowering shoots, narrower lvs, smaller fls and hairless sepals; Aug.

4* SPANISH CATCHFLY *Silene otites*. Medium perennial, stickily hairy near the base. Lvs broadest at tip, in rosettes. Fls small, 3-4 mm, *yellowish*, apparently in whorls, flhds very unlike other catchflies; stamens and styles often on different plants; June-Sept. Dry, sandy places. B (rare), F, G, south-eastern.

5* MOSS CAMPION *Silene acaulis*. Low, hairless, slightly woody perennial, forming characteristic bright green, moss-like *cushions*, often covered with the solitary, pink fls; June-Aug. Wild azalea (p. 172) has duller green lvs and very different fls, and Cyphel (p. 52) has less pointed lvs and yellowish fls. Bare places in mountains, and near sea-level in north. B, S.

6* SWEET WILLIAM CATCHFLY *Silene armeria*. Short hairless annual with all shoots flowering. Lvs lanceolate, greyish, the upper clasping the stem. Fls bright pink, clustered in rather flat-topped heads, petals only notched; June-Sept. Dry, often shady places. F, G, (B, S).

7 FLAXFIELD CATCHFLY *Silene linicola*. Medium annual; stem *rough*, branching. Lvs linear. Fls pink, with red-purple stripes, petals deeply cleft, sepals clearly veined, inflated, contracted at tip; June-July. Flax fields. F, G.

8 ROCK CATCHFLY *Silene rupestris*. Short, slender, hairless, *greyish* perennial. Lvs lanceolate. Fls white or pink, on long stalks, petals notched, sepal-tube 10-veined; June-Sept. Dry places, often in mountains. S; rare and southern in F, G.

9 NORTHERN CATCHFLY *Silene wahlbergella*. Short, unbranched, slightly hairy perennial with long, narrow lvs and solitary fls. Sepal-tube *inflated* and *longer* than the inconspicuous, purplish petals; June-Aug. Mountain meadows on lime. S. **9a** *S. furcata* is stickily hairy and branched, with smaller fls and less inflated sepal-tube.

10* FORKED CATCHFLY *Silene dichotoma*. Medium/tall annual, all stems flowering. Lvs lanceolate. Fls 15 mm, white, occasionally pink, petals *deeply cleft*; sepals hairy, not sticky; May-Aug. Waste places, (B casual), F, G.

On p. 57

8* CORAL NECKLACE *Illecebrum verticillatum* (see p. 57). Prostrate hairless annual. Resembles 7 but has *white* fls with pointed petals; June-Oct. Damp, sandy ground, not on lime. B (rare), F, G, southern.

9* FOUR-LEAVED ALLSEED *Polycarpon tetraphyllum* (see p. 57). Low hairless branching annual. Lvs oval, apparently in *whorls* of four. Fls in clusters, white, tiny, petals shorter than sepals; June-Aug. Sandy places. B (rare), F, southwestern. **9a*** *P. diphyllum* is unbranched, the often purplish lvs in pairs. Mainly coastal.

Pink Family *(contd.)* CAMPIONS AND CATCHFLIES *(contd.)*

1* RED CAMPION *Silene dioica.* Medium/tall hairy perennial. Lvs pointed oval, the basal ones stalked. Fls bright *pink,* sometimes much paler as a result of hybridisation with 2☐, giving a complete range of shades; petals cleft; calyx-teeth triangular, pointed; stamens and styles in separate fls, styles 5; Mar-Nov. Teeth on seed-pod rolled back. Rich soils, especially in or near woods or shady hedges. T.

2* WHITE CAMPION *Silene alba.* Medium/tall, much branched, stickily-hairy perennial, resembling 1, but fls longer and *white,* calyx-teeth narrow lanceolate, blunt, and teeth on seed-pod upright; May-Oct. Roadsides and more or less bare places, often on cultivated ground. T.

3* NIGHT-FLOWERING CATCHFLY *Silene noctiflora.* Medium, stickily hairy annual. Lvs pointed oval. Fls pink, opening at night, and petals rolling inwards by mid-morning to show their *yellowish* undersides; styles *three,* in same fls as stamens. Teeth on seed-pod turned back. Arable fields. T, eastern.

4* RAGGED ROBIN *Lychnis flos-cuculi.* Medium/tall, rough-stemmed perennial. Lvs lanceolate. Fls bright pink, the characteristic irregularly and narrowly 4-lobed petals giving them a *ragged* appearance; May-Aug. Damp meadows and marshy places. T.

5* STICKY CATCHFLY *Lychnis viscaria.* Medium tufted perennial with lfy flowering shoots, *sticky* just below each lf-junction. Lvs lanceolate. Fls 20 mm, bright rosy red, petals notched, apparently *whorled,* in long spikes; May-Aug. Dry, rock places, rarely on lime. T, upland, rare in B.

6* ALPINE CATCHFLY *Lychnis alpina.* Short tufted perennial with lanceolate lvs and smooth stems ending in a compact cluster of fls, smaller (6-12 mm) and pinker than in 5, and *deeply* notched; June-July. Bare places in mountains. B(rare),S.

7* CORN COCKLE *Agrostemma githago.* Tall, hairy annual. Lvs narrow lanceolate. Fls pale red-purple, petals notched, sepals hairy, forming a tube with five *long,* narrow, spreading teeth, projecting well beyond the petals; May-Aug. Cornfields. T, southern, decreasing.

8* SOAPWORT *Saponaria officinalis.* Medium/tall, often rather straggling, hairless perennial, with runners. Lvs lanceolate. Fls soft pink, the petals not notched, apparently stalked, and *standing clear* of the scarcely inflated sepals; styles 2; June-Sept. Hedges and woods by streams, waysides. T, lowland, but (B). A double-fld form☐ is a frequent escape in B.

1 × 2* Hybrid between Red and White Campions

8* Double-flowered Soapwort

Pink Family (contd.)

1* SMALL-FLOWERED CATCHFLY *Silene gallica.* Short/medium, stickily hairy annual. Lvs lanceolate, the lower broadest near tip. Fls 10-12 mm, mostly pointing in one direction; petals notched, white or pink, sometimes with a deep red spot (var. *quinquevulnera* □); June-Oct. Sandy places. B, F, G.

2* SAND CATCHFLY *Silene conica.* Short, stickily hairy, greyish annual. Lvs narrow lanceolate. Fls 4-5 mm, variably pink, the sepal-tube soon *swelling,* clearly veined but shorter than petals; May-June. Sandy places, mostly near the coast. B, F, G, southern.

3* BERRY CATCHFLY *Cucubalus baccifer.* Medium/tall downy perennial with rather brittle branches. Fls c18 mm, drooping, greenish-white, petals deeply cleft; July-Sept. Fr□ a round, *black berry.* Shady places. (B), F, G, eastern.

4 FASTIGIATE GYPSOPHILA *Gypsophila fastigiata.* Very variable low/tall perennial, hairless below, downy above. Lvs linear, 20-80 mm. Fls 5-8 mm, *white or lilac,* in a dense, *flat-topped* head, petals notched, less than twice as long as sepals; styles 2; June-Sept. Dry rocks. G, S, southern, scattered. **4a** G. *repens* is shorter and completely hairless, with lvs often curved and petals twice as long as sepals. Mountains. G, southern.

5 ANNUAL GYPSOPHILA *Gypsophila muralis.* Low/short, usually hairless annual. Lvs linear, 5-25 mm, greyish. Fls 4 mm, *pink,* in a *loose cluster,* petals slightly notched, twice as long as sepals; styles 2; June-Oct. Damp woods and meadows. F, G, S, southern.

6* COW BASIL *Vaccaria pyramidata.* Medium greyish hairless annual with a regularly branching stem. Lvs oval. Fls pink, 10-15 mm, long-stalked, with an inflated, *winged* sepal-tube; June-July. Arable fields, often on lime. (B, casual), F, G.

7* TUNIC FLOWER *Petrorhagia saxifraga.* Short/medium, usually hairless perennial. Lvs linear. Fls pale pink or white, usually solitary, on long stalks, in a *loose head,* petals notched; June-Aug. Dry sandy places and walls. (B, rare), F, G, southern.

8 PROLIFEROUS PINK *Petrorhagia prolifera.* Short/medium hairless annual. Lvs linear, fused into a sheath at the base, about as long as wide. Fls pink, in *dense clusters,* each surrounded by several large, *brown, papery bracts*; fls opening one or two at a time; petals notched; May-Sept. Dry, open places, usually on lime. F, G, S, southern. **8a* Childing Pink** P. *nanteullii* is very similar with a slightly downy stem and longer leaf-sheaths, twice as long as wide. B (rare), F.

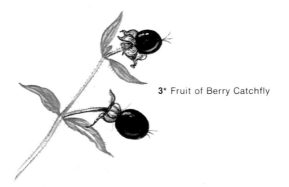

3* Fruit of Berry Catchfly

Pink Family *(contd.)*

PINKS *Dianthus* are perennials (except 7) with stiff, greyish linear lvs. Fls white, pink, or red, with sepals fused into an unribbed tube, surrounded by an epicalyx of 2-6 scales.

1* CHEDDAR PINK *Dianthus gratianopolitanus.* Short, hairless, densely tufted, with *long,* creeping sterile shoots. Lvs *rough-edged.* Fls 20-30 mm, solitary, fragrant, pale pink, petals shortly toothed; epicalyx with 2-4 short segments, pointed; May-July. Sunny rocks, often on lime. B(rare),F,G, scattered. **1a** *D. seguieri* is not greyish and has smaller, deeper pink fls, 2-4 together. G, southern.

2* CLOVE PINK *Dianthus caryophyllus.* Short/medium, hairless, tufted, with *smooth-edged* lvs; sterile shoots short. Fls 35-40 mm, sometimes in loose groups of up to five, fragrant, pink, petals *shortly toothed;* epicalyx as 1; July-Aug. Walls. (T) from S Europe. Also widely cultivated as Carnation.

3* WILD PINK *Dianthus plumarius.* Short, with the habit of 2 but *rough-edged* lvs. Fls 25-35 mm, fragrant, pink or white, the petals *cut to half-way* into narrow, feathery lobes; epicalyx with 2-4 pointed segments up to ⅓ as long as sepal tube; June-Aug. Walls. (T, southern) from S E Europe.

4 JERSEY PINK *Dianthus gallicus.* Short/medium, loosely tufted, *downy.* Lvs stiff, very short, not more than 15 mm. Fls 20-30 mm, mauve pink, the petals as in 3, but less deeply lobed; epicalyx as in 1; June-Aug. Coastal sand dunes. F.

5 LARGE PINK *Dianthus superbus.* Medium/tall, branching, hairless. Fls *30-50 mm,* pink, the petals as in 3 but divided *almost to the base;* epicalyx long, pointed; June-Sept. Dry, often shady places, also by rivers in the north. F,G,S, southeastern. **5a** *D. arenarius* has narrower lvs and white fls. G,S, eastern.

6* MAIDEN PINK *Dianthus deltoides.* Short/medium, loosely tufted, with *short,* creeping sterile shoots; fl-stems *roughly hairy.* Lvs short, roughly hairy at the edges. Fls 15-20 mm, pale pink, spotted, petals toothed; epicalyx usually with two long-pointed scales half as long as the sepal-tube; June-Sept. Dry grassy, usually sandy places. T.

7* DEPTFORD PINK *Dianthus armeria.* Medium stiff hairy *annual,* dark green, not greyish. Lvs shortly sheathed at the base. Fls 8-15 mm, bright pink or red, in dense 2-10-fld clusters, surrounded by *long, green, lfy bracts;* epicalyx with two scales as long as the softly hairy sepal-tube; June-Aug. Dry sandy places and tracksides. T, southern.

8 CARTHUSIAN PINK *Dianthus carthusianorum.* Variable, medium, hairless, with lf-sheaths several times longer than the diameter of the stem. Fls *c.* 20 mm, usually bright pink or red, in dense clusters surrounded by *short brown bracts;* epicalyx half as long as dark sepal-tube; May-Aug. Dry grassy places and open woods. F,G, eastern. **8a* Sweet William** *D. barbatus* □ is the familiar garden plant with very dense, many-fld flat heads and long-pointed epicalyx scales longer than sepal-tube. (T) from S Europe.

8a* Sweet William

Buttercup Family   Ranunculaceae

pp. 66-76. Includes a very mixed bag of rather primitive plants, all with many stamens and normally five, usually very prominent, petals or petal-like sepals. The fls often contain small honey-lvs or nectaries, which secrete nectar.

1* STINKING HELLEBORE *Helleborus foetidus.* Tall foetid perennial with overwintering stems. Lvs palmate, simply divided, all *on the stem,* the uppermost undivided. Fls *bell-shaped,* petal-less, sepals yellow-green, purple-tipped, in clusters; Jan-May. Dry woods and scrub, mainly on lime. B,F,G, southern.

2* GREEN HELLEBORE *Helleborus viridis.* Short perennial with *two root lvs* that die before winter; all lvs palmately divided, except in the south and east where forms with undivided lvs occur. Fls spreading, larger than 1, petal-less, the sepals dull green like the lvs; Feb-Apr. Damp woods in thick leaf-mould. B,F,G, southern.

3* LOVE-IN-A-MIST *Nigella damascena.* Short, branched annual. Lvs divided into *thread-like* segments. Fls solitary, with five pale blue sepals, *surrounded* by feathery lvs; June-July. Dry, open places. (B, casual, F), from Mediterranean. **3a****N. arvensis* is variable with fls often veined green and long-stalked, not surrounded by a web of lvs; June-Sept. Arable fields on lime. (B, casual), F,G.

4* WINTER ACONITE *Eranthis hyemalis.* Low hairless perennial. Lvs palmately lobed, all from roots and appearing after fl-stem has died down, except for three which form a *ruff* immediately below the solitary fl; sepals yellow, usually 6; Jan-Mar. Woods, naturalised from gardens. (B,F,G, scattered), from S Europe. Buttercup Family contd. on p. 68.

Water-lilies   Nymphaeaceae

Hairless perennials of stagnant and slow-flowing fresh water. They root in the mud and have large floating lvs on long stalks, and conspicuous fls.

5* YELLOW WATER-LILY *Nuphar lutea.* Lvs oval, larger than 6, up to 40 cm, and some thin and *submerged.* Fls yellow, much smaller than 6, to 60 mm, held a few cm above the water surface; June-Sept. Fruits roundish, warty. Still or slow-flowing water, sometimes deeper than 6. T, lowland. **5a* Least Water-Lily** *N. pumila* ☐ is smaller with fls to 30 mm and larger gaps between the petals; June-July. Hybridises with 5 and intermediates occur, sometimes in the absence of both parents. Stagnant water, mainly upland.

6* WHITE WATER-LILY *Nymphaea alba.* All lvs *floating,* almost circular, rather broader than the fls, the lobes at the base not overlapping. Fls 50-200 mm, white, fragrant; June-Sept. Fruits carafe-shaped, smooth. Still, shallow water. T, southern in S. **6a** *N. candida* ☐ is rather smaller with basal lobes of lvs overlapping, leaving little or no gap and deeper yellow stamens. F,G,S, scattered.

6a *Nymphaea candida*

Buttercup Family *(contd.)*

1* GLOBE FLOWER *Trollius europaeus.* Short/medium hairless perennial. Lvs palmate, deeply cut. Fls large, almost *spherical,* petal-less, the ten yellow sepals curving in at the top; May-Aug. Damp, grassy places, mainly upland. T.

2* MARSH MARIGOLD *Caltha palustris.* Short stout hairless perennial; dwarf and creeping on mountains. Lvs large, *kidney-shaped,* toothed, dark green, shiny, and often mottled paler on the upper surface. Fls variable, 10-50 mm, with five yellow sepals and no petals; Mar-Aug. Fr rather pod-like, clustered, conspicuous in summer. Wet places. T.

BUTTERCUPS *Ranunculus* mostly have shiny yellow fls (except Large White and Glacier, p. 70) and both petals and sepals, distinguishing them from 1 and 2.

3* MEADOW BUTTERCUP *Ranunculus acris.* Medium/tall hairy perennial with characteristic 2-7-lobed lvs, rather rounded in outline, the end lobe *unstalked* (cf. 4 and 5). Fl-stalks unfurrowed, sepals *erect,* nectaries golden-yellow; Apr-Oct. Meadows, damp grassy places. T. **3a** *R. polyanthemos* is strongly branched, and less hairy, with 5-lobed lvs and narrow lf-segments; May-July. Damp woods on lime. F, G, S, eastern. **3b** *R. nemorosus* is like 3a but has 3-lobed lvs and broader lf-segments; beak on seed longer. Meadows. F, G, S (Oland). **3c Woolly Buttercup** *R. lanuginosus* is very hairy with lvs less deeply divided; nectaries orange-yellow; May-Aug. Damp woods. G.

4* BULBOUS BUTTERCUP *Ranunculus bulbosus.* Short, very hairy perennial with stem-base swollen. Lvs like 3 but less deeply lobed and end lobe *stalked.* Sepals *turned downwards,* fl-stalks furrowed; Mar-June. Drier grassland than 3 and 5, and prefers lime. T. **4a* Hairy Buttercup** *R. sardous* is a hairier annual with stem-base not swollen, smaller, often shiny lvs, and paler fls; May-Oct. Damp arable land, grassy places. T, southern. **4b** *R. illyricus* is medium and less hairy, with very narrow lf-segments. G, S (Gotland).

5* CREEPING BUTTERCUP *Ranunculus repens.* Short/medium, slightly hairy, creeping perennial, with rooting *runners.* Lvs rather triangular in outline. end lobe *stalked.* Resembles 4 more than 3, but sepals erect; May-Sept. Damp, often bare places. T. **5a Jersey Buttercup** *R. paludosus* is hairier and not creeping; seeds with long hooked beak. F. **5b** *R. montanus* agg. has no runners and stem lvs more divided. Mountains. G.

6* GREATER SPEARWORT *Ranunculus lingua.* Tall hairless perennial with long *runners.* Lvs to 25 cm long, *lanceolate,* toothed. Fls *more* than 20 mm, stalks not furrowed; June-Sept. Marshes and fens, becoming rarer. T, scattered.

7* LESSER SPEARWORT *Ranunculus flammula.* Very variable, usually short/ medium, hairless perennial, sometimes creeping and then rooting at some lf-junctions of the often reddish stem, but with *no* runners. Lvs *lanceolate,* short- or unstalked, to 4 cm, more or less toothed. Fls *7-20 mm,* stalks furrowed; June-Oct. Wet places. T. **7a* Creeping Spearwort** *R. reptans* has runners and roots at every lf-junction; lvs long-stalked, broader; fls 5 mm; June-Aug. Hybridises with 7. Gravelly lake-shores. T but scattered, rare in B; mainly in mountains. **7b* Adderstongue Spearwort** *R. ophioglossifolius* is a short erect annual with broader lvs and fls 6-9 mm; May-Aug. Marshes. B (rare), F, south-western. **7c** *R. hyperboreus* is like 7a with broader, 5-lobed lvs and 3 petals. Mountains. S.

8* LESSER CELANDINE *Ranunculus ficaria.* Low/short hairless perennial with rather fleshy, dark green, heart-shaped lvs. In shady places may have bulbils at base of lf-stalk. Usually 8-12 narrow petals, whitening when old; Mar-May. Woods, hedge-banks, bare damp ground. T. **8a** *R. lapponicus* is creeping and rooting at lf-junctions, with 3-lobed lvs, 6-8 petals and down-turned sepals. Mountains. S. **8b** *R. cymbalaria* is smaller with 5-petalled fls. (S).

Buttercup Family *(contd.)*

1* GOLDILOCKS BUTTERCUP *Ranunculus auricomus.* Short, slightly hairy perennial with lowest lvs only slightly lobed. Fls few, sepals purple-tinged, petals 0-5, often *distorted* and sometimes *absent,* otherwise resembling a slender Meadow Buttercup (p. 68); Apr-May. Woods, hedges. T. **1a** *R. nivalis* ☐ is smaller, the 5 petals never distorted. Mountains. S. **1b** *R. sulphureus* is stouter than 1a, with dense brown hairs on sepals. Arctic.

2* CELERY-LEAVED BUTTERCUP *Ranunculus sceleratus.* Medium hairless annual with palmately lobed shiny lvs. Fls numerous, small, petals *no longer* than down-turned sepals; May-Sept. Fr head conical, with several hundred seeds. In or near slow or stagnant water. T, lowland. **2a** *R. pygmaeus* ☐ is very much smaller, its fr-head not conical. Mountains. S.

3* CORN BUTTERCUP *Ranunculus arvensis.* Short/medium hairless annual with small (4-12 mm) pale yellow fls on unfurrowed stalks, and spreading sepals; May-July. Fr *spiny,* strongly ridged. Cornfield weed, often in rather damp places on lime. T, southern.

4* SMALL-FLOWERED BUTTERCUP *Ranunculus parviflorus.* Short hairy, usually sprawling annual, somewhat unbuttercup-like, though most resembling a small Hairy Buttercup (p.68); stem lfy. Root-lvs circular. Fls tiny (3-6 mm), on furrowed stalks, petals no longer than down-turned sepals; May-July. Bare, dry places on lime. B, F, south-western.

5 LARGE WHITE BUTTERCUP *Ranunculus platanifolius.* Tall perennial. Lvs large, shallowly 5-7-lobed. Fls 20 mm, *white,* on long hairless stalks; May-Aug. Damp woods, mainly lowland. F, G. **5a** *R. aconitifolius* is smaller, with 3-5 lobed lvs; fl-stalks hairy. Mountains. F, G.

6 GLACIER BUTTERCUP *Ranunculus glacialis.* Low hairless perennial with thick 3-5-lobed lvs. Fls large, solitary or few, white or pinkish, sepals with purple-brown hairs; June-Aug. Stony ground in mountains near snow-line, not on lime. S.

WATER CROWFOOTS *(Ranunculus* subgenus *Batrachium)* grow in all freshwater habitats from mud to swift water. Lvs palmately lobed when floating and/or much divided and feathery when submerged. Fls white, petals with yellow claw (except 8c). Most are very similar, and often best identified on habitat and lf-type.

7* COMMON WATER CROWFOOT *Ranunculus aquatilis.* Very variable annual/perennial, with submerged lvs branching in three planes and usually also *toothed* floating lvs. Fls 10-20 mm, nectaries circular. Fr-stalk shorter than lf-stalk; Apr-Sept. Still or slow water to 1 m deep. **7a* Pond Water Crowfoot** *R. peltatus* has more shallowly toothed floating lvs, longer (20-30 mm), pear-shaped nectaries and stalk longer than lf-stalk. Western. **7b* Brackish Water Crowfoot** *R. baudotii* is very like 7a but has more rigid lf-segments, and smaller (10-20 mm) fls sometimes with blue-tipped sepals. Brackish water. **7c* Thread-leaved Water Crowfoot** *R. trichophyllus* has no floating lvs and short, stiff, fan-like submerged lvs, repeatedly branching into three; fls 5-10 mm, nectaries half-moon-shaped. Sometimes in brackish water. **7d* Fan-leaved Water Crowfoot** *R. circinatus* is like 7c, but with lvs branching in two dimensions only, and larger (10-20 mm) fls. **7e* Chalk-stream Water Crowfoot** *R. penicillatus* usually lacks floating lvs and is larger and robust, with long trailing stems; fls 20-30 mm, nectaries pear-shaped; fr-stalk longer than lf-stalk. Fast streams, especially in limestone areas. **7f* River Water Crowfoot** *R. fluitans* is like 7e, but never has floating lvs; submerged lvs very dark green with long narrow thread-like segments. Rarely in limestone areas.

8* IVY-LEAVED CROWFOOT *Ranunculus hederaceus.* Creeping annual/perennial, lvs ivy-shaped. Fls 3-6 mm, petals equalling sepals; Apr-Sept. Shallow water, rarely in limestone areas. T, western. **8a* Round-leaved Crowfoot** *R. omiophyllus* has petals twice as long as sepals; fls 8-12 mm. B, F, western. **8b* Three-lobed Crowfoot** *R. tripartitus* is more delicate perennial, with both finely-divided submerged lvs and rounder floating lvs; sepals blue-tipped; Mar-June. B, F, G, western. **8c** *R. ololeucos* is like 8b but petals all white, more than twice as long as sometimes green-tipped sepals; May-July. Shallow water. F, G, western.

Buttercup Family (contd.)

MEADOW-RUES *Thalictrum* have fls in clusters with four inconspicuous petals and no sepals but long stamens which give the fls their colour and rather shaggy appearance. Lvs 2-3-pinnate.

1* COMMON MEADOW-RUE *Thalictrum flavum.* Tall, almost hairless perennial. End lflet *longer* than broad. Fls erect, densely crowded, with whitish petals and long yellow stamens. June-Aug. Wet meadows. T. **1a Small Meadow-rue** *T. simplex* is usually somewhat smaller, with lflets rather narrower, and fls drooping at first. Meadows. G, S, northern. **1b** *T. lucidum* is tufted with narrower lflets. G. **1c** *T. morisonii* has shiny stems, upper lflets narrow and untoothed and fls in more open clusters. G (rare), southern.

2* GREATER MEADOW-RUE *Thalictrum aquilegifolium.* Tall hairless perennial. Lflets about as long as broad. Fls larger than 1 and 3, greenish-white petals and very long, *broad* lilac-coloured stamens; June-July. Woods, meadows. (B), G, S, eastern.

3* LESSER MEADOW-RUE *Thalictrum minus.* Variable medium (in dry places) to tall (in wet places) perennial. End lflet roughly *as long as broad.* Fls yellowish or purplish, in loose clusters; June-Aug. Sand dunes, turf, rocks and mountain ledges, and by streams, mainly on lime. T, scattered.

4* ALPINE MEADOW-RUE *Thalictrum alpinum.* The smallest meadow-rue, a short delicate unbranched perennial. Lvs 2-trefoil, end lflet rounded. Fls purple, in an unbranched stalked spike; stamens violet with yellow anthers; May-July. Damp turf, ledges, usually in mountains. B, S, northern.

5* BANEBERRY *Actaea spicata.* Medium hairless perennial, strong smelling, rather umbellifer-like, with large, long-stalked 2-pinnate or 2-trifoliate root lvs. Fls white, in stalked spikes; petals short, usually 4, stamens long, *white,* conspicuous; May-July. Fr a large, shiny *black berry.* Woods and limestone pavements. T, eastern, rare in B. **5a** *A. erythrocarpa* has more deeply cut lvs and smaller red berries. S, northern.

6* TRAVELLER'S JOY *Clematis vitalba.* Tall clambering woody perennial, sometimes with immensely long stems. Lvs pinnate, often with twining stalks. Fls fragrant, with four greenish petals and conspicuous stamens; July-Sept. Fr □ with long, grey, hairy *plumes,* in dense clusters, persisting through the winter, giving rise to the name Old Man's Beard. Hedges and scrub on lime. B, F, G, southern. **6a** *C. recta* □ looks very different, being erect and not woody; fls white. G (rare), southern.

7 ALPINE CLEMATIS *Clematis alpina.* A much smaller plant than 7, with 2-pinnate lvs and larger, solitary, showy *violet* fls; May-July. Fr with similar hairy plumes. Rocky mountain woods. G, S, rare.

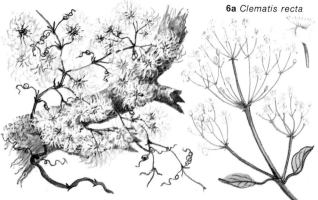

6a *Clematis recta*

6 Traveller's Joy, fr.

73

Buttercup Family *(contd.)*

1* COLUMBINE *Aquilegia vulgaris.* Tall, branching perennial with 2-trifoliate lvs, the lflets rounded, dull green. Fls characteristic, each sepal with a *long spur,* usually violet, but sometimes pink or white, especially when escaped from gardens; May-July. Woods and scrub, on lime. T.

2* MONKSHOOD *Aconitum napellus.* Tall hairless perennial, with lvs palmately divided almost to the midrib. Fls large, *violet or blue,* broadly hooded, in long spikes; May-Sept. Damp woods and by streams. B, F, G, southern. **2a** *A. firmum* has lvs less deeply divided and fewer fls. G. **2b** *A. variegatum* □ has fls blue, white or variegated, the hood taller, almost conical. Mountain woods. G, southern.

3 WOLFSBANE *Aconitum vulparia.* Tall hairless perennial with lvs palmately divided to midrib. Fls pale yellow, narrowly hooded, in branched spikes; June-Aug. Very shady, damp woods. F, G, eastern. **3a Northern Wolfsbane** *A. septentrionale* □ has fls hairy, usually violet, sometimes yellow, the hood broader at the base. S.

4* FORKING LARKSPUR *Consolida regalis.* Short, downy annual; widely branched. Lvs much divided into many very narrow segments. Fls deep purple, *long-spurred* in a loose spike; fl-stalks much longer than their lf-like bracts; June-Aug. Fr hairless. Arable fields on lime. T, but casual in B. **4a* Eastern Larkspur** *C. orientalis* is larger with erect, stickily-hairy branches; fls blue in a dense spike; fl-stalks shorter than bracts; fr hairy. (B, rare, F, G), from S Europe. **4b* Larkspur** *C. ambigua* resembles 4a but is downy with fl-stalks as long as bracts. (B, casual), F.

5 YELLOW PHEASANT'S-EYE *Adonis vernalis.* Short perennial, stem scaly at base. Lvs 2-3-pinnate, feathery. Fls large, 40-80 mm across, *yellow,* with more than ten petals; Apr-May. Dry grassland. F, G, S, south-eastern.

6 LARGE PHEASANT'S-EYE *Adonis flammea.* Short/medium annual, like 7, but fls larger (25-30 mm) and sepals *hairy* at least at base and pressed against petals; June-July. Arable fields on lime. F, G.

7 SUMMER PHEASANT'S-EYE *Adonis aestivalis.* Short hairless annual with feathery 3-pinnate lvs, the lower stalked. Fls *red* (or yellow, var. *citrina*) 15-25 mm, with prominent black centres and black anthers; sepals hairless, touching with 5-8 spreading petals; June-July. Arable fields on lime. F, G. **7a* Pheasant's-Eye** *Adonis annua* has fls always red; June-Aug. Lower lvs unstalked. B, F, G, southern.

8* MOUSETAIL *Myosurus minimus.* Low hairless annual. Lvs linear, slightly fleshy, all in a basal tuft. Fls small, solitary, long-stalked, 5-7-petalled, greenish-yellow, with a greatly elongated plantain-like fr head, fancifully likened to a mouse's tail. Bare, usually damp ground. T.

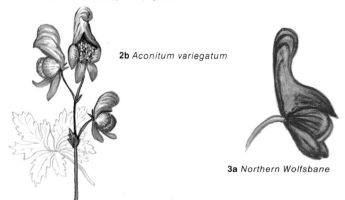

2b *Aconitum variegatum*

3a *Northern Wolfsbane*

1

2

3

4

5

6

7

8

Buttercup Family *(contd.)*

ANEMONES, *Anemone* and *Pulsatilla,* have conspicuous, solitary fls, opening in spring, with brightly coloured sepals and no petals. *Pulsatilla* usually has six, and *Anemone* more than six sepals.

1* PASQUE FLOWER *Pulsatilla vulgaris.* Low hairy perennial. Lvs 2-pinnate, feathery, covered with long hairs. Fls large, purple, *bell-shaped,* 50-80 mm, erect at first, later drooping, anthers bright yellow; Mar-May. Fr☐ with long, silky plumes, sometimes persisting into late summer. Dry grassland on lime. T, southern.

2 PALE PASQUE FLOWER *Pulsatilla vernalis.* Short hairy perennial. Lvs much less divided than in 1, with broader lobes and fewer, shorter hairs. Fls similar to 1, but *white* inside and violet or pink outside, drooping at first, later erect; Apr-June. Stem lengthening in fr. Meadows, especially in mountains. F, G, S, eastern.

3 SMALL PASQUE FLOWER *Pulsatilla pratensis.* Low hairy perennial, much resembling 1, but fls smaller (40 mm), always *drooping;* dark purple in the north, paler southwards; May. Root lvs very feathery, stem lengthening in fr. Meadows. G, S.

4 EASTERN PASQUE FLOWER *Pulsatilla patens.* Differs from 1 in having broadly lobed root lvs, and *spreading* petals, so that fl is not bell-shaped; Apr-May. Meadows. S.

5* WOOD ANEMONE *Anemone nemorosa.* Low/short hairless perennial. Stem lvs 3, long-stalked, deeply palmately lobed, in a whorl just below the fl; root lvs similar, appearing after fls, few, sometimes absent. Fls solitary, white, 20-40 mm, sometimes pinkish, especially on the underside of the 6-12 hairless sepals; Mar-May. Deciduous woods, except on very acid soils, mountain meadows. T.
5a* Blue Anemone *A. apennina*☐ has blue-purple fls with 8-18 sepals; Mar-Apr. (B), F, G. **5b** *A. narcissiflora* ☐ has smaller fls 3-8 together in an umbel. Mountains. G (rare), southern.

6* YELLOW ANEMONE *Anemone ranunculoides.* Short hairy perennial. Lvs like 5, those on the stem very short-stalked. Fls bright *yellow,* buttercup-like, 15-20 mm, usually solitary and with 5-8 sepals, hairy beneath; Mar-May. Woods. T, but (B).

7 SNOWDROP WINDFLOWER *Anemone sylvestris.* Short/medium hairy perennial. Lvs palmately lobed. Fls white, to 70 mm, with only *five* sepals, hairy beneath, anthers yellow; Apr-June. Dry woods on lime. F, G, S, southern.

8 HEPATICA *Hepatica nobilis.* Low/short hairless perennial. Lvs *3-lobed,* untoothed, evergreen and purplish below. Fls solitary, 15-25 mm, usually blue-purple, occasionally pink or white, with 6-9 sepals and three sepal-like bracts immediately below them; anthers whitish; Mar-Apr. Woods and scrub on lime. F, G, S.

5a* Blue Anemone

5b *Anemone narcissiflora*

1* Pasque Flower

Barberry Family   Berberidaceae

1* BARBERRY *Berberis vulgaris.* Deciduous shrub to 4 m; stem with 3-pointed *spines.* Lvs oval, sharply toothed. Fls yellow, in drooping spikes; May-June. Fr a bright red berry. Hedges, dry woods, often on lime. T, southern. **1a* Oregon Grape** *Mahonia aquifolium* □ is much shorter, with spineless stem, lvs pinnate and spine-toothed, fls fragrant, in erect spikes, and fr blue-black; Mar-May.

Fumitory Family   Fumariaceae

Mostly rather floppy, hairless plants with lvs several times pinnately divided, and characteristic tubular, two-lipped, spurred fls. Fumitories *Fumaria* are mostly annual weeds of arable fields and waste places.

2* YELLOW CORYDALIS *Corydalis lutea.* Short perennial with lfy stems and lvs ending in a lflet. Fls bright *yellow,* in dense spikes opposite upper lvs; May-Nov. Fr drooping. Walls. (B,F,G), from S Europe. **2a*** *C. ochroleuca* □ has lf-stalks narrowly winged, cream-coloured fls, and an erect fr. Walls and rocks.

3* BULBOUS CORYDALIS *Corydalis solida.* Short perennial with only *1-3 lvs* on the stem and a scale below the lowest lf; lflets in threes. Fls 18-30 mm long, pink-purple, more than ten to each rather lax, *terminal* spike; bracts lobed; Apr-May. Woods, hedges. T but (B). **3a*** *C. bulbosa* is rather larger with no scale below the lowest lf; bracts not lobed, fl-spikes dense. **3b** *C. intermedia* has fls 10-15 mm long, only 2-8 to a spike, and unlobed bracts. F,G,S. **3c** *C. pumila* is like 3b with lobed bracts. G,S.

4* CLIMBING CORYDALIS *Corydalis claviculata.* Medium/tall, *climbing* annual, with lvs ending in a tendril. Fls 5-6 mm long, creamy, 6-8 in dense spikes opposite lvs; June-Sept. Rocky, often wooded places, not on lime. T, western.

5* COMMON FUMITORY *Fumaria officinalis* agg. Much the commonest species. Weak scrambling annual. Lf-segments flattened. Fls pink, tipped darker, the largest *7-9 mm* long, sepals more than ¼ the length of the petals; Apr-Oct. Bracts shorter than the fr-stalks. T.

6* WALL FUMITORY *Fumaria muralis.* Rather robuster than 5, with larger fls 9-12 mm, 12 on each spike: spikes shorter than their stalk; lower petals turned up at the edges. Fr stalks *not* curved back. Often on hedge-banks. T, western. **6a*** *F. bastardii* has fl-spikes longer than their stalks, fls 9-11 mm, 15-25 to a spike, and upper petals compressed, the lower with flat edges. B,F, western. **6b*** *F. martinii* is like 6a with fls 11-13 mm. B,F.

7* RAMPING FUMITORY *Fumaria capreolata.* Robust climbing annual, up to 1 m long. Fls white, tipped blackish-pink, the largest 10-14 mm, on short stalks, c 20 in each spike; upper petals compressed, lower with turned-up edges; May-Sept. Fr stalks *turned back.* Often in hedges.T,southern. **7a*** *F. purpurea* has purple fls, upper petal not compressed. B(rare). **7b*** *F. occidentalis* has lower petal with broad flat margin. B (Cornwall, rare).

8* SMALL FUMITORY *Fumaria parviflora.* Slender annual. Lf-segments chan-nelled. Fls *5-6 mm,* white or pinkish, tipped darker pink in *dense 20-fld* spikes on very short stalks; sepals less than ¼ the length of the petals; June-Sept. Bracts as long as fr-stalks. Usually on chalk. B,F,G, southern. **8a*** *F. densiflora* has fls 6-7 mm, always pink; bracts longer than fr-stalk. T, southern. **8b*** *F. vaillantii* agg. has lf-segments flat, fls always pink, 6-12 on each spike; and bracts slightly shorter than fr-stalks. T, southern. **8c** *F. schleicheri* is like 8b, but has deep pink fls and bracts much shorter than fr-stalks. F,G.

1a* Oregon Grape

Poppy Family   Papaveraceae

POPPIES *Papaver* are mostly weeds of cultivated ground with large, solitary, brightly-coloured fls, whose four petals have a rather creased, silky appearance. They have two sepals which fall off soon after the buds open. Juice usually white. Lvs 1-2-pinnate.

1* COMMON POPPY *Papaver rhoeas.* Medium, usually roughly hairy annual. Fls 70-100 mm, deep scarlet, often with a *dark centre,* anthers blue-black; fl-stalks with spreading hairs; June-Oct. Pod almost *round,* hairless. Arable fields and disturbed ground. T, southern.

2* PRICKLY POPPY *Papaver argemone.* Short, stiffly hairy annual with hairs closely pressed to stem. Fls 20-60 mm, generally smaller than 1 and 4, petals *pale scarlet,* not overlapping; May-July. Pod long and narrow, with numerous bristles. Arable fields, especially on sandy soils. T, southern.

3* ROUGH POPPY *Papaver hybridum.* Resembles 2 but with a *rounded* pod, densely covered with stiff yellow bristles. Fls darker, crimson; June-Aug. Arable fields, mainly on lime. T, southern.

4* LONG-HEADED POPPY *Papaver dubium.* Medium annual, with hairs closely pressed to stems; juice white. Fls paler than 1, 30-70 mm, *without* dark centres; anthers violet; June-Aug. Pod much *longer* than wide, hairless. Arable fields and disturbed ground. T. **4a* Babington's Poppy** *P. lecoqii* has juice turning yellow on exposure to air, petals more orange-red, and anthers brown or yellow. On lime. B,F.

5* OPIUM POPPY *Papaver somniferum.* Medium/tall, almost hairless perennial with large, wavy, greyish lvs, irregular in outline. Fls up to 180 mm, from *lilac* with dark centres to white; June-Aug. Arable fields or waste ground, mainly an escape from cultivation. (T, southern). Grown since early times as a source of opium.

6 ARCTIC POPPY *Papaver radicatum* agg. Low/short tufted hairy perennial with deeply lobed lvs; juice white or yellow. Fls yellow, sometimes very pale, 20-40 mm; June-Aug. Pod *softly hairy,* less than twice as long as wide. Bare, rocky ground in mountains. S.

7* YELLOW HORNED-POPPY *Glaucium flavum.* Medium/tall, hairless, rather greyish perennial, with wavy edged lvs clasping the branching stem. Root lvs more deeply lobed, all lvs rough to the touch. Fls 60-90 mm, yellow, variable; June-Sept. Pods *very long,* up to 300 mm, sickle-shaped, hairless. Sea shingle and waste places inland. T, western. **7a* Red Horned-Poppy** *G. corniculatum* □ has fls 30-50 mm, orange to red; stem and pod hairy. Bare and waste places. (T, casual) from S Europe.

8* WELSH POPPY *Meconopsis cambrica.* Medium, slightly hairy perennial with long-stalked, deeply lobed root lvs and similar, but short-stalked stem lvs. Fls usually *more than 50 mm,* yellow; June-Aug. Pod hairless, longer and narrower than in 6. Damp woods and rocks. B,F, upland and as a garden escape in B and G.

9* GREATER CELANDINE *Chelidonium majus.* Medium/tall, almost hairless perennial, its branched stems easily broken, revealing an orange juice. Lvs rather greyish, pinnately lobed. Fls *20-25 mm,* smaller than other poppies, in clusters of 3-8; Apr-Oct. Hedges, walls, waste places, often near houses. T.

7a* Red Horned Poppy

Key to Crucifers

The Crucifer or Cabbage Family, Cruciferae (pp. 84-99), is one of the most distinctive families, with its four petals arranged crosswise. Some other superficially similar four-petalled flowers are Greater Celandine (p. 80), Tormentil and other potentillas (pp. 112-14), Willowherbs (p. 154) and Bedstraws (pp. 188-90). These are included in the Main Key on p. 13. The shape of the fruits is one of the most important clues to crucifer identity.

Plant

foetid: Wall Rockets 84; Narrow-leaved Pepperwort, Field Pennycress, Swinecress 98

garlic-smelling: Garlic Mustard 90; Garlic Pennycress 98

grey or white: Alisons, Ball Mustard 84; Woad 86; Wild Cabbage, Rape, Black Mustard, Bastard Cabbage 88; Sea Kale. Stocks 92; Garden Rockcress, Tower Mustard 94; Alisons 96; Field Pepperwort, Perfoliate Pennycress 98

with bulbils: Coralroot Bittercress 90

with runners: Creeping Yellowcress 86; Cuckoo Flower, Large Bittercress 90; Sea Kale, Horse-radish, Dittander 92; Alpine Rockcress 94; Hoary Cress 98

Stems

woody below: Golden Alyssum 84; Wallflower 86; Wild and Wallflower Cabbages 88; Dame's Violet 90; Sea Kale, Stocks 92; Hoary Cress 98

leafless: Yellow Whitlow-grass 84; Isle of Man Cabbage 88; Shepherd's Cress, Rock and Common Whitlow-grasses, *Draba nivalis, D. fladnizensis* 96

angled: Wallflower, Common Wintercress 86; Large Bittercress, Watercress 90

unbranched: Yellow Whitlow-grass, Ball Mustard 84; Cuckoo Flower, Coralroot Bittercress, Garlic Mustard 90; Hairy Rockcress 94; Whitlow-grasses 96

zig-zag: Wavy Bittercress 94

Flowers: fragrant: Wallflower 86; Dame's Violet 90; *Erysimum odoratum* 299.

yellow: all on pp.84-8; *Cardamine enneaphyllos* 90; Tower Mustard, Towercress 94; *Draba nemorosa* 96

white: Wild Radish 88; all on pp.90-2

pink/purple: Cuckoo Flower, Coralroot Bittercress, Dame's Violet, Honesties 90; Sea Rocket, Stocks, Early Scurvy-grass 92; Northern and Tall Rockcresses 94; Candytufts 96; Alpine Pennycress 98

Petals: notched: Hoary Alison, Common and Rock Whitlow-grasses 96

unequal: White Ball Mustard 92; Candytufts, Shepherd's Cress 96

absent: Narrow-leaved Bittercress 90; Hairy Bittercress 94; Narrow-leaved Pepperwort, Lesser Swinecress 98

veined: Bastard Cabbage, Wild Radish 88.

Sepals

erect (fls yellow): Yellow Whitlow-grass, Small Alison, Ball Mustard 84; Hedge and Treacle Mustards, Common Wintercress, Creeping Yellowcress 86; Wild and Wallflower Cabbages, Hairy Rocket, Wild Radish 88

spreading (fls yellow): Wall Rockets 84; Tall Rocket, Warty Cabbage, Yellowcresses 86; Wild Turnip, Black Mustard, Charlock, Bastard Cabbage 88

persistent: Small Alison 84

Seed-pods

long and narrow: Wall Rockets 84; Wallflower, Tall Rocket, Hedge and Treacle Mustards, Common Wintercress 86; all on pp. 88-90; Stocks 92; all on p. 94

oblong: Yellow Whitlow-grass, Gold of Pleasure 84; Woad, Great Yellowcress 86; Perennial Honesty 90; White Ball Mustard, Early Scurvy-grass 92; all on pp. 96-8

spherical: Alisons, Ball Mustard 84; Warty Cabbage 86; Austrian Yellowcress 299; Sea Kale, Horse-radish, Dittander, Common scurvy-grass 92; Common Whitlow-grass 96; Garden Cress, several Pepperworts 98

circular: Honesty 90; Wild Candytuft 96; Field Pennycress 98

2-lobed: Buckler Mustard 84; Swinecresses 98

notched: Wild Candytuft, Shepherd's Purse 96

wrinkled: Ball Mustard 94

warty: Warty Cabbage 96; Swinecress 98

jointed or segmented: Bastard Cabbage, Wild Radish 88; Sea Rocket, Sea Kale 92

flattened: Buckler Mustard 94; Honesties 90; Dittander 92; Northern and Tall Rockcresses 94; Wild Candytuft, Shepherd's Purse and Cress, Hutchinsia 96

triangular: Shepherd's Purse 96; Alpine, Perfoliate and Mountain Pennycress 98

twisted: Tower Mustard 94; Twisted Whitlow-grass 96

4-angled: Treacle Mustard, Common Wintercress 86; Black Mustard, Hairy Rocket 88

overtopping fls: London Rocket 86; Hairy Bittercress 94

spreading: Small Alison, Ball Mustard 84; Creeping Yellowcress 86; Wallflower Cabbage 88; Dame's Violet, Honesties, Watercress 90; Common Scurvy-grass 92; Shepherd's Purse, Wall Whitlow-grass 96; Austrian Yellowcress 299

closely pressed to stem: Hedge Mustard 86; Black Mustard 88; Stocks 92; Hairy Rockcress, Tower Mustard 94

hanging: Woad 86

Cabbage Family   Cruciferae

Crucifers get their name from the characteristic cross made by their four, usually well separated petals; sepals 4; fls usually in stalked spikes. Seed pods are important characters and are either long and thin (siliquae), or less than three times as long as broad (siliculae) and of various shapes.

1* YELLOW WHITLOW-GRASS *Draba aizoides.* Low tufted hairless perennial with stiff narrow undivided lvs but a *lfless* stem. Petals yellow, much longer than sepals; Mar-May. Pod elliptical. Rocks and walls on lime. B(rare),F,G, southern, scattered. **1a** *D. alpina* agg. has broader lvs with a few branched hairs; June-July. S, northern. **1b** *D. crassifolia* is like 1a, but has unbranched hairs, still broader, thicker lvs, and smaller paler fls.

2* ANNUAL WALL ROCKET *Diplotaxis muralis.* Short/medium, hairless annual/biennial, much branched, with stems *lfless* or almost so. Root lvs stalked, deeply pinnately lobed. Fls 10-15 mm, deep yellow; June-Sept. Pods longer, exceeding their stalks, held at an angle to the stem. Walls, rocks, also a weed of bare and waste places. T, southern. **2a* Perennial Wall Rocket** *D. tenuifolia* is a taller perennial, woody at the base, with greyish lvs on the stem, fls 15-30 mm, and pods held stiffly erect, parallel to the stem, shorter than their stalks. B,F,G, southern. **2b** *D. viminea* is smaller, with smaller, paler fls. (G), from S Europe.

3* SMALL ALISON *Alyssum alyssoides.* Low/short, downy annual. Lvs small, lanceolate. Fls tiny, *pale* yellow, 3-4 mm, in dense spikes; style *minute;* Apr-June. Pod almost *round,* the hairy sepals persisting with it. Sandy fields, often on lime. T, eastern, but (B, rare).

4 MOUNTAIN ALISON *Alyssum montanum.* Low/short, hairy, often rather whitish perennial, with broader lvs and slightly larger fls than 3. Fls *bright* yellow, style longer than in 3; Apr-June. Pod round, sepals not persisting. Sandy places. F,G, southern. **4a* Golden Alyssum** *A. saxatile* is woody at base, with lvs often lobed and pods longer. Stony places on lime. (B),G(rare).

5* GOLD OF PLEASURE *Camelina sativa.* Short/medium, usually hairless annual. Lvs narrowly arrow-shaped, clasping the stem. Fls 3 mm, yellow; May-July. Pod *oblong,* becoming yellowish. A weed of arable and waste places. T, a rare casual in the north and west. **5a** *C. microcarpa* ☐ is hairy, the grey-green pods as long as wide. **5b** *C. alyssum* ☐ has a squashy, more rounded pod. Flax fields. **5c** *C. macrocarpa* ☐ is larger with larger pods. Flax fields. 5a-c are not in B.

6* BALL MUSTARD *Neslia paniculata.* Medium downy annual with lanceolate lvs, sometimes slightly toothed, the lower stalked, the upper clasping the stem. Fls yellow, 3-4 mm; June-Sept. Pod rounded, often rather flattened, *wrinkled.* Waste places. (B,casual,S),F,G, southern. **6a Mitre Cress** *Myagrum perfoliatum* ☐ is hairless and greyish, with more toothed lvs; pod larger, broad at the tip, very variable, more or less tapering to the thicker stalk. (T, casual) from S Europe.

7 BUCKLER MUSTARD *Biscutella laevigata* agg. Short, slightly hairy perennial, with very variable, more or less deeply lobed root lvs. Fls 5-10 mm, yellow; May-July. Pod much wider than long, flattened, with two almost *circular* lobes. Dry limestone rocks, occasionally in waste places. F,G.

On p. 87
9* CREEPING YELLOWCRESS *Rorippa sylvestris* (see p. 87). Straggling, hairless perennial with runners and short/medium flg shoots. Lvs pinnately lobed, the upper sometimes unlobed. Fls 5 mm, yellow, petals longer than sepals; June-Sept. Pods long, 6-18 mm, equalling stalk. Damp and bare places. T. **9a* Marsh Yellowcress** *R. palustris* is erecter and stronger-stemmed; fls 3 mm, petals not longer than sepals; June-Oct; pod 4-9 mm, oblong, shorter than stalk. **9b* Iceland Yellowcress** *R. islandica* ☐ is like 9a but always prostrate and has pod longer than stalk. B,S. **9c** *R. pyrenaica* is downy at base, with upper lvs clasping the stem, smaller fls and shorter fr. F,G, southern.

1

2

3

4

5

5a

5b

5c

6

6a

7

Cabbage Family (contd.)

1* WALLFLOWER *Cheiranthus cheiri*. Medium/tall perennial with narrow lanceolate untoothed lvs covered with *flattened* hairs. Fls *25 mm*, orange-yellow or orange-brown. fragrant: Mar-June. Pods long, flattened. Old walls, dry limestone rocks. (B, F, G), from the Mediterranean. **1a** *Erysimum crepidifolium* has hairs not flattened, lvs often lobed or toothed, smaller (18 mm) yellow fls than Wallflower and pods 4-angled; June-Sept. Dry grassland, often on lime. G. **1b** *E. odoratum* is like 1a, but taller, with more deeply toothed lvs and fls brighter yellow. F, G, eastern. **1c* Hare's-ear Cabbage** *Conringia orientalis* is like 1b but lvs clasp stem and fls smaller, greenish. Pods to 140 mm. (T), from E Europe.

2* HEDGE MUSTARD *Sisymbrium officinale*. Medium/tall, *stiff,* roughly hairy (occasionally hairless) annual with spreading branches and rosette lvs deeply pinnately lobed. Fls *3 mm*, pale yellow; May-Sept. Pods 6-20 mm, much shorter than 2a-2f, erect and closely *pressed* to stem, unlike 2a-2f. Waste places and hedge banks. T. **2a** *S. strictissimum* has lvs not or shallowly-toothed, not lobed; fls bright yellow, 4-6 mm, and pods hairless. (B), F, G. **2b* London Rocket** *S. irio* is often hairless, with lvs more narrowly lobed and pods hairless, overtopping fls when young. (T, rare in B). **2c* False London Rocket** *S. loeselii* is like 2b but hairy with larger fls and young pods not overtopping fls. **2d* Eastern Rocket** *S. orientale* has topmost lvs not lobed, fls 7 mm, and pods hairy at first. (T). **2e*** *S. austriacum* is biennial/perennial with some lvs not lobed, fls golden yellow, 7-10 mm and pods and their stalks twisted. (B, S), F, G. **2f*** *S. volgense* is like 2b with larger fls.

3* TALL ROCKET *Sisymbrium altissimum*. Tall, branched annual, hairless above. Root lvs pinnate, roughly hairy, soon dying; stem lvs unstalked, with very *narrow, deeply cut* lobes. Fls 10 mm, pale yellow; June-Aug. Pods long, to 100 mm. Waste places. (T), from E Europe. **3a* Flixweed** *Descurainia sophia* ☐ has lvs 2-pinnate with even narrower lobes, fls 3 mm, petals shorter than sepals, and pods 10-40 mm. T.

4* WOAD *Isatis tinctoria*. Medium/tall, mostly hairless biennial/perennial. Lower lvs narrow, downy; upper greyish, hairless, *arrow-shaped,* clasping the stem. Fls 4 mm, yellow; June-Aug. Pods dark brown, oblong, *hanging*. Dry places and rocks. (T, southern), from S Europe.

5* WARTY CABBAGE *Bunias orientalis*. Medium/tall, hairless perennial; stems warty. Lvs pinnately lobed, the topmost unlobed. Fls sulphur-yellow, 15 mm; May-Aug. Pods rounded, *warty,* shiny, long-stalked. Waste places. (T), from E Europe. **5a*** *B. erucago* is similar with pods 4-winged.

6* TREACLE MUSTARD *Erysimum cheiranthoides*. Short/tall, *square-stemmed* annual covered with flattened branched hairs. Rosette lvs lanceolate, *shallowly* and irregularly toothed, soon dying. Fls 6 mm, yellow, fl-stalks longer than sepals; June-Sept. Pods long, 10-30 mm, 4-angled. Waste places. T. **6a** *E. repandum* has fls 10 mm with a longer style, and pods 45-100 mm, rounded. G. **6b** *E. hieracifolium* is biennial/perennial, with fls 10 mm on stalks equalling sepals, and pods 30-35 mm, parallel to stem. Grassland. G.

7* COMMON WINTERCRESS *Barbarea vulgaris*. Medium/tall *hairless* perennial. Lvs with 2-5 pairs of coarse lobes, end lobe shorter than rest of lf; upper lvs undivided, toothed. Fls 7-9 mm, bright yellow, buds hairless; May-Aug. Pods long, 15-30 mm, stiffly *erect*. Damp places and roadsides. T. **7a* Small-flowered Wintercress** *B. stricta* has end lobe longer than rest of lf and 1-2 pairs of side lobes; fls 5-6 mm, buds downy. **7b* Medium-flowered Wintercress** *B. intermedia* has all stem lvs lobed. (T, southern) from Mediterranean. **7c* American Wintercress** *B. verna* has all lvs lobed, the lower with 6-10 pairs; Apr-July; pods 30-60 mm. (B, F, G, western), from Mediterranean.

8* GREAT YELLOWCRESS *Rorippa amphibia*. Tall hairless perennial, with runners. Lvs lanceolate, toothed or pinnately lobed. Fls 6 mm, bright yellow; June-Aug. Pods *oblong, long-stalked*. Wet places. T. **8a* Austrian Yellowcress** *R. austriaca* has lvs clasping stem and fls 3 mm; pod spherical, shorter-stalked. (B, F, S), G.

Cabbage Family *(contd.)*

1, 2 and 8 are cultivated forms and their wild relatives, so much variability must be expected.

1* WILD CABBAGE *Brassica oleracea.* Medium/tall hairless perennial with a thick, straggling stem, woody at the base and clearly marked with old *lf scars.* Lvs *greyish,* pinnately lobed, somewhat clasping the stem. Fls 30-40 mm, in a long spike, the unopened *buds above* the fls; May-Sept. Sea-cliffs in B and F, western; also (T), widely cultivated as cabbage, cauliflower, etc.

2* WILD TURNIP *Brassica rapa.* Tall slender annual/biennial. Root lvs pinnately lobed, roughly hairy, bright green, stem lvs clasping. Fls 20 mm, in a more or less flat-topped spike, the *fls above* the unopened buds; Apr-Aug. Pods long. Cultivated, and on bare ground, especially by streams. (T). **2a* Rape** *B. napus* has greyish lvs and fls dull yellow in a spike like 1, but less elongated. Fls buff in var. *napobrassica* (Swede). **2b*** *B. juncea* is hairless, with greyish lvs, the lower very large and compound, the end lflet much larger than the others; fls level with the buds. (B, G), from Asia.

3* BLACK MUSTARD *Brassica nigra.* Tall, rather greyish annual. Fls yellow, 15 mm; June-Aug. All lvs *stalked,* the lowest pinnately lobed, bristly and bright green. Pods long, pressed against stem. Waste places, sea cliffs, stream banks, woods; also cultivated. T. **3a* Hoary mustard** *Hirschfeldia incana* is whitish with coarse hairs below, and has paler fls. Waste places. (B, G), F.

4* CHARLOCK *Sinapis arvensis.* Medium/tall, roughly hairy annual. Lower lvs large, irregularly lobed and toothed, the upper narrow, with pointed teeth, *unstalked* but not clasping the stem. Fls 15-20 mm, yellow; Apr-Oct. Pods long, markedly beaded. Arable fields, waste places, most often on lime. T. **4a* White Mustard** *S. alba* □ has all lvs deeply pinnately-lobed, sepals longer. Often cultivated and as a weed. (T, southern), from Mediterranean.

5* HAIRY ROCKET *Erucastrum gallicum.* Short/medium, densely hairy annual. Lvs pinnately lobed, the lower stalked, the upper not clasping the stem. Fls pale yellow, 10-15 mm, sepals erect; May-Sept. Pods *less than 40 mm long,* beaded. Waste places, dry rocks. T, southern, but (B, rare). **5a*** *E. nasturtiifolium* has upper lvs not clasping, fls bright yellow and sepals spreading.

6* WALLFLOWER CABBAGE *Rhynchosinapis cheiranthos.* Medium biennial with base of lfy stem hairy. Lvs deeply pinnately lobed. Fls yellow, 20-25 mm; June-Aug. Pods long, *more than 40 mm.* Dry and waste places, rarely on lime. (B), F, G, south-western. **6a* Isle of Man Cabbage** *R. monensis* is entirely hairless, with an almost lfless stem. Sandy places by coast. B, western. **6b* Lundy Cabbage** *R. wrightii* is densely hairy. Cliffs. B, rare (Lundy Is.).

7* BASTARD CABBAGE *Rapistrum rugosum.* Medium hairy annual whose stalked lvs have three pairs of lobes. Fls pale yellow, 10 mm; May-Sept. Pods with *two segments,* the upper large and rounded, with an abrupt beak, the lower narrower, closely pressed to the stem. Waste ground. (B, F, G), from the Mediterranean. **7a*** *R. perenne* is a perennial, stiffly hairy below, with six pairs of lf-lobes; fls bright yellow; and upper pod-segment narrowing gradually into beak.

8* WILD RADISH *Raphanus raphanistrum.* Medium, roughly hairy annual, slightly resembling 4 but with the teeth of the upper lvs blunt. Fls larger (25-30 mm), pale yellow or white with lilac veins; May-Sept. Pods long, markedly *beaded,* and breaking easily at the joints. Bare and waste ground. T. **8a* Sea Radish,** ssp. *maritimus,* is taller and bushier, with fls always yellow, and pods not breaking easily. Coasts. B, F, G. **8b* Garden Radish** *R. sativus* has an edible tuberous root, fls white or lilac, and pods not beaded.

4a* White Mustard

Cabbage Family (contd.)

BITTERCRESSES *Cardamine* have pinnate lvs (except 1a) and long pods.

1* CUCKOO FLOWER *Cardamine pratensis* agg. Medium hairless perennial, sometimes tufted, with root lvs forming a *rosette;* stem lvs dissimilar. Fls variable, 12-20 mm, lilac or white, anthers yellow; Apr-June. Pods to 40 mm. Damp places. T. **1a** *C. bellidifolia* □ is dwarf and tufted, with all lvs undivided, few or none on stems, and very small (7-10 mm) white fls. S, mountains.

2* LARGE BITTERCRESS *Cardamine amara.* Medium hairless perennial with basal lvs *not* in a rosette; stem lvs with *broader* lflets than 1. Fls white, 12 mm, anthers violet; Apr-June. Damp or wet places, often in woods. T. **2a*** *C. raphanifolia* □ is taller with a large, rounded terminal lflet, deep lilac fls and yellow anthers. Streamsides. (B, rare), from S Europe.

3* CORALROOT BITTERCRESS *Cardamine bulbifera.* Medium hairless perennial with *no root lvs;* upper stem lvs narrow lanceolate, with purple-brown *bulbils* at their base. Fls 12-18 mm, lilac to purple; Apr-June. Pods 20-35 mm. Woods, often of beech, on lime. T, southern. **3a** *C. heptaphylla* has all lvs pinnate and no bulbils; fls *c.* 20 mm, often white, pods up to 80 mm. F, G, southern. **3b** *C. pentaphyllos* is like 3a, but with lvs palmate or trefoil. **3a** × **3b** occurs. Mountain woods. G, southern. **3c** *C. enneaphyllos* □ has 2-trefoil lvs, no bulbils and drooping pale yellow or white fls. Mountain woods. G, southern.

4* NARROW-LEAVED BITTERCRESS *Cardamine impatiens.* Medium hairless annual/biennial. Lvs with up to 18 *3-lobed lflets;* lf-stalks clasping the stem. Fls white, but petals inconspicuous or absent; anthers greenish-yellow; May-Aug. Pods explosive when ripe. Damp, shady, often rocky places, on lime in the north. T, mainly western. **4a** *C. parviflora* is a short annual, each lf with fewer, unlobed lflets; petals always present. F, G, S, scattered.

5* GARLIC MUSTARD *Alliaria petiolata.* Medium hairless biennial with long-stalked lvs, heart-shaped at the base, smelling of *garlic* when crushed. Fls white; Apr-Aug. Pods long. Hedges and open woods. T.

6* DAME'S VIOLET *Hesperis matronalis.* Medium/tall hairy perennial, with toothed lanceolate short-stalked lvs. Fls 15-20 mm, white or violet, very *fragrant;* May-Aug. Pods long, to 100 mm, curving upwards. Hedges, roadsides, a frequent garden escape. (B, G, S), F.

7* PERENNIAL HONESTY *Lunaria rediviva.* Tall perennial with a hairy stem. Lvs pointed oval, sharply toothed, the upper stalked. Fls fragrant, purple, *c.* 20 mm; May-July. Pods long, very *flattened,* elliptical with pointed tip. Damp woods, mainly on lime. T but (B), eastern. **7a* Honesty** *L. annua* is biennial with unstalked, coarsely toothed upper lvs, and larger, scentless fls, 25-30 mm; Apr-June. Pods almost round. A frequent garden escape. (T), from S E Europe.

8* WATERCRESS *Nasturtium officinale.* Short/medium *creeping* hairless perennial, with pinnate lvs, green in autumn. Fls white; May-Oct. Pods less than 18 mm long, with two rows of seeds inside. In and beside shallow, flowing water, often on lime. T, southern. **8a* One-rowed Watercress** *N. microphyllum* is extremely like 8 but with lvs turning purplish-brown in autumn, and pods more than 16 mm long, containing only one row of seeds; fls two weeks later. Less often on lime. **8** × **8a*** is intermediate, with lvs turning purplish-brown, but deformed pods. Occurs with parents, and often cultivated.

2a* *C. raphanifolia*

3c *Cardamine enneaphyllos*

1a *C. bellidifolia*

4a *C. parviflora*

Cabbage Family (contd.)

1* SEA ROCKET *Cakile maritima* agg. Short/medium greyish hairless *succulent* annual. Lvs linear to pinnately lobed. Fls lilac, sometimes very pale; June-Sept. Pods egg-shaped, angled. Sandy coasts. T.

2* SEA KALE *Crambe maritima*. Stout medium greyish hairless perennial, making large clumps, with thick stems, woody at the base. Lvs large, rounded, fleshy, with *crinkly* lobes or teeth, but narrower up the stem. Fls 10-15 mm, white, in a rather flat-topped head; June-Aug. Pods globular. Sea shores, on sand and shingle. T.

3* WHITE BALL MUSTARD *Calepina irregularis*. Medium/tall hairless annual. Stem lvs *clasping*, toothed, arrow-shaped. Fls 2-4 mm, with unequal petals; May-June. Pods more or less spherical, beaked. Waste places, mainly on lime; not a coastal plant. (B, rare), F, G.

4* HOARY STOCK *Matthiola incana*. The garden plant, a medium/tall, grey, stickily hairy annual/perennial; stem lfless and usually woody at the base. Lvs narrow lanceolate, untoothed. Fls *30 mm, purple,* sometimes very pale, fragrant; May-July. Pods long, stout, hairy. Sea cliffs. B(rare), F, western. **4a* Sea Stock** *M. sinuata* □ is a less bushy, whitish biennial, lfy and not woody at the base; root lvs lobed or wavy-edged. Pods stickily hairy. Sand dunes.

5* HORSE-RADISH *Armoracia rusticana*. Tall hairless perennial with *large* (to 50 cm), very long-stalked, wavy-edged, toothed root lvs and *narrow,* short-stalked stem lvs. Fls 8-9 mm, white, densely clustered on long-stalked lfy spikes; May-Aug. Pods almost spherical. Cultivated, and naturalised in damp places and waysides. (T, southern) from E Europe.

6* DITTANDER *Lepidium latifolium*. Tall greyish hairless perennial resembling 5 but with smaller shorter-stalked lvs, not wavy-edged. Fls 2-3 mm, sepals *white-edged;* July-Aug. Pods rounded, rather flattened. Damp places, usually near the coast. T, southern.

7* COMMON SCURVY-GRASS *Cochlearia officinalis* agg. Very variable, low/medium hairless perennial with *fleshy,* heart-shaped lvs, the upper loosely clasping the stem. Fls white, 8-10 mm; Apr-Aug. Pod *swollen,* almost spherical. Dry saltmarshes and banks, occasionally inland. T. **7a* English Scurvy-grass** *C. anglica* □ has upper lvs clearly clasping the stem, lowest lvs not heart-shaped, tapering to the base, and usually larger fls; Apr-July. Hybridises with 7. Muddy shores. B, G, S. **7b** *C. aestuaria* has pod with blunt or notched tip. F. **7c** *C. fenestrata* has pod 3-4 times as long as broad. S. **7d** *Kernera saxatilis* has lvs broad to narrow lanceolate, often well toothed, not fleshy, the upper sometimes clasping the stem. Mountains, on lime. G, southern, rare.

8* EARLY SCURVY-GRASS *Cochlearia danica*. Very variable, low/short overwintering annual, differing from 7 especially in its oval pods, often lilac fls, and *stalked* upper lvs; Jan-Sept. Usually sandy places by the sea. T. **8a Greenland Scurvy-grass** *C. groenlandica* is perennial with usually lfless stems. Iceland.

4a* Sea Stock

7a* English Scurvy-grass

93

Cabbage Family (contd.)

1* HAIRY BITTERCRESS *Cardamine hirsuta*. Low/short, rather hairy annual, with a compact rosette of pinnate lvs, each with 3-7 pairs of lflets. Fls less than 5 mm, with four stamens, less obvious than *long, erect* pods; Feb-Nov. Bare open ground. T. **1a* Wavy Bittercress** *C. flexuosa* is usually biennial and has wavy stems, a loose rosette of fewer lvs, each with 13-15 lflets, six stamens and pods not overtopping stem; Apr-Sept. Damp woods and rocks. T.

2* NORTHERN ROCKCRESS *Cardaminopsis petraea*. Short, slightly hairy perennial with *lobed* lower lvs and narrow, toothed stem lvs. Fls 5-8 mm, white, sometimes lilac; June-Aug. Pods long, strongly *flattened*. Rocks and river shingle in mountains. B,G,S.

3 TALL ROCKCRESS *Cardaminopsis arenosa*. Annual/perennial, taller than 2 and all lvs more deeply lobed. Fls larger than 2, white in the north, lilac further south; Apr-June. Pods like 2. Sandy soils, usually lime-rich. F,G,S. **3a** *C. halleri* is smaller with runners and rounded lvs; Apr-May; pod stalks longer. Mountains, naturalised by roadsides. F,G, scattered. **3b** *Arabidopsis suecica* has lvs less deeply lobed, fls always white and pods not flattened. Gravel. S.

4* THALE CRESS *Arabidopsis thaliana*. Short/medium annual with roughly hairy, *elliptical* toothed root lvs, forming a rosette; stem lvs narrower, not stalked or clasping the stem. Fls 3 mm, white; Mar-Oct. Pods long, only slightly flattened. Dry sandy soils and walls. T. **4a** *Sisymbrium supinum* is larger, with pinnately cut lvs, fls slightly larger and pods not flattened. F,G,S. **4b** *Braya linearis* is more tufted, with untoothed linear lvs. Arctic gravels on lime. S.

5* HAIRY ROCKCRESS *Arabis hirsuta* agg. Short/medium hairy biennial/perennial with both stem lvs and long pods *tightly pressed* against stem; stem lvs tapering at base; root lvs in a rosette, long oval, only slightly toothed. Fls white, 3-5 mm, numerous; May-Aug. Dry grassland, dunes, rocks, walls, on lime. T. **5a* Bristol Rockcress** *A. stricta* is a low/short perennial, with stem sparsely lfy and root lvs more deeply toothed; fls creamy; Mar-May. B, western, rare.

6* TOWER MUSTARD *Arabis glabra*. Tall, almost hairless, *grey-green* biennial with softly hairy, toothed, short-lived root lvs and arrow-shaped stem lvs *clasping* the stem. Fls 5-6 mm, creamy; May-July. Pods long, numerous, closely pressed to stem. Dry banks, roadsides, open woods. T, eastern. **6a** *A. pauciflora* has hairless root lvs which persist after flowering, and guitar-shaped stem lvs; pods fewer. Dry rocky meadows and woods, on lime. F,G, south-western.

7* ALPINE ROCKCRESS *Arabis alpina*. Low/short creeping hairy perennial, with long *runners*, forming conspicuous grey-green *mats*. Stem lvs deeply toothed, bases bluntly arrow-shaped. Fls 6-10 mm, white, petals spreading; May-July. Mountain rocks. B(rare),S. **7a* Garden Rockcress** *A. caucasica* is greyish-white; stem lvs with fewer shallower teeth, sharply arrow-shaped at base; fls 15 mm; Mar-May. Garden escape, on rocks and by roads, on lime. (T), from S Europe.

8* TOWERCRESS *Arabis turrita*. Medium, softly hairy biennial/perennial with an often reddish stem. Root lvs long-stalked, densely grey-hairy. Fls yellowish-white; Apr-July. Pods very *long,* up to 150 mm, all *twisted* to one side and downwards. Rocks, walls. (B,F,G), southern, from S Europe.

9 ANNUAL ROCKCRESS *Arabis recta*. Short, softly hairy annual with *untoothed* lvs, the oval root ones soon withering, the stem ones clasping with arrow-shaped points. Fl white, less than 5 mm, petals erect; Apr-June. Pods long, 10-35 mm, spreading. Mountain rocks. F,G.

Cabbage Family *(contd.)*

1* WILD CANDYTUFT *Iberis amara.* Short, slightly downy annual with more or less deeply toothed, hairless lvs scattered up the stem. Fls 6-8 mm, white or mauve, petals *unequal*; May-Nov. Pods rounded, less than 5 mm long. Bare ground and arable fields, on lime. B', F, G, southern. **1a** *I. intermedia* is larger with narrower, pointed, untoothed upper lvs and pods more than 5 mm. F, G, southern. **1b** *I. umbellata* □ has purplish fls in a flat-topped head. Garden escape. (B, F, G), from the Mediterranean.

2* SHEPHERD'S PURSE *Capsella bursa-pastoris* agg. Very variable, low/ medium annual/biennial, hairy or hairless. Lvs lanceolate, pinnately lobed, deeply toothed or untoothed, mainly in a basal rosette. Fls 2-3 mm, white; all year. Pod an inverted, notched *triangle.* Extremely common weed. T. **2a*** *C. rubella* □ has reddish petals, as long as sepals, and pods with rounded lobes. (B, rare), F.

3* SHEPHERD'S CRESS *Teesdalia nudicaulis.* Low/short, hairless annual with an almost *lfless* stem, and a rosette of pinately lobed root lvs. Fls 2 mm, petals unequal, not notched; June-Oct. Open sandy places, not on lime. T, mainly southern.

4* HOARY ALISON *Berteroa incana.* Medium hairy annual, grey with hairs. Lvs lanceolate, untoothed. Fls 5-8 mm, white, petals deeply *cleft;* June-Oct. Pods oval. Casual on waste ground. (B, F, S), G, eastern.

5* SWEET ALISON *Lobularia maritima.* Short, densely hairy, greyish annual or short-lived perennial, with *narrow* untoothed lvs. Fls 5-6 mm, fragrant, white, petals not notched; June-Sept. A garden escape, especially near the sea. (T), from S Europe.

6* WALL WHITLOW-GRASS *Draba muralis.* Short hairy annual with sparsely lfy stem. Lvs rounded, toothed, *partly clasping* the stem. Fls 2-3 mm, white, petals not notched; Apr-June. Pods elliptical, 5 mm, spreading. Rocks and walls, mainly on lime. T, scattered. **6a** *D. nemorosa* has narrower lvs, smaller pale yellow fls and longer fr stalks. G, S. **6b* Twisted Whitlow-grass** *D. incana* □ is a taller biennial/perennial with lfier stems, narrower lvs, larger fls, notched petals and longer pods twisted when ripe; May-July. Rocks, dunes, mountains. B, G, S, northern. **6c** *D. daurica* agg. is like 6b, but with creamy fls and straight pods. S. 6d-6f are all low perennials with stems lfless or almost so; June-July. Mountain rocks. **6d* Rock Whitlow-grass** *D. norvegica* agg. □ has lvs not always toothed and petals notched. B, S. **6e** *D. nivalis* is smaller than 6c, with stems always lfless and lvs densely hairy, usually untoothed. S. **6f** *D. fladnizensis* is like 6d, with stem hairless and lvs sparsely hairy. S.

7* COMMON WHITLOW-GRASS *Erophila verna* agg. Very variable, low, slightly hairy annual with a lfless stem. Lvs lanceolate, in a basal rosette. Fls white, petals *cleft to base,* 3-6 mm; Mar-May. Pods elliptical or rounded, on long stalks. Bare, often sandy ground, rocks. T.

8* HUTCHINSIA *Hornungia petraea.* Low hairless annual with a rosette of pinnate lvs and a lfy stem. Fls minute, greenish-white, petals not notched; Mar-May. Pod elliptical' 2 mm. Bare sandy ground, rocks, mainly on lime. T, southern, scattered. **8a** *Hymenolobus procumbens* is taller, with undivided stem lvs and fls slightly larger; pods kidney-shaped, flat, to 5 mm. F, G, scattered.

1b *Iberis umbellata*

2a* *Capsella rubella*

6b* Twisted Whitlow-grass

6d* Rock Whitlow-grass

1

2

3

4

5

6

7

8

Cabbage Family (contd.)

1* HOARY CRESS *Cardaria draba.* Medium, usually hairless perennial with *greyish,* pointed, toothed lvs clasping the stem. Fls 5-6 mm, white, stalked, in flat-topped, *umbel-like* clusters; May-June. Pods kidney-shaped, beaked. Roadsides and disturbed ground, forming clumps. T, southern.

2* FIELD PEPPERWORT *Lepidium campestre.* Medium greyish annual/biennial, with many spreading hairs. Lvs lanceolate, the upper rather triangular and clasping the stem. Fls 2-3 mm, white, petals *longer* than sepals, anthers yellow; May-Aug. Pods roundish, flattened, notched, rough, *as long* as their stalks. Dry banks, disturbed ground. T. **2a* Smith's Pepperwort** *L. heterophyllum* is an often shorter, more branched perennial, with longer petals, violet anthers, and smooth pods. Also in grassland. T, south-western. **2b* Poor Man's Pepperwort** *L. virginicum* has downward directed hairs, toothed lvs not clasping stem, and pods shorter than stalks. (T) from N America. **2c*** *L. divaricatum* has toothed lvs not clasping stem lvs; petals equalling or shorter than sepals. (B, G, S) from Africa. **2d*** *L. perfoliatum* has root lvs 2-pinnate, stem lvs rounded and enclosing the stem, pale yellow petals longer than sepals, and round pods. (T) from E Europe. **2e* Tall Pepperwort** *L. graminifolium* is perennial with lvs not clasping, sepals white-edged and pods not notched. (B), F, G.

3* GARDEN CRESS *Lepidium sativum.* Short hairless strong-smelling annual, the cress used for salads. Lvs *linear,* not clasping the stem, the lower pinnate. Fls sometimes reddish, 4-5 mm, petals longer than sepals; June-July. Pod roundish, flattened *longer* than its stalk. Disturbed ground. (T), from W Asia.

4* NARROW-LEAVED PEPPERWORT *Lepidium ruderale.* Short/medium, almost hairless annual/biennial, smelling like 3, with pinnate root lvs and narrow undivided stem lvs. Fls green, usually petalless; May-Sept. Pods oval, flattened, *shorter* than their stalks. Waste places, commonest near coasts. T. **4a*** *L. densiflorum* is not cress-scented and has hairs pointing upwards, stem lvs slightly toothed, fls sometimes with minute petals and pods as long as stalks. (T) from N America. **4b* Least Pepperwort** *L. neglectum* is like 4a with spreading hairs, undivided stem lvs and round pod. **4c** *L. bonariense* is hairy, lvs pinnate. (B, G, S).

5* FIELD PENNYCRESS *Thlaspi arvense.* Short/medium hairless, foetid annual with broad lanceolate toothed lvs clasping the stem, the lower not in a rosette. Fls 4-6 mm, white, anthers yellow; May-Aug. Pods rounded, with *broad,* almost translucent *wings,* the very short style persisting in a deep notch. Cultivated and waste ground. T. **5a* Garlic Pennycress** *T. alliaceum* smells of garlic and has narrower, less winged pods; Apr-June. (B), F, G.

6* ALPINE PENNYCRESS *Thlaspi alpestre.* Very variable short hairless perennial with non-flowering rosettes. All lvs *untoothed,* somewhat clasping the stem. Fls white or mauve, 4-8 mm, anthers violet; Apr-July. Pods heart-shaped, winged, the style usually projecting beyond the notch. Limestone rocks and lead-mine debris, and in mountain woods away from lime. T, scattered, eastern.

7* PERFOLIATE PENNYCRESS *Thlaspi perfoliatum* is low/short and rather greyish, with all lf rosettes flowering. Lvs lanceolate, deeply clasping stem. Fls white, anthers yellow; Apr-May. Pods heart-shaped, winged. Stony and waste ground, on lime. T, southern, rare in B.

8 MOUNTAIN PENNYCRESS *Thlaspi montanum.* Short hairless mat-forming perennial, with *sparsely lfy* stems. Root lvs long-stalked. Fls 10-12 mm, anthers yellow; Apr-June. Pods heart-shaped, flattened, broadly winged, style projecting beyond notch. Rocky grassland, on lime. F, G, southern.

9* SWINECRESS *Coronopus squamatus.* Prostrate annual/biennial. Lvs 1-2-pinnate, lflets narrow. Fls 2-3 mm, white, in small clusters opposite lvs; June-Sept. Pods longer than stalks, rounded, clearly 2-lobed, warty, *not* notched, with prominent style. Characteristic weed of paths and *trodden* ground. T, southern.

10* LESSER SWINECRESS *Coronopus didymus.* More delicate than 9, often *semi-erect,* with a pungent smell and more feathery lvs. Fls 1-2 mm, white or petalless; June-Sept. Pods shorter than stalks, more obviously 2-lobed than 9, and *notched,* with obscure style. Waste places. T, southern.

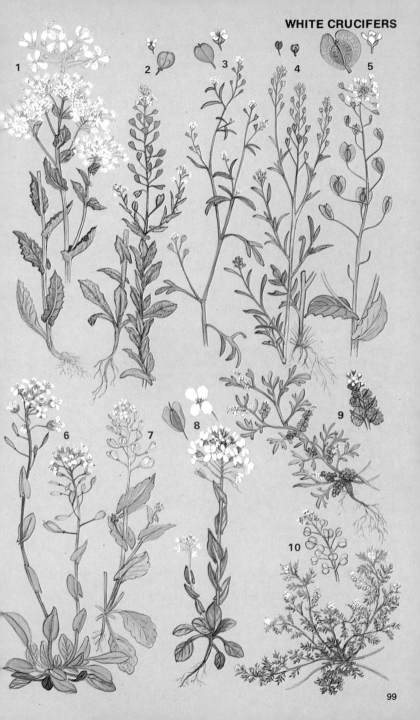

Mignonette Family   Resedaceae

1* WELD *Reseda luteola*. Tall hairless unbranched biennial. Lvs lanceolate, *untoothed,* wavy-edged. Fls 4-5 mm, yellow-green, with *four* sepals and petals; June-Sept. Pods globular, erect. Disturbed ground, often on lime. T, southern. **1a** *Sesamoides canescens* has 6-petalled fls. F, western, rare.

2* WILD MIGNONETTE *Reseda lutea*. Medium branched biennial/perennial, the stem clothed with small *pinnate* lvs. Fls 6 mm, yellow-green, with *six* sepals and petals; June-Sept. Pods oblong, erect. Disturbed ground, often on lime. T, southern. **2a* Corn Mignonette** *R. phyteuma* is a short downy annual with sometimes unlobed lvs, fls whitish and pods drooping. (B, F, G). from S Europe. **2b* White Mignonette** *R. alba* is medium, the white fls with five sepals and petals. (B, F, G), from S Europe.

Sundew Family   Droseraceae

Insectivorous plants, whose lvs are covered with long sticky hairs which curve inwards to trap insects.

3* COMMON SUNDEW *Drosera rotundifolia*. Low/short perennial with a lfless fl-stem more than twice as long as the neat rosette of *round* red lvs with long stalks. Fls 5 mm, white, 5-petalled, in a shortly stalked spike, arising from the rosette; June-Aug. Hybridises with 4a. Wet heaths and moors, sphagnum bogs, never on lime. T.

4* OBLONG-LEAVED SUNDEW *Drosera intermedia*. Differs from 3 in its long *narrow* lvs, tapering to their shorter hairless stalks. Fl-stem arising from *below* rosette and curving upwards. Habitat similar. T, rather scattered. **4a* Great Sundew** *D. anglica* □ has larger longer lvs and fl-stem arising from rosette. Sometimes on lime. Northern.

Stonecrop Family   Crassulaceae   see p. 102

5* NAVELWORT *Umbilicus rupestris*. Low/medium hairless perennial, the *round* fleshy lvs with a *"navel"* in the centre, above the lf-stalk; stem lvs less rounded. Fls bell-shaped, toothed, greenish or pinkish-white; June-Aug. Rocks, banks, walls, not on lime. B, F, western.

6* ROSEROOT *Rhodiola rosea*. Short hairless grey perennial, often purple-tinged. Lvs thick, stiff, *succulent,* more or less oval, often hiding the stem. Fls yellow, 4-petalled, in rather flat-topped heads, stamens and styles on separate plants; anthers purple; May-June. Fr orange. Mountain rocks, sea-cliffs. B, S, northern, and rare in F.

Saxifrage Family   Saxifragaceae   see pp. 104-106

7* OPPOSITE-LEAVED GOLDEN SAXIFRAGE *Chrysosplenium oppositifolium*. Low creeping perennial, slightly hairy; stem square. Lvs roundish, bluntly toothed, opposite; root lvs as long as stalks. Fls 3-4 mm, greenish, *without* petals, sepals 4, stamens 8, anthers bright yellow; March-July. Wet places. T, but western in S. **7a* Alternate-leaved Golden Saxifrage** *C. alternifolium* □ has a triangular stem, alternate, broader, more toothed lvs, root lvs on long stalks, and fls 5-6 mm. Eastern. **7b** *C. tetrandrum* is like 7a, but completely hairless and with 4 stamens. S, northern.

Grass of Parnassus Family   Parnassiaceae

8* GRASS OF PARNASSUS *Parnassia palustris*. Short hairless tufted perennial with a rosette of heart-shaped, untoothed, stalked lvs and a stem with a *single* lf and a *solitary* white fl, 15-30 mm, with five clearly veined petals; June-Sept. Wet, marshy places. T.

1
2
4a
4
3
5
6
7a
7
8

Stonecrop Family   Crassulaceae

See also p. 100. Mostly hairless perennials with untoothed, fleshy, un- or short-stalked lvs. Fls star-like, usually conspicuous, with five petals and sepals. Characteristic of dry rocky places and walls, except 7.

1* ORPINE *Sedum telephium.* Very variable, short/tall sometimes greyish perennial, with clusters of erect, often red-tinged lfy stems. Lvs up to 80 mm long, oblong, *flattened*, slightly toothed, alternate. Fls pinkish-red, whitish, greenish or yellowish, petals spreading, in flattened heads; July-Sept. Shady places and woods. T. **1a* Caucasian Stonecrop** *S. spurium* □ is shorter and creeping, forming mats, with erect, downy flg stems; lvs to 30 mm, more toothed, and larger fls, usually pinkish-red, rarely white, petals erect. Garden escape. (T), from Caucasus.

2* BITING STONECROP *Sedum acre.* Low creeping evergreen with *peppery* taste. Lvs 3-6 mm, cylindrical, not spreading, yellowish, broadest at base. Fls 12 mm, bright yellow; May-July. Dry, bare places, often on lime. T. **2a* Tasteless Stonecrop** *S. sexangulare* is not peppery, has lvs 3-9 mm, broadest at middle, and paler 8-10 mm fls; July-Sept. T, eastern, but (B).

3* WHITE STONECROP *Sedum album.* Bright green, often red-tinged, mat-forming evergreen, creeping with short, erect flg stems. Lvs *6-12 mm,* rarely to 25 mm, alternate, cylindrical, rather flattened on the upper surface. Fls 6-9 mm, *white,* in a dense, much-branched, *flat-topped* head; June-Aug. Rocks, walls. T, southern.

4* REFLEXED STONECROP *Sedum reflexum.* Short creeping evergreen, with short, erect flg shoots and low lfy non-flg ones. Lvs.8-20 mm, *cylindrical,* not persisting when dead. Fls 15 mm, *yellow,* usually with seven petals and sepals; June-Aug. Rocks, walls. T, southern, but (B). **4a* Rock Stonecrop** *S. forsteranum* has narrower flattened lvs, crowded into terminal rosettes on the non-flg shoots; dead lvs persistent; fls 12 mm. Sometimes in damp places. B, F, G, western.

5 ALPINE STONECROP *Sedum alpestre.* Low evergreen, somewhat resembling 2 but not peppery. Lvs rather flattened, parallel-sided. Fls 6-8 mm, dull yellow; June-July. Mountain rocks, not on lime. F, G.

6* ENGLISH STONECROP *Sedum anglicum.* Low mat-forming evergreen, greyish but soon reddening. Lvs *3-5 mm,* cylindrical, alternate. Fls 12 mm, *white* (pink below), the flhead much less branched than in 3; June-Sept. B, F, S, western, rarely on lime. **6a* Thick-leaved Stonecrop** *S. dasyphyllum* is smaller and tufted with opposite, slightly sticky or downy lvs, and fls 5-6 mm. (B), F, G, southern.

7* HAIRY STONECROP *Sedum villosum.* Low, reddish, *downy* biennial. Lvs 6-12 mm, alternate, flattened on the upper surface. Fls 6 mm, *pink,* long-stalked; stamens 10; June-Aug. Wet places and by streams, in mountains. T, scattered. **7a** *S. rubens* is annual with lvs 10-20 mm and unstalked fls; stamens 5; May-July. (F, G), from S Europe. **7b** *S. hirsutum* is very hairy; fls usually white. Dry rocks. F. **7c** *S. andegavense* is a hairless annual with more rounded lvs; fls usually white. F. **7d** *S. hispanicum* has 6 white petals with pink veins. (G).

8 ANNUAL STONECROP *Sedum annuum.* Low hairless annual/biennial. Lvs flattened on both surfaces, 6 mm. Fls *yellow,* short-stalked; June-Aug. G, S, only on mountains in the south.

9* MOSSY STONECROP *Crassula tillaea.* Very low moss-like annual, with tiny crowded broad opposite lvs pressed against stem; red lvs and stem conspicuous in winter. Fls white, 1-2 mm, petals usually three, shorter than sepals; June-Sept. Damp sandy places, wet in winter. B, F, G, western. **9a* Water Tillaea** *C. aquatica* has narrower, less crowded lvs and short-stalked 4-petalled fls. B (rare), G, S. **9b** *C. vaillantii* is like 9a with long-stalked fls. F.

10 HEN-AND-CHICKENS HOUSELEEK *Jovibarba sobolifera.* Succulent perennial with lvs in roundish rosettes, often tipped red. Fls pale yellow, bell-shaped, 6-petalled; July-Aug. Dry, often sandy grassland. G. **10a** *Sempervivum tectorum* has fls flat, the 12 or more petals with smooth edges. (G), scattered.

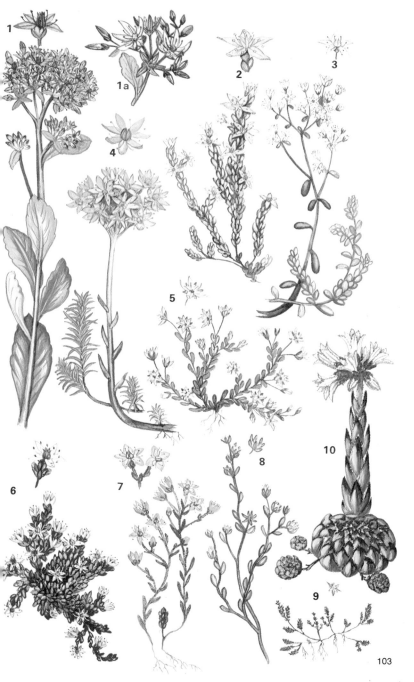

Saxifrage Family   Saxifragaceae

See also p. 100. Mostly rather short perennials. Lvs alternate (except Purple Saxifrage, p. 106). Stamens 10, petals, sepals, styles 5. Fr a capsule in two parts, often joined at the base.

1* STARRY SAXIFRAGE *Saxifraga stellaris*. Low/short, sparsely hairy, with lfless stem and a *rosette* of very short-stalked, *toothed* lvs. Fls white, in a lax head, 10-15 mm, with conspicuous red anthers; sepals turned down; June-Aug. The commonest white-fld mountain saxifrage; in wet places. T. **1a** *S. foliolosa* is smaller and has red bulbils. S.

2* ST PATRICK'S CABBAGE *Saxifraga spathularis*. Loosely tufted, with short, lfless fl-stem. Lvs spoon-shaped, mostly in rosettes, hairless, with pointed teeth and a narrow *translucent* margin; stalks long, flattened. Petals 4-5 mm, white with several *red spots*, sepals turned down; June-July. Rocks, not on lime. Ireland. **2a* Pyrenean Saxifrage** *S. umbrosa* □ is very tufted, the lvs rounded with blunt teeth, on short, hairy stalks. (B, rare.) **2b* Kidney Saxifrage** *S. hirsuta* □ has hairy, kidney-shaped lvs on long cylindrical stalks. **2** × **2a* London Pride** *S.* × *urbium* □ (B, F), **2** × **2b*** *S.* × *polita* (Ireland), and **2a** × **2b*** *S.* × *geum* (B, F, G) are all intermediate between the parents. **2c* Round-leaved Saxifrage** *S. rotundifolia* has larger, rounder lvs and yellow spots on the fls. (B, F), G.

3* MEADOW SAXIFRAGE *Saxifraga granulata*. Short/medium, downy, with *bulbils* at lf-bases, but no barren shoots, Lvs kidney-shaped, shallowly lobed, mainly basal. Fls 20-30 mm, white, not spotted, in loose clusters; Apr-June. Grassy places, lowland. T.

4 LIVELONG SAXIFRAGE *Saxifraga paniculata*. Short, tufted with long, oblong, finely-toothed lvs, lime-encrusted, in hemispherical rosettes. Fl-stem long, mostly lfless, with a *loose panicle* of white fls; June-Aug. Mountain rocks. F, G. **4a** *S. cotyledon* is larger with broader, pointed lvs. Fl-stem much denser, with flg branches from the base. G, S.

5* FRINGE CUPS *Tellima grandiflora*. Stout, medium, hairy. Lvs heart-shaped. lobed, dark *green*. Fls greenish, *reddening*, in *nodding* stalked spikes, the petals jaggedly *toothed*; Apr-Aug. Damp shady places. (B) from N America.

6* PICK-A-BACK PLANT *Tolmiea menziesii*. Creeping, medium, hairy. Lvs *yellowish-green*, heart-shaped, lobed, with buds at their base in autumn. Fls green with *purple-veined* sepals, in stalked spikes, the petals very *narrow*; Apr-Aug. Damp shady places. (B) from N America.

Saxifrage Family (contd.)

1* ARCTIC SAXIFRAGE Saxifraga nivalis. Low, with basal rosette of small, thick lvs, 15-30 mm, rounded, coarsely toothed, purple beneath, and sticky at the edges. Fl-stalks lfless, very hairy. Fls whitish, unstalked, in a crowded head, petals unspotted; sepals spreading; July-Aug. B, S, mountains. **1a** S. tenuis is slenderer and less hairy, with its fls stalked, in a looser head. S, northern.

2* RUE-LEAVED SAXIFRAGE Saxifraga tridactylites. Low, erect, stickily hairy annual, often reddish. Upper lvs spoon-shaped, 3-5-lobed, the lower untoothed; all lvs on stem. Fls c. 5 mm, white, with notched petals; Apr-May. Walls, bare places, lowland, mainly on lime. T, southern. **2a** S. adscendens is larger and not stickily hairy, with all lvs toothed, the lower in a rosette; June-July. Mountains, S. **2b** S. osloensis is intermediate between 2 and 2a. S, rare.

3* YELLOW SAXIFRAGE Saxifraga aizoides. Low/short, very lfy, slightly hairy. Lvs 10-20 mm, narrow, unstalked, fleshy. Fls in loose heads, petals yellow, well spaced; June-Sept. Streams and wet places in mountains. B, G, S.

4* MARSH SAXIFRAGE Saxifraga hirculus. Low, downy, loosely tufted or mat-forming. Lvs narrower than 3, with long red-brown hairs. Fls bright yellow, often red-spotted, much larger than 3, sometimes solitary; sepals turned down in fr; Aug, rather shy-flowering. Marshy and boggy places on moors. T, scattered; rare in B.

5* HIGHLAND SAXIFRAGE Saxifraga rivularis. Very low, hairless, with runners from bulbils at stem base; stem lfy. Lvs stalked, with 3-5 blunt lobes. Fls few, 6-10 mm, dowdy, white; July-Aug. Wet mountain rocks. B (rare), S.

6* DROOPING SAXIFRAGE Saxifraga cernua. Low/short, slender, almost hairless; lvs with 3-7 blunt lobes. Fls surprisingly large, white, solitary, very often replaced by red bulbils at base of lvs. Mountain rocks. B (rare), S.

7* MOSSY SAXIFRAGE Saxifraga hypnoides. Low/short, hairy, with many long, lfy, barren shoots forming a mat. Lvs long-hairy, narrow, usually 3-lobed, clustered at the end of each stem. Fls 10-15 mm, white, buds nodding; bulbils often present; May-July. Wet, often grassy places in hills. B, F, S, western. **7a* Irish Saxifrage** S. rosacea agg. may have habit of 7 or 8, otherwise as 7 with erect fl-buds, broader lf-lobes and no bulbils. B (Ireland, rare), F, G.

8* TUFTED SAXIFRAGE Saxifraga cespitosa. Low, tufted, hairy. Lvs on rosettes, like 5 but more deeply lobed. Fls whitish, often petalless, smaller than 7. Rocks. B (rare), S.

9* PURPLE SAXIFRAGE Saxifraga oppositifolia. Creeping, with trailing stems covered with small, unstalked, opposite lvs, often lime-encrusted. Fls 10-20 mm, purple, almost unstalked. Apr-May, July. Damp places in mountains, often on lime. B, S.

10 HAWKWEED SAXIFRAGE Saxifraga hieracifolia. Low/short, with a rosette of hairy oval lvs. Fl stem lfless with a spike of small reddish-green fls. Mountains, S.

Rose Family Rosaceae

Lvs alternate, usually pinnate, with stipules at base of lf-stalk. Fls usually open with five conspicuous petals, except for those on this page; often with an epicalyx – an outer ring of sepals below the true sepals.

1* MEADOWSWEET *Filipendula ulmaria*. Tall hairless perennial with lfy stems. Lvs long-stalked, pinnate, with 2-5 pairs of toothed lflets, each more than 20 mm, and small lflets in between, *stipules* green above, downy and pale beneath. Fls in *dense clusters*, creamy, fragrant, with 5 or 6 petals, 2-5 mm; June-Sept. Marshes, fens, swamps, wet meadows and woods. T. **1a* Dropwort** *F. vulgaris* □ is short/medium, with sparsely lfy stem, the shorter lvs with up to 20 pairs of lflets, each less than 20 mm. Fl-head rather flattened, less dense, petals 5-9 mm. Dry grassland, on lime. T, southern.

2 GOATSBEARD SPIRAEA *Aruncus dioicus*. Tall perennial with very large lvs, like 1 but *without* stipules, the flheads with many long, *finger-like* branches, held more or less horizontal; June-Aug. Mountain woods. F, G.

3* AGRIMONY *Agrimonia eupatoria*. Medium downy perennial with pinnate lvs, not at all sticky. Fls 5-8 mm, *yellow*, in a spike; June-Aug. Fr grooved, covered with small erect *hooks* at its apex. Dry grassy places, often on lime. T. **3a* Fragrant Agrimony** *A. procera* is aromatic and altogether larger with lvs stickily hairy beneath; fr hardly grooved, some of the hooks bent back. Rarely on lime. Southern. **3b* Bastard Agrimony** *Aremonia agrimonioides* is shorter with lfy clusters of fls, 7-10 mm; May-June. Mountain woods, waysides. (B), G.

4* GREAT BURNET *Sanguisorba officinalis*. Medium/tall hairless perennial. Lvs pinnate, with 3-7 pairs of lflets, 20-40 mm long. Fls tiny, on dense *oblong* heads, 10-20 mm long, stamens and styles in the same fl; petals absent, sepals deep red; June-Sept. Damp grassland. T, southern.

5* SALAD BURNET *Sanguisorba minor*. Short, rather greyish, almost hairless perennial, similar to 4, but with 4-12 pairs of lflets all less than 20 mm. Fls in *round* heads, the *upper with red styles*, the lower with yellow stamens; sepals green, no petals; May-Sept. Dry, grassy places, mainly on lime. T, southern. **5a* Pirri-pirri Bur** *Acaena anserinifolia* □ is a creeping under-shrub, with fls whitish, stamens and styles in the same fl, and fr softly spiny. Not on lime. (B), from Australia.

6* LADY'S MANTLE *Alchemilla vulgaris* agg. A variable collection of similar microspecies: low/medium, usually rather densely hairy, but sometimes hairless perennials. Lvs palmately lobed, *not* cut to the base, *green* on both sides. Fls 3-5 mm, in loose clusters with an epicalyx but no petals, sepals green, anthers yellow; May-Sept. Grassy places. T.

7* ALPINE LADY'S MANTLE *Alchemilla alpina*. Smaller and hairier than 6, with lvs divided *to the base* into 5-7 narrow lobes, *silvery grey* with hairs beneath. Fls pale green, 3 mm; June-Aug. Mountain grassland. T. **7a*** *A. conjuncta* has lvs divided to about ⅔ into 7-9 broader segments. River shingle. T, but (B, northern, rare), from the Alps.

8* PARSLEY PIERT *Aphanes arvensis* agg. More or less prostrate hairy, pale green annual. Lvs 3-lobed, toothed. Fls minute, green, in clusters *opposite* the lvs, surrounded by the toothed lf-like stipules; Apr-Oct. Bare ground. T, southern.

1a* Dropwort

5a* Pirri-pirri Bur

109

Rose Family *(contd.)*

ROSES *Rosa* are shrubs with thin stems armed with thorns. Lvs pinnate, with stipules. Fls large, white, pink or red, with large, floppy petals, easily detached. Fr a berry-like hip.

1* DOG ROSE *Rosa canina* agg. Variable; stems arching, to 3 m. Thorns *curved.* Lvs with 2-3 pairs of toothed lflets, *hairless* or sometimes downy beneath. Fls 45-50 mm, pink or white, styles not joined; June-July. Sepals with narrow lobes, falling before the hips turn red. Hedges, scrub. T, commoner in the south. **1a*** *R. stylosa* has styles joined in a column. B, F, G, western.

2* FIELD ROSE *Rosa arvensis.* Scrambling, forming bushes not more than 1 m tall, unless on other shrubs; stems green, thorns curved. Lflets hairless, 2-3 pairs on each lf. Fls 30-50 mm, *always* white, the styles *joined* into a column; sepals scarcely lobed, soon falling; July-August. Hips red. Woods, hedges, scrub. B, F, G, southern.

3* BURNET ROSE *Rosa pimpinellifolia.* Suckering stems with *straight* thorns and stiff bristles, forming short bushes to 50 cm. Lvs with 3-5 pairs of small, rounded lflets. Fls 20-40 mm, cream, sometimes pink; sepals not lobed; May-July. Ripe hips dark purple or *black.* Dry, open places, often on coastal dunes. T, rare in S. **3a Provence Rose** *R. gallica* ☐ has larger, more leathery, bluish-green lflets, larger deep pink fls and hips red. F, G.

4* DOWNY ROSE *Rosa tomentosa* agg. To 2 m; resembling 1 but lflets *densely downy;* thorns may be straight. Fls 30-40 mm, deep pink, sometimes pale; June-July. Sepals pinnate, sometimes persisting till red hips fall or decay. Hedges, scrub, especially in the hills. T, northern. **4a* Sweet Briar** *R. rubiginosa* agg. has lflets densely coated with sweet-smelling brown sticky hairs. Scrub, usually on lime. T, southern.

5* BRAMBLE *Rubus fruticosus* agg. Very variable scrambling shrub armed with easily detached thorns; over 2000 microspecies have been described. Rather woody, with long biennial, usually angled stems, rooting where they touch the ground. Lvs prickly, with 3-5 lflets. Fls 20-30 mm, white or pink; May-Nov. Fr (blackberry) with several fleshy segments, red at first, *purplish-black* when ripe. Woods, scrub, open and waste ground. T. **5a* Dewberry** *R. caesius* has weak, round, less prickly stems; lvs always with three lflets; fls always white; fr bluish, with a waxy bloom, fleshier, with fewer and larger segments. Also in damp places, preferring lime. T, southern.

6* RASPBERRY *Rubus idaeus.* Unbranched *erect* perennial with woody, biennial stems, armed with weak prickles. Lvs with 3-7 lflets. Fls smaller than in 5, the petals *erect,* well spaced, *always* white, rather inconspicuous; May-Aug. Fr *red* when ripe. Shady places. T, and widely cultivated.

7* STONE BRAMBLE *Rubus saxatilis.* Resembles 5 but smaller, slenderer, prostrate, and spineless; stems annual. Lvs with three lflets. Fls smaller than 5 or 6, 8-10 mm, the narrow petals *no longer than* the sepals; June-Aug. Fr red when ripe, the segments fewer and larger. Shady, rocky places, often on limestone. T, north-eastern.

8* CLOUDBERRY *Rubus chamaemorus.* Low/short creeping downy perennial with *rounded,* palmately lobed lvs. Fls always white, conspicuous, solitary, stamens and styles on separate plants; June-Aug, but a shy flowerer in B. Fr *orange* when ripe. Upland bogs and damp moors. B, G, S, northern.

9 ARCTIC BRAMBLE *Rubus arcticus.* Short spineless creeping perennial. Lvs with three lflets, smaller than 7. All stems flg; fls 15-25 mm, *bright red,* petals sometimes toothed; June-July. Moors. S, probably extinct in B.

2a Provence Rose

Rose Family (contd.)

1* WILD STRAWBERRY *Fragaria vesca.* Low/short perennial, with long *runners,* rooting at intervals. Lvs trefoil, bright green, with pointed teeth, the lflets paler beneath and hairy with *flattened* silky hairs, less than 80 mm long; hairs on stalks *not* spreading. Fls 12-18 mm, white, on a short stalk no longer than the lvs; Apr-July. Fr the familiar strawberry, but much smaller, covered all over with seeds, the sepals bent back from it. Dry, grassy places and woods, often on lime in the north. T. **1a* Hautbois Strawberry** *F. moschata* □ is larger, usually without runners; lflets more than 60 mm; fl-stalks longer than lvs; fls 15-25 mm; fr without seeds at base. T, but (B), eastern. **1b** *F. viridis* is smaller with short runners, downy or hairless lflets and sepals pressed against fr; fr not easily detached from stalk. F, G, S. **1c* Garden Strawberry** *F.* × *ananassa* □ has much larger fls and fr. (T), a frequent escape.

2* BARREN STRAWBERRY *Potentilla sterilis.* Low hairy perennial, usually with runners, and with *spreading* hairs on stem and lvs. Lvs smaller than 1, blunt-toothed, and bluish; lflets 5-25 mm, toothed. Fls 10-15 mm, white, with gaps between the slightly notched petals; Feb-May. Fr dry, *not* strawberry-like. Dry grassland and open woods. T, southern. **2a White Cinquefoil** *P. alba* has lvs with five very narrow, scarcely toothed lobes, green above and silvery below. Fls 15-20 mm; Apr-May. F, G, south-eastern. **2b Pink Barren Strawberry** *P. micrantha* is shorter, with no runners and often pinkish fls, 6-8 mm. G. **2c** *P. montana* has lflets untoothed in lower part, green above and grey beneath, and larger fls; May-June. Woods. F.

3* ROCK CINQUEFOIL *Potentilla rupestris.* Short/medium hairy perennial without runners. Root lvs *pinnate* with rather rounded lflets, and simpler, usually trefoil lvs up the stem. Fls 15-25 mm, *white;* May-June. Rocky slopes. T, scattered, rare in B.

4* MARSH CINQUEFOIL *Potentilla palustris.* Short/medium hairless perennial with pinnate lvs. Fls *star-shaped,* erect; petals narrow, deep purple, shorter than the broader spreading, *maroon* sepals, 10-15 mm long; May-July. Lime-free wet places. T.

5* WATER AVENS *Geum rivale.* Medium downy perennial with pinnate root lvs, becoming trefoil up the stem, and small stipules, 5 mm, at base of stem lvs. Fls nodding, *bell-shaped,* petals dull pink, 8-15 mm; sepals purple; Apr-Sept. Fr with persistent hairy style, hooked. Damp, often shady places. T. **5** × **6*** *G.* × *intermedium* is found where the parents grow together.

6* HERB BENNET *Geum urbanum.* Medium downy perennial. Lvs pinnate, the smaller with lf-like stipules up stems. Fls 8-15 mm, *yellow,* petals erect, sepals green; style as in 5; May-Sept. Woods and shady places on fertile soils. T. **6a Alpine Avens** *G. montanum* is low/short with fls 25-40 mm, solitary, 6-petalled, on lfless stems; styles in fr long, feathery, not hooked. Rocky mountain pastures. G. **6b** *G. hispidum* is hairier with smaller fls and stipules. S.

7* MOUNTAIN AVENS *Dryas octopetala.* Low creeping downy perennial with long stems. Lvs *oblong,* blunt-toothed, 5-20 mm, green above, *grey* beneath. Fls 20-40 mm, white, with *8 or more* petals, and many conspicuous yellow stamens; May-July. Fr with long, feathery styles. Mountain rocks and to sea-level in the north, on lime. B, S.

1* Wild Strawberry

1a* Hautbois Strawberry

1c* Garden Strawberry

2* Barren Strawberry

Rose Family *(contd.)*

CINQUEFOILS *Potentilla* (pp. 112-14) are usually creeping perennials with compound lvs, conspicuous yellow (p. 114) or white (p. 112) fls, five petals (except 2), numerous stamens, and an epicalyx.

1* SHRUBBY CINQUEFOIL *Potentilla fruticosa.* Small, branched, downy *shrub* up to 1 m high. Lvs deciduous, pinnate; lflets untoothed, greyish, hairy, narrow, usually five. Fls *c.* 20 mm, in loose clusters; May-July. Wet hollows, river banks, upland rocks. T, scattered, rare in B.

2* TORMENTIL *Potentilla erecta.* Creeping and patch-forming, with short, downy flg stems, not rooting at lf-junctions. Root lvs in rosette, trefoil, soon withering; stem lvs unstalked, appearing to have five lflets from conspicuous toothed lf-like stipules. Fls 7-11 mm, with *four* petals and sepals, numerous in loose heads; May-Sept. Moors and grassy places, not on lime. T. **2a* Trailing Tormentil** *P. anglica* □ has stems longer, rooting at lf-junctions, some lvs with five lflets, rosette persistent, and short-stalked stem lvs with untoothed stipules; fls 14-18 mm, solitary, petals 4-5. Often in woods. T, southern. **2b* Creeping Cinquefoil** *P. reptans* □ has stems to 1 m, rooting at lf-junctions, all lvs with 5-7 lflets, stem lvs stalked with untoothed stipules; fls 17-25 mm, solitary, petals 5. Waste places, bare ground, hedges, also on lime. **2, 2a** and **2b** all frequently hybridise.

3* SPRING CINQUEFOIL *Potentilla tabernaemontani.* Low/short, rather hairy, mat-forming, with creeping, *rooting* branches. Root lvs with 5-7 lflets, each 10-40 mm, and narrow stipules; upper lvs trefoil, unstalked. Fls 10-20 mm; epicalyx segments blunt, lanceolate; Apr-June. Dry, often rocky limestone grassland. T, southern. **3a* Alpine Cinquefoil** *P. crantzii* is not mat-forming, rarely rooting; fls 10-25 mm, often orange-spotted; epicalyx segments oblong; June-July. Mountain rocks. T. **3b** *P. heptaphylla* is taller and more erect, not mat-forming; lflets 5-7; fls 10-15 mm, numerous. G, S. **3c** *P. thuringiaca* is medium, erect, very hairy, with 5-9 lflets each up to 60 mm or more; epicalyx segments pointed. G, S. **3d** *P. pusilla* has star-shaped hairs and smaller fls. G. **3e** *P. aurea* is like 3a, but has appressed silky hairs and end tooth of lflets always smaller than neighbours. Not on lime. G, southern. **3f** *P. hyparctica* is like 3a with unspotted petals. S, Arctic.

4* HOARY CINQUEFOIL *Potentilla argentea* agg. Short/medium, with woolly, usually erect flg stems. Lvs with five lflets, dark green above, silvery *white* with dense hairs beneath; lflets pinnately lobed, except at base, margins inrolled. Fls 10-15 mm, style rounded at tip; June-Sept. Dry grassy places. T, southern. **4a** *P. collina* has 5-7 lflets, white or grey-green beneath; fls 8-10 mm, style club-shaped. F, G, S. **4b** *P. inclinata* has lvs grey beneath. F, G. **4c** *P. nivea* agg. □ usually has trefoil lvs. S.

5* SULPHUR CINQUEFOIL *Potentilla recta.* Short/medium, hairy, *stiffly* erect, lower lvs with 5-7 lflets to 100 mm. Fls 20-25 mm, petals longer than sepals, pale yellow, in clusters; June-Sept. Dry grassy and waste places. (B, S), F, G, southern. **5a* Norwegian Cinquefoil** *P. norvegica* is shorter with trefoil lvs; fls 10-15 mm, brighter yellow, petals no longer than sepals, which enlarge in fr. (B, F), G, S. **5b*** *P. intermedia* is like 5a with 5 lflets. (T, eastern).

6 GREY CINQUEFOIL *Potentilla cinerea.* Mat-forming with short, woody stems, as 4 but all creeping, *densely* covered with grey hairs. Lvs with 3-5 lflets, grey-green above, green beneath; stipules very narrow. Fls 10-15 mm, few; Apr-May. Dry grassy and rocky places, mainly upland. F, G, S, south-eastern.

7* SILVERWEED *Potentilla anserina.* Creeping with long runners, rooting at lf-junctions. Lvs *pinnate*, often silvery, at least beneath, lflets toothed. Fls 15-20 mm, *solitary*, petals twice as long as sepals; May-Aug. Damp grassy places. T. **7a** *P. supina* does not root at lf-junctions; lvs green; petals shorter than sepals. Dry grassy places. F, G, eastern. **7b** *P. multifida* does not creep; lvs 2-pinnate with narrow segments, silvery beneath only; fls smaller, in loose clusters. S, northern.

8* SIBBALDIA *Sibbaldia procumbens.* Very low, tufted and covered with stiff hairs. Lvs trefoil, lflets 3-toothed at tip. Fls 5 *mm,* few; petals shorter than sepals, sometimes absent; July-Aug. Mountain grassland. T.

Rose Family *(contd.)*

1* CRAB APPLE *Malus sylvestris*. Small tree to 10 m, rather spiny. Lvs toothed, pointed oval, hairless when mature. Fls 30-40 mm, white or pink, in loose clusters; May. Fr the familiar green apple, sometimes turning red. Woods and hedges. T. **1a* Cultivated Apple** *M. domestica* ☐ is unarmed; mature lvs downy. (T).

2 WILD PEAR *Pyrus pyraster*. Deciduous tree to 20 m, usually *spiny*; twigs grey to brown. Lvs roundish to elliptical, slightly toothed, hairless when mature. Fls white, 5-petalled, in clusters; Apr. Fr roundish to pear-shaped, not fleshy, yellow. brown or black, surmounted by dead sepals. Woods, scrub. ?B,F,G. **2a* Cultivated Pear** *P. communis* may not be spiny and has twigs red-brown and fr fleshy and sweet-tasting. Widely cultivated, sometimes naturalised in hedges. (T). **2b** *P. salvifolia* is smaller, with narrower untoothed lvs. grey-downy beneath. F, G. **2c* Plymouth Pear** *P. cordata* may be a shrub to 8 m, and has purplish twigs, oval lvs, young lvs and sepals thickly downy, smaller fls, often pink on the back, and rounder fr with no dead sepals. B (rare), F.

3* WHITEBEAM *Sorbus aria* agg. Very variable tree, to 25 m. Lvs to 120 mm, more or less pointed oval, widest below the middle, *silvery white* with short hairs beneath, the teeth pointing to the tip. Fls 10-15 mm, dull white, in dense, rather flat heads; May-June. Fr 8-15 mm long, scarlet. Woods and scrub on lime. B, F, G. **3a* Swedish Whitebeam** *S. intermedia* ☐ has lvs with much more deeply cut lobes. (B), F, G, and planted T. **3b* Broad-leaved Whitebeam** *S. latifolia* agg. ☐ has lvs more deeply and sharply toothed than 3, but shallower and more tri-angular than 3a; fls 20 mm and fr orange or brown. Woods, limestone rocks, sometimes bird-sown from gardens. Several hybrids occur between 3, 3a, 3b and 4.

4* ROWAN or MOUNTAIN ASH *Sorbus aucuparia*. A smallish tree, to 15 m, occasionally taller. Lvs *pinnate* with 5-7 pairs of lflets, green on both sides. Fls clear white, 8-10 mm, in an umbel-like head; May-June. Fr orange-scarlet, 8-15 mm long. Woods, heaths, moors, rocky places, often upland; *rarely on lime*.

5* WILD SERVICE TREE *Sorbus torminalis*. Deciduous tree to 25 m. Lvs much more *deeply cut* than 3a, to 90 mm, green on *both* sides. Fls white, 10-15 mm, in loose clusters. Fr 12-18 mm long, brown. Woods, especially on neutral or slightly basic soils. B, F, G.

6* HAWTHORN *Crataegus monogyna*. Hairless *thorny* deciduous shrub or small tree, 2-10 m. Lvs deeply *3-5 lobed,* more than half-way to midrib. Fls 8-15 mm, white, in broad dense flattened clusters; one style; May-June. Fr a *crimson* berry (haw). Hedges, scrub. T. **6a* Midland Hawthorn** *C. laevigata* ☐ has lvs lobed less than half-way to midrib; styles 2. More tolerant of shade. T, southern. Hybrid 6 × 6a is abundant and variable. **6b** *C. calycina* has minutely toothed lf-lobes, narrower sepals and brighter haws. F, G (eastern), S.

7 AMELANCHIER *Amelanchier ovalis*. Deciduous tree to 12m, with toothed hairless oval lvs, 30-70 mm. Fls white in short erect spikes, petals 10-13 mm, *narrow*; Apr. Fr a small purple-black berry. Woods, scrub. F, G, southern. **7a* Juneberry** *A. grandiflora* has young lvs purplish and fl spikes drooping; some-times bird-sown. **7b** *A. spicata*, also occasionally bird-sown, has lvs thickly white-downy when young.

1

1a

2

3

3a

3b

4

5

6

6a

6

7

Rose Family (contd.)

1* BLACKTHORN Prunus spinosa. Deciduous *thorny* shrub, 1-4 m, suckering to form *dense thickets*; twigs blackish, usually downy when young. Lvs oval, toothed, 20-40 mm, *matt*. Fls white, solitary, in short open spikes, usually appearing *before* the lvs; late Mar-May. Fr (sloe) globular, *bluish-black*, 10-15 mm, very sour. Scrub, hedges, woodland. T. **1a* Cherry-Plum** or **Myrobalan** P. cerasifera may be a tree, is almost hairless and spineless, and has glossy green twigs and lvs (40-70 mm), larger fls appearing with the lvs in Feb-Mar, and larger red or yellow fr. Widely planted, especially in hedges. (B, F, G) from S E Europe.

2* WILD PLUM Prunus domestica. Small deciduous tree to 10 m; often suckering, *thornless*, the twigs *grey-brown*, downy or not when young. Lvs oval, toothed, 30-80 mm, usually down beneath. Fls greenish-white, 15-25 mm, in small clusters, *appearing with the lvs*; Apr-May. Fr *egg-shaped*, 20-75 mm, blue-black, purple or red. Woods, hedges. (T) from Caucasus. **2a Bullace** spp. insititia produces damsons and greengages, so fr more globular, purple, red, yellow or green. May be a shrub and have spines, and has twigs densely downy and smaller, purer white fls.

3* WILD CHERRY or GEAN Prunus avium. Deciduous tree to 25 m, with peeling red-brown bark. Lvs pointed oval, 60-150 mm, toothed, *matt* above, somewhat downy beneath, with two conspicuous *red glands* at the top of the stalk. Fls white, 15-30 mm, 2-6 together in loose clusters; Apr-May. Fr yellow, then bright red, finally dark red, often sharply acid. Woods, hedges. T. **3a* Dwarf Cherry** P. cerasus is usually a shrub, or a tree to 10 m, with lvs dark green and glossy above and often narrower and with no red glands. Often near houses. (T) from S W Asia. **3b Ground Cherry** P. fruticosa is a low hairless shrub, to 1.5 m, with lvs like 3a and much smaller fls (10-15 mm). Scrub. G. **3c St Lucie's Cherry** P. mahaleb □, a shrub or tree to 10 m, has lvs with conspicuous marginal glands, small fragrant white fls and black fr. F, G, S.

4* BIRD CHERRY Prunus padus. Deciduous tree or shrub to 15 m with lvs 50-100 mm, almost hairless, its long crowded *drooping* fl-spikes quite unlike 3; fls 10-15 mm; May. Fr *black*. Woods, hedges, mainly upland. T. **4a* Rum Cherry** P. serotina is always a tree, to 25 m, with aromatic bark, shiny lvs to 120 mm, fls 5-10 mm and fr rather purplish. (T) from N America. **4b Choke Cherry** P. virginiana is like 4a, but usually a shrub, with bark not aromatic, matt lvs and fr dark red.

5* CHERRY LAUREL Prunus laurocerasus. Usually a tall shrub, with twigs *green*. Lvs *large*, 200 mm × 60 mm, rather *leathery*. Fls white, c. 10 mm, in short spikes, *shorter* than the lvs. Fr purple-black. Woods, scrub; widely planted and so often bird-sown near houses. (B, F) from S E Europe. **5a* Portugal Laurel** P. lusitanica may be a tree and has twigs dark red, thinner, more pointed, darker green lvs, and fl spikes longer than lvs. (B, F) from S W Europe.

6* HIMALAYAN COTONEASTER Cotoneaster simonsii. Erect shrub to 4 m, deciduous or *half-evergreen*; twigs *downy*. Lvs roundish, untoothed, pointed, 10-30 mm, dark green above and downy when young, *sparsely hairy* beneath. Fls *white* with red markings, 2-4 together. Fr orange-red. Scrub, often on limy soils. (B, F) from the Himalayas. **6a* Wild Cotoneaster** C. integerrimus □ is always deciduous and lower-growing to 1 m, with all but the youngest twigs hairless, lvs larger (20-50 mm) and densely grey-downy beneath, and fls pink. Limestone rocks, stony ground, mainly upland. T, eastern; rare in B. **6b** C. niger is like 6a, but taller, with lvs white-downy beneath and fr black. S. **6c** C. nebrodensis is also like 6a, but taller, with twigs and lvs much downier and fls in clusters of 3-12. F, G, rare.

7* SMALL-LEAVED COTONEASTER Cotoneaster microphyllus. Prostrate evergreen shrub with stiff, spreading branches; twigs downy. Lvs *very small*, 5-8 mm, dark glossy green above and grey-downy beneath. Fls solitary, *white*. Fr red. Bird-sown on *limestone rocks*. (B) from Himalayas. **7a* Wall Cotoneaster** C. horizontalis is deciduous, with stiffer and straighter branches on a horizontal plane, twigs brown-woolly, lvs almost hairless below and fls pink and sometimes in pairs. Also bird-sown. (B) from China. For more Cotoneasters, see p. 304.

Key to Peaflowers

(Pages 122-35)

The members of the Peaflower Family, Leguminosae (pp. 122-35), have the most distinctive flower-shape in the whole British flora. Some of the smaller species, however, such as the clovers, have to be fairly closely examined for this distinctive shape to become apparent. At first sight a clover appears to have a single globular flower, but in fact it is a tight head of small peaflowers. The family includes several shrubs and undershrubs.

Trees, shrubs or undershrubs

All on p. 124; Rest-harrow, Large Yellow Rest-harrow 130

Stems

ridged or angled: Broom, False Acacia 124; Bithynian Vetch 126; Spring, Sea and Tuberous Peas, Vetchlings 128

winged: Winged Broom 124; Everlasting and Marsh Peas, Hairy and Bitter Vetchlings 128

spiny: Gorse, Petty Whin, False Acacia 124; Rest-harrow 130

Leaves

grasslike: Grass Vetchling 128

palmate: Lupins 124

trefoil: Broom, Laburnum 124; pp. 130, 132 and 134

pinnate: Bladder Senna, False Acacia 124, all on pp. 122 and 126; Spring and Sea Peas, Meadow and Bitter Vetchling 128; Sainfoin, Kidney Vetch 130; Horseshoe Vetch, Birdsfoot 132

lanceolate: Broom, Dyer's Greenweed, Petty Whin 124

1 pair: Everlasting and Tuberous Peas, Meadow and Hairy Vetchlings 128

as spines: Gorse 124

as tendrils: Yellow Vetchling 128

with dark spot: Spotted Medick 132

Tendrils present

p. 126; Everlasting, Marsh and Tuberous Peas; Hairy, Meadow and Yellow Vetchlings 128

Flowers

yellow: Wild Lentil, Yellow Alpine Milk-vetch, Yellow Milk-vetch 122; all on p. 124; *Vicia pisiformis*, Bush and Yellow Vetches 126; Meadow and Yellow Vetchlings 128; Large Yellow Rest-harrow, Kidney Vetch, Classical Fenugreek, Lucerne 130; all on p. 132; Fenugreek, Mountain Clover 134

white/greenish: Goat's Rue, Wild Liquorice, Yellow Milk-vetch 122; Tree and Garden Lupins, False Acacia 124; Wood Vetch 126; Meadow and Hairy Vetchlings, Sea and Tuberous Peas 128; Red, White, Crimson, Haresfoot and Suffocated Clovers 134

pink/purple: Purple Milk-vetch, Goat's Rue 122; Wild Lupin 124; pp. 126 and 128; Sainfoin, Rest-harrow and Small Rest-harrow, Kidney Vetch 130; Birdsfoot 132; p. 134

blue/violet: Purple Milk-vetch 122; Wild Lupin 124; Tufted and Bush Vetches 126; Spring and Sea Peas, Hairy and Bitter Vetchlings 128; Lucerne 130

parti-coloured: Crown Vetch 122; Wild Lupin 124; Tufted, Fine-leaved, Danzig, Yellow and Bithynian Vetches 126; Large Yellow Rest-harrow, Kidney Vetch, Classical Fenugreek, Lucerne 130

in spikes: Milk-vetches (*Astragalus*) 122; p. 124; pp. 126, 128 and 130

in clusters: Scorpion Senna 124; all on p. 122 except Milk-vetches (*Astragalus*); Kidney Vetch 130; all on pp. 132 and 134

solitary: Hairy Tare, Common, Yellow and Bithynian Vetches 126; Vetchlings 128; Classical Fenugreek 130; Dragon's Teeth, Spotted Medick 132

Pods

Inflated: Bladder Senna 124

angled: Dragon's Teeth 132

hanging: Small Rest-harrow 130

curved: Classical Fenugreek, Lucerne 130; Horseshoe Vetch, Black Medick 132

buried: Burrowing Clover 134

warty: Sainfoin 130

spirally twisted: Lucerne 130; Spotted Medick 132

spreading: Birdsfoot Trefoil, Birdsfoot 132

Pea Family Leguminosae

pp. 122-35. Has highly distinctive 5-petalled flowers, the broad and often erect "standard" at the top, the two narrower "wings" at the sides, and the two lowest joined at the "keel" which conceals the stamens and styles. Fruit a usually elongated pod.

MILK-VETCHES *Astragalus*. Lvs pinnate, usually with a terminal lflet. Fls in spikes or heads at base of lvs, keel *blunt*-tipped; calyx-teeth short.

1* PURPLE MILK-VETCH *Astragalus danicus*. Low/short downy perennial. Fls *violet;* May-July. Pods whitely hairy. Mountain and lime-rich grassland, dunes. T. **1a* Alpine Milk-vetch** *A. alpinus* □ has much paler fls and brown hairs on pods. Mountain rocks and grassland. T, but rare in B. **1b** *A. arenarius* has fewer, narrower lflets and paler fls on shorter stalks. G,S. **1c** *A. norvegicus* □ has fewer lflets and paler violet fls. S.

2 WILD LENTIL *Astragalus cicer*. Medium/tall hairy perennial. Fls pale *yellow,* sepal-tube and pod black-haired; June-July. Grassy places. F,G.

3 YELLOW ALPINE MILK-VETCH *Astragalus frigidus*. Short perennial, almost hairless. Lvs greyish. Fls yellowish-white; teeth of reddish sepal-tube tipped with black hairs; July-Aug. *Mountain* grassland. F,G,S, northern. **3a** *A. penduliflorus* has narrower lflets, yellower fls and sepal-tube green. S.

4* WILD LIQUORICE *Astragalus glycyphyllos*. Medium/tall straggling perennial, slightly hairy; stems *zigzag*. Fls greenish-cream; June-Aug. Grassland, scrub, open woods. T.

5* YELLOW MILK-VETCH *Oxytropis campestris*. Short hairy perennial. Lvs pinnate, silky, with terminal lflet. Fls in rounded heads on lfless stems, creamy white or pale *yellow,* often tinged purple, keel finely *pointed;* June-July. Mountain rocks and grassland, rarely by the sea. B (rare), S. **5a Hairy Milk-vetch** *O. pilosa* is very hairy and has pale yellow fls on stems lfy at base. Grassland, rocks. F, G, eastern.

6* MOUNTAIN MILK-VETCH *Oxytropis halleri*. Like 5 but fls *purple*. B(rare), S. **6a Northern Milk-vetch** *O. lapponica* □ is slenderer and has smaller fls on stems lfy at base. S. **6b** *O. deflexa* is like 6a with whitish fls. S, Arctic, rare.

7* CROWN VETCH *Coronilla varia*. Medium/tall straggling hairless perennial. Lvs pinnate, with a terminal lflet. Fls in rounded heads, *particoloured* (pink, lilac, keel tipped purple); June-Aug. Pod 4-angled. Grassy places. (B,S),F,G.

8* GOAT'S RUE *Galega officinalis*. Medium/tall *erect* hairless perennial. Lvs pinnate, with a terminal lflet. Fls in spikes, white or pinkish-lilac. 5 *bristle-like* calyx-teeth; July-Sept. Pod rounded. Damp and waste ground. (B),F,G.

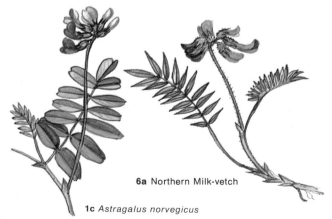

6a Northern Milk-vetch

1c *Astragalus norvegicus*

123

Pea Family *(contd.)*

1* GORSE *Ulex europaeus.* Evergreen shrub to 2½ m, hairy. Lvs as rigid furrowed *spines.* Fls rich golden-yellow, in lfy stalked spikes, almond-scented, wings longer than keel and standard longer still; sepals yellowish, with spreading hairs as long as petals; all year. Heaths, grassland. B,F,G, western. **1a* Dwarf Gorse** *U. minor* is smaller, often prostrate, with shorter, less stout spines, smaller and paler fls, wings as long as keel, and sepals nearly as long as petals; July-Nov. B, F. **1b* Western Gorse** *U. gallii* □ is intermediate between 1 and 1a, with harder yellow fls. July-Nov. B, F.

2* BROOM *Cytisus scoparius.* Tall deciduous *spineless shrub,* sometimes prostrate by the sea; almost hairless; stems ridged. Lvs lanceolate and *trefoil.* Fls yellow in lfy stalked spikes; Apr-June. Heaths, open woods. T. 2a and 2b are shorter, with clover-like lvs. **2a Clustered Broom** *Chamaecytisus supinus* has fls in terminal clusters. On lime. F, G. **2b** *C. ratisbonensis* has one-sided spikes of red-spotted fls. Dry places. G, eastern.

3* DYER'S GREENWEED *Genista tinctoria.* Medium deciduous *spineless* undershrub, sometimes prostrate, slightly hairy. Lvs *lanceolate.* Fls yellow, standard equalling keel, in lfy stalked spikes; June-Aug. Pods hairless. Grassland, heaths, open woods. T. **3a* Hairy Greenweed** *G. pilosa* is half-evergreen and more often prostrate, with small dark green oval lvs, smaller fls and downy pods; May-June. Drier habitats. **3b Winged Broom** *Chamaespartium sagittale* has young stems broadly winged. F, G. **3c Black Broom** *Lembotropis nigricans* has trefoil lvs and fls in longer lfless spikes, wings shorter than keel. Dry places. G, eastern.

4* PETTY WHIN *Genista anglica.* Slender short/medium undershrub, almost hairless, *spiny* (rarely spineless). Lvs small, pointed oval. Fls yellow in lfy stalked spikes, standard shorter than keel; Apr-June. Heaths, moors. T, western. **4a German Greenweed** *G. germanica* has branched spines and smaller hairy flowers with standard much shorter than keel. F, G, S, eastern.

5* LABURNUM *Laburnum anagyroides.* Deciduous *tree* to 7 m; bark smooth. Lvs trefoil. Fls yellow in drooping stalked spikes; May-June Woods and scrub in mountains; widely planted. (B), F, G.

6* BLADDER SENNA *Colutea arborescens.* Deciduous shrub to 3 m, hairy. Lvs *pinnate.* Fls in stalked spikes, deep yellow, sometimes marked red; June-Aug. Pods papery, inflated. Open woods, scrub, also naturalised on waste ground in N and W. (B), F, G.

7* TREE LUPIN *Lupinus arboreus.* Evergreen shrub to 3 m, hairy. Lvs *palmate.* Fls in a stalked spike, *yellow* or white, sometimes tinged mauve, scented. Sandy soils. (B), from California. May-Aug. **7a* Sweet Lupin** *L. luteus* is a hairy annual widely grown for fodder. (B), F, G.

8* WILD LUPIN *Lupinus nootkatensis.* Medium/tall perennial, hairy. Lvs *palmate,* lflets 6-8. Fls in stalked spikes, *blue* and purple; May-July. River shingle, moors. (B,S), from N America. **8a* Garden Lupin** *L. polyphyllus* has larger lvs with 10-17 lflets and longer spikes of sometimes also pink, white or yellow fls. Widespread garden escape. (T), from N America. **8b* Annual Lupin** *L. angustifolia,* smaller, slenderer and always annual, is widely grown as a crop, as are the shorter Bitter Blue Lupin *L. micranthus,* covered with brown hairs, and White Lupin *L. albus,* whose white fls are tipped blue.

9* FALSE ACACIA *Robinia pseudacacia.* Deciduous tree to 25 m, suckering; bark grey-brown, deeply furrowed; shoots *thorny.* Lvs pinnate. Fls in drooping stalked spikes, white; June. Sandy soils. (T), from N America.

10 SCORPION SENNA *Coronilla emerus.* Deciduous shrub to 3-4 m. Lvs pinnate, *greyish.* Fls in heads, pale yellow often tipped red; Apr-June. Pods 50-100 mm. Scrub. F, G, S. **10a Small Scorpion Vetch** *C. vaginalis* is shorter, with shorter rounder green lflets and pods less than 35 mm. G. **10b** *C. minima* is like 10a, with lflets unstalked. F. **10c Scorpion Vetch** *C. coronata* is shorter and not woody. Dry places on lime. G. 10a-10c have much shorter pods.

125

Pea Family *(contd.)*

VETCHES *Vicia*. Mostly climbing or clambering plants. Lvs pinnate ending (except 4) in a tendril, usually branched. Fls usually in heads, on stalks from base of lvs. Pods long, more or less flattened. All except 8 differ from Peas *Lathyrus* (p. 128) in not having stems winged or angled.

1* TUFTED VETCH *Vicia cracca*. Clambering perennial to 1½ m, slightly downy. Lflets 8-12 pairs. Fls *10-40* in one-sided spikes, *blue*-violet, 10-12 mm; June-Aug. Pods brown. Hedges, bushy places. T. **1a* Fine-leaved Vetch** *V. tenuifolia* has much narrower lflets and larger, often white-winged fls. T, eastern, but (B). **1b* Danzig Vetch** *V. cassubica* has shorter lflets and 4-15 pinker fls with whitish wings and keel. T but (B). **1c* Fodder Vetch** *V. villosa* □ is a variable annual with 6-8 pairs of lflets, often larger fls, sometimes with white or yellow wings, and sepal-tube swollen at base; June-Nov. Bare and waste ground. T but (B).

2* WOOD VETCH *Vicia sylvatica*. Clambering hairless perennial to 2 m. Lflets 5-12 pairs. Fls *5-20* in one-sided spikes, white or *pale* lilac, with purple veins, 12-20 mm; June-Aug. Pods black. Open woods, rarely on sea cliffs and shingle. T. **2a** *V. dumetorum* has blue-purple fls. **2b** *V. pisiformis* □ has yellow fls. Both 2a and 2b have 3-5 pairs of lflets and brown pods. F, G, S, eastern.

3* BUSH VETCH *Vicia sepium*. Medium/tall clambering downy perennial. Lflets 3-9 pairs. Fls 2-6 in a short spike, *blue*-purple, rarely yellow, 12-15 mm; Apr-Nov. Pods black. Hedges, bushy places. T. **3a*** *V. pannonica* has narrower lflets and longer, sometimes yellow fls. Grassy places. F, (B, G).

4* UPRIGHT VETCH *Vicia orobus*. Medium *erect* perennial, slightly downy; stems *unwinged*. Lvs ending in a minute *point*, not a tendril. Fls pink-lilac; May-June. Pods yellowish-brown. Scrub, rocks. T, western.

5* COMMON VETCH *Vicia sativa* agg. Variable short/medium clambering downy annual. Lflets 3-8 pairs, varying from almost linear to oval, tendrils sometimes unbranched; usually a *dark spot* on toothed stipules at base. Fls 1-2, reddish-purple, *10-30 mm;* Apr-Sept. Pods downy or hairless, black to yellow-brown. Bare or grassy ground; large forms are cultivated. T. **5a* Spring Vetch** *V. lathyroides* □ is a prostrate downy annual, with 2-4 pairs of narrow lflets, tendrils always unbranched, untoothed unspotted stipules, solitary deep lilac 5-8 mm fls and black hairless pods. Apr-May. Sandy ground, especially near sea.

6* HAIRY TARE *Vicia hirsuta*. Slender short annual, almost hairless. Lflets 4-10 pairs, tendrils sometimes unbranched. Fls 1-8 in spike, pale *lilac, 4-5 mm;* sepal-teeth longer than tube; May-Aug. Pods downy. Grassy places. T. **6a* Smooth Tare** *V. tetrasperma* □ has 3-6 lflet pairs, 1-2 larger (4-8 mm) deep lilac fls and hairless pods. **6b* Slender Tare** *V. tenuissima* □ has 2-5 lflet pairs, 2-5 still larger (6-9 mm) deep lilac fls, and pods downy or hairless; June-Aug. B, F, G, southern. 6a and 6b both have calyx-teeth shorter than tube and tendrils usually unbranched.

7* YELLOW VETCH *Vicia lutea*. More or less prostrate downy annual. Lflets 3-10 pairs, tendrils branched or not. Fls 1-3, *yellowish-white*, often tinged purple, 20-35 mm; June-Sept. Pods usually downy. Bare and sparsely grassy places, often on shingle by the sea. B,F,G.

8* BITHYNIAN VETCH *Vicia bithynica*. Medium clambering hairy annual; stems 4-angled. Lflets 2-3 pairs, large, *broad*. Fls 1-3, purple, wings and keel *creamy white*, 16-20 mm; May-June. Pods brown or yellow, downy. Grassy places, hedges. B, F, western. **8a** *V. narbonensis* has broader lflets, longer, dark purple fls and black pods. Woods. F, (G).

2b *Vicia pisiformis*

127

Pea Family (contd.)

PEAS *Lathyrus*. Differ from all Vetches (p. 126), except Bithynian Vetch, in their winged or angled stems, but cf 4a. Most have fewer lflets, but some have more. Pods usually brown.

1* BROAD-LEAVED EVERLASTING PEA *Lathyrus latifolius*. Climbing perennial to 3 m, hairless or downy; stems winged. Lflets *one* pair, large, *broad*. Fls 5-15, bright magenta-pink, 20-30 mm; July-Sept. Scrub, widespread garden escape. (B),F,G. **1a* Narrow-leaved Everlasting Pea** *L. sylvestris* □ has narrower lvs and stipules, and smaller muddy pink fls; June-Aug. Woods, scrub, hedges. T. **1b** *L. heterophyllus* has 2-3 pairs of lflets and smaller fls. F,G,S.

2* MEADOW VETCHLING *Lathyrus pratensis*. Short/tall clambering perennial, hairless or downy; stems angled. Lflets one, occasionally two, pairs. Fls 4-12, *yellow;* May-Aug. Pods black. Grassy places. T.

3 SPRING PEA *Lathyrus vernus*. Short *erect* hairless perennial, stems angled. Lflets 2-4 pairs, *no* tendril. Fls 3-10, red-purple, fading blue, 13-20 mm; April-June. Pods brown. Woods, scrub. F,G,S. **3a* Black Pea** *L. niger* □ is sparsely downy, with 3-6 lflet pairs, smaller bluish-purple fls and black pods; May-July. T, rare in B.

4* BITTER VETCHLING *Lathyrus montanus*. Short erect perennial, hairless; stems *winged*. Lflets 2-4 pairs, sometimes very narrow (var. *tenuifolius* □), lvs ending in a *point* not a tendril. Fls 2-6, red-purple fading blue, 10-16 mm; Apr-July. Pods red-brown. Woods, scrub, heaths. T. **4a** *L. pannonicus* has stems not or narrowly winged, lflets very narrow, fls creamy tinged red-purple, and pale brown pods. F,G, southern.

5* SEA PEA *Lathyrus japonicus*. Prostrate fleshy perennial, hairless or downy; stems angled. Lflets 2-5 pairs, greyish, sometimes no tendril. Fls 2-12, purple fading blue, 14-22 mm; June-Aug. Shingle and dunes *by sea*. B,G,S.

6* MARSH PEA *Lathyrus palustris*. Medium/tall clambering perennial, slightly downy. Lflets 2-5 pairs, tendril branched. Fls 2-8, bluish-purple, 12-20 mm; June-July. Pods black. *Marshes*, fens. T, northern.

7* TUBEROUS PEA *Lathyrus tuberosus*. Medium/tall clambering perennial, more or less hairless; stems angled. Lflets one pair. Fls 2-7, bright *crimson*, slightly fragrant, 12-20 mm; June-July. Grassy and cultivated ground. (B),F,G.

8* HAIRY VETCHLING *Lathyrus hirsutus*. Medium/tall clambering annual, slightly downy; stems winged. Lflets one pair, tendrils branched. Fls 1-3, pale red-purple, fading blue, keel *creamy*, 7-15 mm; June-Aug. Pods brown, downy. Grassy and bare places. (B),F,G, eastern. **8a** *L. sphaericus* has stems angled, solitary orange-red fls and hairless pods. F,G,S, southern.

9* GRASS VETCHLING *Lathyrus nissolia*. Short/tall erect annual, hairless or slightly downy. Unique among peaflowers of region for long narrow *grasslike* 'lvs', (actually modified stems) with no tendrils, very hard to detect among the grass. Fls 1-2, long-stalked, crimson; May-July. Pods pale brown. Grassy places. B,F,G.

10* YELLOW VETCHLING *Lathyrus aphaca*. Medium/tall hairless clambering annual. Unique among peaflowers for its lvs having become unbranched *tendrils* direct from the stems and its stipules developed into *triangular* greyish apparent leaves, *joined* around the stem. Fls usually solitary, yellow; May-Aug. Dry grassy places. B,F,G.

3a* Black Pea

4* Bitter Vetchling (var. *tenuifolius*)

129

Pea Family (contd.)

1* SAINFOIN *Onobrychis viciifolia.* Short/medium erect, rarely prostrate, downy perennial. Lvs *pinnate.* Fls in stalked spikes, bright *pink;* June-Sept. Pods warty, pitted. Dry grassy and waste places; widely cultivated. (T), from E Europe. **1a** *O. arenaria* has lflets narrower and smaller paler fls with purple veins. F, G, southern. **1b** *Hedysarum hedysaroides* has purplish fls and pods constricted between seeds. G.

2* REST-HARROW *Ononis repens.* Short/medium semi-erect hairy undershrub; stem hairy all round, sometimes with soft *spines.* Lvs *trefoil* or oval, blunt or slightly notched. Fls in lfy stalked spikes, pink, wings equalling keel; July-Sept. Dry grassy places. T. **2a* Spiny Rest-harrow** *O. spinosa* ☐ is erect with sharper spines, stem with two lines of hairs, lflets never notched and sometimes pointed, and reddish-pink fls with wings shorter than keel. **2b** *O. arvensis* is taller, with fls in pairs. G, S.

3 LARGE YELLOW REST-HARROW *Ononis natrix.* Medium erect undershrub, stickily hairy. Lvs trefoil, variable. Fls in branched spikes, *deep yellow,* often veined red or purple; May-Aug. Pods short. Dry open places. F, G, southern. **3a** *O. pusilla* has broader lflets and fls half the size. F.

4* SMALL REST-HARROW *Ononis reclinata.* Low semi-prostrate annual, stickily hairy, with shiny dots. Lvs trefoil. Fls in lfy spikes, pink, 5-10 mm; May-June. Pods hang straight down when ripe. Sandy turf and cliffs by sea. B (rare), F.

5* KIDNEY VETCH *Anthyllis vulneraria.* Very variable, low/short, semi-prostrate/erect, annual/perennial, silkily hairy. Lvs *pinnate,* the lowest sometimes lanceolate. Fls in heads, sometimes paired, yellow, orange, red, purple, whitish or particoloured; two lfy bracts at base; Apr-Sept. Dry grassland, often by sea and on mountains. T.

6* RIBBED MELILOT *Melilotus officinalis.* Medium/tall hairless biennial. Lvs trefoil. Fls in stalked *spikes,* yellow, keel shorter than wings and standard; June-Sept. Pods hairless, brown when ripe. Bare and waste ground. T, but (B). **6a* Tall Melilot** *M. altissima* has deeper yellow fls, keel equalling wings and standard, and pods downy, black when ripe. **6b** *M. dentata* has smaller fls, wings shorter than standard, longer than keel, and hairless net-veined pods, blackish-brown when ripe. Saline soils. G, S. **6c* Small Melilot** *M. indica* is a short/medium annual, with smaller fls, wings and keel both shorter than standard, and pods hairless, net-veined, olive-green when ripe. (B), F, G, southern.

7* WHITE MELILOT *Melilotus alba.* Tall hairless biennial. Lvs trefoil. Fls in stalked spikes, *white;* July-Oct. Pods hairless, net-veined, brownish when ripe. Bare and waste ground; grown as Bokhara Clover. T, but (B).

8 CLASSICAL FENUGREEK *Trigonella foenum-graecum.* Short/medium annual, almost hairless. Lvs trefoil, toothed. Fls *1-2* at base of upper lvs, yellowish-white, tinged purple at base; Apr-May. Pods hairless, slightly curved, beaked. Bare and waste ground; widely cultivated. F, G, southern. **8a** *T. monspeliaca* is hairy, with smaller lvs and fls; fls yellower, not purple at base, 4-14 in a spike. F.

9* LUCERNE *Medicago sativa.* Medium hairless perennial. Lvs trefoil. Fls in short spikes, of varying shades of violet (spp. *sativa),* yellow (**9a** ssp. *falcata* ☐), green and livid black (hybrids between the two); June-Oct. Pods sometimes curved or spiral. Grassy, waste and cultivated places; widely grown as Alfalfa. T.

5* Kidney Vetch, colour varieties

9* Hybrids of Lucerne

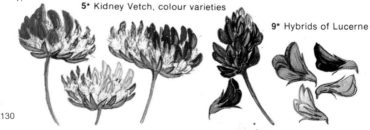

131

Pea Family (contd.)

1* BIRDSFOOT TREFOIL *Lotus corniculatus.* More or less prostrate perennial, downy or hairless. Lvs with five lflets, lowest pair bent back so that lvs appear *trefoil.* Fls 2-7 in a head, yellow, often tinged orange or red, 10-16 mm; sepal-teeth erect in bud, an obtuse angle between the two upper; May-Sept. Pods straight, in a head resembling a bird's foot, 15-30 mm. Grassy places. T. **1a* Narrow-leaved Birdsfoot Trefoil** *L. tenuis* □ is slenderer, with narrower lflets, fewer and smaller yellow fls, and rear sepal-teeth converging; June-Aug. **1b* Greater Birdsfoot Trefoil** *L. uliginosus* is larger and erect, medium/tall, with broader lflets greyish beneath, 5-12 fls in each head and sepal-teeth spreading in bud, the two upper at an acute angle; June-Aug. Damp grassland, marshes.

2* HAIRY BIRDSFOOT TREFOIL *Lotus subbiflorus.* Prostrate *hairy* annual. Lvs as 1 (above). Fls 2-4 in a head, on stalks much longer than lvs, orange-yellow, 5-10 mm; June-Sept. Pods 6-12 mm. Grassy places. B, F. **2a* Slender Birdsfoot Trefoil** *L. angustissimus* is less hairy, with 1-2 fls in a head, on stalks usually shorter than lvs, yellow, and much longer (15-30 mm) pods; May-Aug.

3* DRAGON'S TEETH *Tetragonolobus maritimus.* Prostrate patch-forming perennial, hairless or downy. Lvs trefoil. Fls *solitary,* pale yellow, sometimes red-veined, 25-30 mm; May-Sept. Pods 4-angled, dark brown, 30-60 mm. Grassy places, often damp. T, but (B, rare).

4* HORSESHOE VETCH *Hippocrepis comosa.* Prostrate downy perennial. Lvs *pinnate,* ending in a lflet. Fls 5-12 in a head, yellow, 6-10 mm; May-July. Pods curved, with horseshoe-like segments, 15-30 mm. Dry grassland, often on limy soils. B, F, G.

5* BIRDSFOOT *Ornithopus perpusillus.* Prostrate downy annual. Lvs *pinnate,* ending in a lflet. Fls 3-8 in a head, yellowish-white or pink, veined red, 3-5 mm; bracts pinnate, lf-like; May-Aug. Pods beaded, 10-18 mm, in a head resembling a bird's foot. Bare and grassy places. T, southern. **5a* Orange Birdsfoot** *O. pinnatus* □ has 1-5 orange-yellow fls with no lf-like bract. B(rare), F.

6* BLACK MEDICK *Medicago lupulina.* Low, often prostrate annual, usually downy. Lvs trefoil, lflets with a minute *point.* Fls yellow, 10-50 in a short-stalked rounded head, 3-8 mm; Apr-Oct. Pods curved, *black* when ripe, not covered by dead fls. Bare and grassy places; widely cultivated. T. **6a* Lesser Trefoil** *Trifolium dubium* □ is almost hairless; lflets slightly notched, not pointed, the middle one longer-stalked; fls 3-15 in head; pods straight, covered by dead brown fls; May-Sept. Cf. 8. **6b* Slender Trefoil** *Trifolium micranthum* □, is like 6a with all lflets unstalked, fls 1-6 in head, with long hairlike stalks, 2-3 mm. Southern.

7* SPOTTED MEDICK *Medicago arabica.* Prostrate annual, hairless or slightly downy. Lvs trefoil, usually with a *dark spot* on each lflet; stipules toothed. Fls 1-4 in a head, yellow, 5-7 mm; Apr-Sept. Pods spiny, spirally coiled, faintly netted. Bare and grassy places. B, F, G, southern. **7a* Toothed Medick** *M. polymorpha* is smaller and hairier, with unspotted lvs, stipules more deeply toothed, and strongly netted but more shortly spiral pods; May-Aug. **7b* Bur Medick** *M. minima* is smaller still, and downy, with unspotted lvs, untoothed pointed stipules, and two rows of hooked spines on the pods; Apr-July. T.

CLOVERS: *Trifolium.* pp. 132-34. lvs trefoil. Fls small, usually in a dense rounded head, wings longer than keel, sepal-tube 5-toothed. Pods covered by dead fls. Grassy places.

8* HOP TREFOIL *Trifolium campestre.* Low/short annual, hairy. Lvs trefoil, lflets with *no* point, the middle one longer-stalked. Fls *20-30* in rounded stalked 10-15 mm heads, yellow; May-Sept. Pods covered by pale brown dead fls. Grassy places. T. **8a* Large Hop Trefoil** *T. aureum* is a stouter biennial, with narrower lflets, all unstalked, larger, more golden-yellow, unstalked fls, and pods with a hooked beak; June-Aug. T, but (B). **8b** *T. spadiceum* is slenderer and almost hairless, with all lflets unstalked, and much larger, often paired heads of larger golden-yellow fls, turning dark brown when dead; June-Aug. Not on lime. F, G, S, eastern. **8c** *T. patens* is less hairy, with larger lvs and longer fls. Damp places. F. **8d** *T. badium* has fls turning bright chestnut brown. G.

Pea Family (contd.)

1* RED CLOVER *Trifolium pratense*. Variable erect low/tall hairy perennial. Lflets often with a whitish crescent; stipules triangular, bristle-pointed. Flhds globular or egg-shaped, sometimes paired, usually unstalked, with two lvs closely beneath, 20-40 mm; fls of varying shades of *pink-purple*, rarely creamy or white; May-Nov. Widely cultivated. T. **1a* Zigzag Clover** *T. medium* is usually short/medium, with stems often slightly zigzag, stipules narrow, not bristle-pointed, and flhds longer-stalked, fls a uniform hard reddish-purple; June-Aug. Also in open woods and scrub. **1b Mountain Zigzag Clover** *T. alpestre* has narrower lflets and stipules, and unstalked flhds. Uplands. F, G, eastern.

2 MOUNTAIN CLOVER *Trifolium montanum*. Short/medium hairy perennial. Flhds often paired, well stalked, 15-30 mm; fls *yellowish-white*, rarely pink, turning brownish; May-July. Upland. F, G, S. **2a* Sulphur Clover** *T. ochroleucon* □ has two lvs closely below the larger (20-40 mm) un- or short-stalked flhds. Lowland. B, F, G. **2b** *Dorycnium pentaphyllum* may be woody; keel dark red. G.

3* WHITE CLOVER *Trifolium repens*. Creeping perennial, more or less hairless, stems rooting at lf-junctions. Lflets usually with a whitish mark. Flhds globular, long-stalked, 15-30 mm; fls *white*, pink, or rarely purple, turning brown, scented, angles between sepal-teeth acute; May-Nov. Widely cultivated. T. **3a* Western Clover** *T. occidentale* has unmarked lvs, almost circular lflets, red stipules and small heads of scentless fls; Mar-May. Near the sea. B, F, western. **3b* Alsike Clover** *T. hybridum* □ is short/medium and erect, with unmarked lvs and shorter-stalked heads of more often pink fls, angles between sepal-teeth blunt; June-Sept. Often on waste ground; also widely cultivated. (B, S), F, G. **3c** *T. michelianum* is annual with pink fls. Wet places. F, rare.

4* STRAWBERRY CLOVER *Trifolium fragiferum*. Creeping hairy perennial; stems often rooting at lf-junctions. Flhds globular, 10-14 mm, *swelling* in fr to appear like miniature pale pink strawberries; fls pink; June-Sept. T. **4a* Sea Clover** *T. squamosum* is a short, often erect annual with two lvs closely beneath the unstalked flheads and fr heads not swollen but with conspicuously star-like spreading sepal-teeth; June-Oct. Damp grassland, especially by the sea. B, F. **4b* Reversed Clover** *T. resupinatum* is a hairless annual with fls twisted upside down so that heads appear flatter. Bare ground. (B, casual), F, G, southern.

5* KNOTTED CLOVER *Trifolium striatum*. Low-spreading or erect downy annual. Lvs downy on *both* sides. Flhds egg-shaped, unstalked at base of lvs, 10-15 mm; fls pink, sepal-teeth swollen in fr, with erect teeth; June-July. Dry places. T, southern. **5a* Clustered Clover** *T. glomeratum* is hairless and often semi-prostrate, and has lflets often with a pale patch, flheads globular, and calyx-teeth turned back in fr. B, F. **5b* Twin-flowered Clover** *T. bocconei* has pale pink fls in often paired but unequal heads. B (rare), F. **5c* Rough Clover** *T. scabrum* is more often prostrate, with white fls and calyx-teeth curving outwards in fr; May-June. B, F, G, southern.

6* HARESFOOT CLOVER *Trifolium arvense*. Low/short hairy annual/biennial. Lflets narrow. Flhds *egg-shaped*, stalked, 10-20 mm; fls whitish or pale pink, much shorter than sepal-tube; June-Sept. Scarce on lime. T. **6a* Upright Clover** *T. strictum* □ has larger but still narrower lflets, and smaller (7-10 mm), more globular, longer-stalked heads of larger bright pink fls, much longer than calyx; June. B (rare), F.

7* CRIMSON CLOVER *Trifolium incarnatum*. Short/medium hairy annual. Flhds elongated, 10-40 mm; fls *bright* crimson, pink or creamy white, sepal-teeth spreading in fr; June-July. Widely cultivated. Pale-fld subsp. *molinerii*, Long-headed Clover, is native on cliff-tops by the sea. B, F(G, S), southern. **7a** *T. rubens* is hairless with longer heads of purpler fls. G.

8* FENUGREEK *Trifolium ornithopodioides*. Prostrate hairless annual. Fls 2-4 in short-stalked heads, white or pale pink, 6-8 mm; June-July. Pods much longer than sepal-tube. Bare dry places, often damp in winter. B, F, G, southern. **8a* Burrowing Clover** *T. subterraneum* □ has larger (8-14 mm) fls and fr heads buried in the ground; May-June.

9* SUFFOCATED CLOVER *Trifolium suffocatum*. Prostrate hairless annual. Flhds unstalked, tightly packed, 5-6 mm; fls white, almost *hidden* by green calyx-teeth, recurved in fr; July-Aug. Dry places. B, F.

Wood-Sorrel Family Oxalidaceae

Downy, lvs trefoil, often closing up at night. Fls 5-petalled,cup-shaped.

1* WOOD-SORREL *Oxalis acetosella*. Low creeping perennial. Fls *white*, usually veined mauve, sometimes purplish; Apr-May. Woods, also on mountains. T.

2* YELLOW OXALIS *Oxalis corniculata*. Low/short creeping annual/perennial, stems rooting at lf-junctions. Lvs often purplish, stipules tiny. Fls *yellow*, 4-7 mm; May-Oct. Fr stalks bent right back. Bare, especially cultivated, ground. T, southern, but (B). **2a* Upright Yellow Oxalis** *O. europaea* is erect, not rooting at lf-junctions, with lvs rarely purplish, no stipules and fr stalks not bent back. (T) from N America. **2b*** *O. stricta* has tiny stipules and bent fr stalks. (B, F, G).

3* PINK OXALIS *Oxalis articulata*. Low/short tufted perennial. lvs spotted pale orange beneath. Fls in a cluster, *rose-pink*; May-Sept. Garden escape. (B, F), from S America. **3a*** *O. corymbosa* is smaller, with a cluster of bulbils at the stem base, spots mainly near lf edges and fewer, more purplish-pink fls. **3b*** *O. latifolia* is almost hairless, with lvs unspotted but sometimes purplish beneath, and pale to deep pink fls.

4* BERMUDA BUTTERCUP *Oxalis pes-caprae*. Low/short tufted perennial. Lflets sometimes with a pale brown blotch. Fls in an umbel-like cluster, *yellow*, 20-25 mm, Mar-June. Naturalised weed. (B, rare, F), from S Africa.

Geranium Family Geraniaceae: see p. 138

5* COMMON STORKSBILL *Erodium cicutarium*. Low/medium annual/biennial, hairy, sometimes stickily so; often strong-smelling. Lvs *2-pinnate,* stipules lanceolate. Fls in umbel-like heads with five, easily shed, often unequal petals, purplish-pink to white, sometimes with a blackish spot at base of two upper petals, 4-11 mm; Apr-Sept. Fr with a long twisted beak. Bare and grassy places, frequent by the sea. T. **5a* Musk Storksbill** *E. moschatum* smells musky, is usually stickily hairy, and has 1-pinnate lvs, broader stipules and larger (15 mm) always purplish-pink fls with no blotches. **5b Caltrop** *Tribulus terrestris* (Zygophyllaceae), creeping on dry ground, has smaller pale yellow fls, and tiny star-like fr. F.

6* SEA STORKSBILL *Erodium maritimum*. Low, often prostrate, downy annual/biennial. Lvs oval, toothed, often in a rosette. Fls usually solitary, pale pink or white, the five petals dropping very soon; May-Aug. Fr with a long twisted beak. Bare ground, short turf, usually near the sea. B, F.

Flax Family Linaceae

Slender, hairless. Lvs undivided, untoothed; no stipules. Fls 5-petalled (except 10), open. Fr a capsule.

7* PERENNIAL FLAX *Linum perenne* agg. Short/medium hairless perennial. Lvs more or less linear, alternate, 1-veined or obscurely 3-veined. Fls in a cluster, on erect stalks, pale to bright blue, 25 mm; anthers and stigmas unequal; inner sepals often blunt, much shorter than fr; May-Aug. Dry grassland. B, F, G. **7a* Pale Flax** *L. bienne* □ has 1-3 veined lvs, pale bluish-lilac fls that drop early, and all sepals pointed, nearly as long as the more pointed fr. B, F. **7b* Common Flax** *L. usitatissimum* □ is annual with lvs always 3-veined and sepals like 7a; June-Oct. Widely cultivated. (T). **7c** *L. viscosum* is downy, has broader lvs and pink fls, blue when dry. G, southern. **7d** *L. tenuifolium* has smaller pink or white fls with all sepals pointed. F, G. **7e White Flax** *L. suffruticosum* may be woody at base and has white fls with all sepals pointed. F.

8* PURGING FLAX *Linum catharticum*. Low annual. Lvs oblong, opposite, 1-veined. Fls white, 4-6 mm; May-Sept. Grassland, fens, mainly on lime. T.

9 YELLOW FLAX *L. flavum*. Variable medium hairless perennial, sometimes woody at base. Lvs lanceolate, the lower broader at tip. Fls yellow, *c* 30 mm, in a cluster. Dry grassy places. G.

10* ALLSEED *Radiola linoides*. A tiny (rarely above 5 cm) annual, with repeatedly branching thread-like stems. Lvs oblong, opposite, 1-veined. Fls 4-petalled, white, 1 mm, equalling sepals; July-Aug. Bare damp places, not on lime. T, western.

Geranium Family   Geraniaceae

pp. 136-38. Hairy or downy. Lvs deeply palmately lobed or cut, with stipules. Fls 5-petalled, sepals often bristle-tipped, stamens prominent. Fr ends in long straight pointed beak, whence "cranesbill".

1* MEADOW CRANESBILL *Geranium pratense*. Medium/tall perennial. Lvs cut nearly to base. Fls *blue*, 25-30 mm; petals not notched; June-Sept. Fl stalks bent down after flg, becoming erect when fr ripe. Grassy places, usually on lime. T. **1a* Wood Cranesbill** *G. sylvaticum* □ has less deeply cut lvs, generally smaller, more reddish-mauve fls and fl stalks always erect. Also in woods and on mountains. Northern.

2* BLOODY CRANESBILL *Geranium sanguineum*. Short/medium tufted perennial. Lvs cut nearly to base. Fls *bright* purple-crimson, very rarely pink, 25-30 mm. June-Aug. Dry grassy places, usually on lime. T.

3* FRENCH CRANESBILL *Geranium endressii*. Medium perennial. Lvs cut to more than half-way. Fls *pink*, unveined, 25 mm; petals not notched; June-Aug. Garden escape. (B,F), from S W France. The hybrid **3** × **3a***, with dark-veined, slightly notched pink petals, is more frequent in B than either. **3a* Pencilled Cranesbill** *G. versicolor* □ has paler pink fls, pencilled with violet veins, and notched petals. From Italy and Balkans. **3b* Knotted Cranesbill** *G. nodosum* has stems swollen at lf-junctions, smaller purple-veined mauve fls and petals deeply notched. (B,F,G), from S Europe. **3c* Dusky Cranesbill** *G. phaeum* □ has less deeply cut lvs, smaller drooping maroon fls and unnotched wavy-edged petals. Damp and shady places; garden escape in the West. (B),F,G.

4 MARSH CRANESBILL *Geranium palustre*. Medium/tall perennial. Lvs cut to more than half-way. Fls pale *purple* or lilac, 30 mm; petals not or very slightly notched; July-Aug. Fr stalks bent down. Damp grassy places. F,G,S, eastern.

5* HEDGEROW CRANESBILL *Geranium pyrenaicum*. Medium perennial. Lvs cut to about half-way. Fls pink-purple to lilac, often *mauve-pink,* 15 mm; petals well notched; May-Sept. Fr stalks bent down. Grassy places, open woods. T, southern, but (B). **5a** *G. bohemicum* agg. □ is annual/biennial, with smaller, bright violet-blue fls, less notched petals and erect fr stalks. Eastern. **5b** *G. divaricatum* is annual; petals pink, less notched, shorter. G.

6* HERB ROBERT *Geranium robertianum*. Medium annual/biennial, strong-smelling, often reddening. Lvs triangular, 1-2-pinnate. Fls *pink,* 20 mm; petals scarcely notched, pollen orange; Apr-Nov. Fr slightly wrinkled. Shady places, rocks, shingle. T. **6a* Little Robin** *G. purpureum* □ has much smaller fls, yellow pollen and strongly wrinkled fr. B,F.

7* SHINING CRANESBILL *Geranium lucidum*. Short annual, almost hairless, often reddening. Lvs *glossy,* rounded, cut to half-way. Fls pink, 15 mm; petals not notched; May-Aug. Fr wrinkled. Shady rocks and hedge-banks. T.

8* DOVESFOOT CRANESBILL *Geranium molle*. Short semi-prostrate annual, with long hairs on stems. Root lvs cut to less than half-way, stem lvs to half-way or more, all rounded. Fls dark or pale pinkish-purple, 10 mm; petals well notched; Apr-Sept. Fr hairless, usually wrinkled. Bare and grassy places. T. **8a* Round-leaved Cranesbill** *G. rotundifolium* is taller, with rounder, scarcely cut, wavy-edged lvs, pink fls, unnotched petals and downy unwrinkled fr. B,F,G. **8b* Small-flowered Cranesbill** *G. pusillum* □ has only short hairs on stem, stem lvs more narrowly cut to more than half-way, smaller lilac fls, half the stamens without anthers, and fr downy, not wrinkled.

9* CUT-LEAVED CRANESBILL *Geranium dissectum*. Short/medium straggling annual/biennial. Lvs cut almost to base. Fls pink-purple, 8 mm; petals notched; May-Sept. Fr downy. Bare and sparsely grassy places. T. **9a* Long-stalked Cranesbill** *G. columbinum* □ sometimes reddens and has larger longer-stalked fls, petals not notched and fr hairless. Dry grassy places.

8a* Round-leaved
Cranesbill

8b* Small-flowered Cranesbill

CRANESBILLS

Spurge Family Euphorbiaceae

SPURGES *Euphorbia*. Stems with milky juice; mostly hairless. Lvs undivided, usually alternate and untoothed. Fls in usually broadening umbel-like clusters, yellowish-green; no petals or sepals, but a solitary 3-styled female f., surrounded by several 1-anthered male fls is cupped by a calyx-like bract, with 4-5 glands on its lip, and 2-3 small lflike bracts at its base. Fr rounded, stalked.

1* WOOD SPURGE *Euphorbia amygdaloides*. Medium/tall unbranched perennial, downy, often tinged red. Lvs lanceolate. Fls yellow, glands horned, bracts joined together; *Apr-May. Woods*, scrub. B,F,G. **1a* Cypress Spurge** *E. cyparissias* □ is shorter and hairless, with narrower lvs, turning bright red, and bracts free; May-Aug. Bare and grassy places. T, but (B). **1b* Leafy Spurge** *E. esula* forms patches, has bracts free and glands long- or short-horned; June-Aug. Disturbed ground. T, but (B). **1c** *E. lucida* is like 1b with lvs shiny above. G, rare.

2* SEA SPURGE *Euphorbia paralias*. Short/medium greyish perennial; unbranched. Lvs fleshy, oval, midrib *not* prominent beneath, overlapping up the stem. Fls with short-horned glands; June-Sept. Sands by the *sea*. B,F,G, southern. **2a** *E. seguieriana* has narrower, more pointed, less fleshy lvs and glands not horned; Apr-Aug. Woods, grassy places, on lime. F,G.

3* PORTLAND SPURGE *Euphorbia portlandica*. Short greyish hairless perennial; branched from the base, often reddening. Lvs oval, *midrib* prominent below. Fls with long-horned glands; Apr-Sept. Sand-dunes and cliff-tops. B,F.

4* CAPER SPURGE *Euphorbia lathyris*. Tall greyish biennial. Lvs narrow, opposite. Fls with horned glands; June-Aug. Fr large, *3-sided*, resembling a green caper. Bare and cultivated ground. B (rare), F, G.

5* IRISH SPURGE *Euphorbia hyberna*. Medium little-branched perennial; often in clumps. Lvs untoothed, downy beneath, reddening. Fls with yellow bracts and *kidney-shaped* yellow glands; Apr-July. Fr with long and short warts. Damp or shady places. B,F, southern. **5a Hairy Spurge** *E. villosa* is taller, with lvs downy on both sides and fr scarcely warty; June-July. F,G, southern. **5b Marsh Spurge** *E. palustris* is taller still than 5a, with hairless greyish lvs and fr with short warts only. F,G,S. **5c* Sweet Spurge** *E. dulcis* has lvs almost hairless and glands green, turning purple. (B),F,G. 5a, 5b and 5c all have lvs sometimes finely toothed at tip, green bracts and oval glands. **5d** *E. brittingeri* is woody at base and downier, with lvs all toothed and fr slightly curved. F,G, southern.

6* SUN SPURGE *Euphorbia helioscopia*. Short, usually unbranched, annual. Lvs oval, finely toothed. Fls in a broad umbel, bracts often yellowish, glands green, oval; all year. Fr smooth. Common weed. T.

7* PETTY SPURGE *Euphorbia peplus*. Low/short, often branched, annual. Lvs oval, untoothed. Fls green, glands horned; all year. Fr with wavy ridges. Common weed of cultivation. T. **7a* Broad-leaved Spurge** *E. platyphyllos* □ is taller and may be downy, with longer lvs finely toothed at the tip, larger yellower fls with rounded glands and fr with rounded warts; June-Sept. B,F,G. **7b* Upright Spurge** *E. serrulata* is slenderer than 7a and always hairless, with stems often red and long warts on fr. Open woods. T, rare in B.

8* DWARF SPURGE *Euphorbia exigua*. Low slender greyish annual. Lvs linear. Fls yellowish, glands horned; May-Oct. Fr smooth. Weed, especially of cornfields. T.

9* PURPLE SPURGE *Euphorbia peplis*. Prostrate fleshy annual; stems crimson. Lvs oblong, greyish, opposite. Fls stalked, along stems, 1-2 mm; July-Sept. Sandy shores. B (rare), F, southern.

10* DOG'S MERCURY *Mercurialis perennis*. Low/short unbranched downy foetid perennial. Lvs broad lanceolate, opposite. Fls in stalked tassels, green, 4-5 mm; stamens and styles on separate plants; Feb-May. Woods, shady places, often carpeting the ground; also on mountains. T. **10a* Annual Mercury** *M. annua* is paler, almost hairless and often branched, with female tassels scarcely stalked and sometimes on same plant as male; May-Nov and through mild winters. Bare ground. **10b** *M. ovata* has very short fl-stalks. G, eastern, rare.

141

Balsam Family   Balsaminaceae

BALSAMS *Impatiens*. Hairless annuals, stems fleshy. Lvs oval, stalked, slightly toothed. Fls in long-stalked spikes from base of lvs, 5-petalled, with a broad lip, a small hood and a spur, often curved. Fr cylindrical, explosive.

1* HIMALAYAN BALSAM *Impatiens glandulifera*. Tall, stems often reddish. Lvs opposite or in threes, with small red teeth. Fls large, pale to dark pink-purple, with short bent spur; July-Oct. Bare places, especially by streams. (T).

2* TOUCH-ME-NOT BALSAM *Impatiens noli-tangere*. Medium/tall, lvs alternate. Fls *yellow*, sometimes spotted red-brown. spur curved downwards; July-Sept. Damp shady places. T.   **2a* Orange Balsam** *I. capensis* □ has red-spotted orange fls with spur bent into a crook; July-Sept. By rivers and canals. (B, F), from N America.

3* SMALL BALSAM *Impatiens parviflora*. Short/medium, lvs alternate. Fls *pale* yellow, spur short and straight, 3-5 mm; June-Sept. Shady or bare places. (T), from C Asia.

Milkwort Family   Polygalaceae

Hairless perennials. Lvs unstalked, usually alternate, no stipules. Fls 5-sepalled, the two inner large and petal-like on either side of the three true petals, which are joined at the base. Stamens also petal-like, in two Y-shaped bundles of four. Fr flat, heart-shaped, slightly winged.

4* COMMON MILKWORT *Polygala vulgaris*. Low perennial. Lvs lanceolate, pointed, broadest at or below the middle, the lower shorter and broader than the upper. Fls in stalked spikes, usually blue, also often mauve, pink, white or white tipped mauve or blue, 4-7 mm; May-Sept. Fr shorter than inner sepals. Grassy places. T.   **4a* Chalk Milkwort** *P. calcarea* has blunt lvs broadest above the middle, the lower stalked and in a rosette; fls bright gentian blue or white; and fr broader than inner sepals; Apr-June. Only on lime. B, F, G, western.   **4b Tufted Milkwort** *P. comosa* has lower lvs blunt and fls usually pink. F, G, S, eastern.   **4c* Heath Milkwort** *P. serpyllifolia* □ has at least the lower lvs opposite and more crowded, and shorter spikes of usually dark blue or dark pink fls. Not on lime. Southern.   **4d* Dwarf Milkwort** *P. amarella* □ has the lower lvs in a rosette, the stem ones blunt, broadest near the tip, and fls 2-4 mm; May-July. Northern; rare in B.

5 SHRUBBY MILKWORT *Polygala chamaebuxus*. More or less prostrate evergreen undershrub. Lvs oval to lanceolate, rather leathery. Fls usually yellow tipped purple, also red, pink or white; Apr-Sept. Woods, grassy places, especially on mountains. F, G, south-eastern.

Daphne Family   Thymelaeaceae

Hairless or almost so. Lvs untoothed, shortly stalked, usually alternate, no stipules. Fls in clusters at base of lvs, petalless with 4 petal-like sepals joined at base.

6* MEZEREON *Daphne mezereum*. Deciduous shrub to 2 m. Lvs lanceolate, pale green. Fls strongly scented, *pinkish-purple;* Feb-Apr, before the lvs. Fr a *red* berry. Woods on lime. T.

7* SPURGE LAUREL *Daphne laureola*. Evergreen shrub to 1 m. Lvs lanceolate, dark green, rather leathery. Fls slightly scented, *yellow-green;* Jan-Apr. Fr a *black* berry. Open woods, often on lime. B, F, G.

8 ANNUAL THYMELAEA *Thymelaea passerina*. Medium annual. Lvs almost linear, pointed. Fls 1-3 together, greenish, tiny; July-Oct. Fr a downy nut. Dry bare uncultivated places. F, G.

Rue Family   Rutaceae

9 BURNING BUSH *Dictamnus albus*. Medium/tall bushy hairy perennial, pungent from aromatic oil, which may burst into flame if a match is held near it in still summer weather. Lvs pinnate. Fls 5-petalled, white or pink with purple streaks and prominent purple stamens, in stalked spikes. Dry places. G.

143

Currant Family   Grossulariaceae

Small deciduous shrubs, to 1·5 m. Lvs alternate, palmately lobed. Fls 5-petalled, green. Fr a berry.

1* RED CURRANT *Ribes rubrum* agg. Stems erect. Lvs hairless or downy beneath, *not* aromatic. Fls purple-edged, in drooping stalked spikes; Apr-May. Fr red, acid. Damp woods, fens, by fresh water; widely cultivated. T. **1a* Black Currant** *R. nigrum* □ has aromatic lvs and black fr. **1b* Mountain Currant** *R. alpinum* □ is bushier, with smaller, more deeply lobed lvs, fls in erect clusters, male and female on different bushes, each with a conspicuous green bract, and tasteless fr. Upland woods, often on lime; rarely cultivated. **1c** *R. petraeum* may be taller, with larger wrinkled lvs, pinkish fls and purpler fr. F,G, southern.

2* GOOSEBERRY *Ribes uva-crispa.* Stems spreading, *spiny.* Lvs smaller than 1. Fls purple-edged, petals turned back, 1-3 together; Mar-May. Fr egg-shaped, green, often tinged red or yellow, usually hairy. Woods, scrub, hedges; widely cultivated. T.

Spindle-Tree Family   Celastraceae

3* SPINDLE-TREE *Euonymus europaeus.* Deciduous shrub, rarely a small tree, to 6 m, with 4-sided green twigs. Lvs elliptical, opposite,scarcely toothed. Fls small, greenish-white, 4-petalled, in small clusters; May-June. Fr a bright *coral-pink* berry. Woods, scrub, hedges, often on lime. T. **3a** *E. latifolius* has grey-brown, more rounded twigs, larger, more obviously toothed lvs, and browner fls, usually 5-petalled. (F), G.

Box Family   Buxaceae

4* BOX *Buxus sempervirens.* Evergreen shrub or small tree, to 5 m, twigs green. Lvs oval, shiny, *leathery,* opposite. Fls petalless, greenish-yellow, in small clusters at base of upper lvs; Mar-May. Fr dry. Dry hillsides, widely planted. B (rare), F, G.

Buckthorn Family   Rhamnaceae

5* BUCKTHORN *Rhamnus catharticus.* Deciduous shrub or small tree to 6 m, often thorny. Lvs broad elliptical, finely toothed, *opposite,* with *conspicuous* veins. Fls small, green, 4-petalled, in small clusters at base of upper lvs; male and female separate; May-June. Fr a black berry. Scrub, hedges, fens, usually on lime. T. **5a** *R. saxatilis* has smaller, rounder lvs on short stalks. G, southern, rare.

6* ALDER BUCKTHORN *Frangula alnus.* Deciduous shrub or small tree to 5 m, thornless. Lvs broad elliptical, untoothed, *alternate.* Fls small, green, usually 5-petalled, in small clusters at base of upper lvs; May-June. Fr a black berry, red at first. Damp heaths and woods, fens; usually not on lime. T.

Oleaster Family   Elaeagnaceae

7* SEA BUCKTHORN *Hippophae rhamnoides.* Deciduous shrub or small tree to 11 m, *thorny,* suckering freely and forming thickets. Brown twigs and untoothed narrow lanceolate alternate lvs both covered with *silvery scales,* the older lvs brownish. Fls tiny, green, petalless, male and female on different plants; Apr-May, before lvs. Fr a bright orange berry. Coastal dunes, sea cliffs, mountain river beds. T.

Tamarisk Family   Tamaricaceae

8* TAMARISK *Tamarix gallica.* Evergreen shrub to 3 m, hairless; twigs reddish. Lvs minute, *scale-like,* overlapping, alternate, often greyish, producing a feathery foliage. Fls pink, 5-petalled, in stalked spikes; July-Sept. By salt and fresh water, often planted. (B),F. **8a** *Myricaria germanica* has longer fatter spikes of larger fls. River gravels. F, G, S, eastern.

Bladder-Nut Family   Staphyleaceae

9* BLADDER-NUT *Staphylea pinnata.* Deciduous shrub to 5 m. Lvs pinnate, opposite. Fls white, 5-petalled, in hanging stalked clusters; May-June. Fr *inflated.* Woods on lime; often planted. (B, F), G.

145

Mallow Family   Malvaceae

Downy or softly hairy. Lvs rounded, palmately lobed, toothed, stalked, alternate, with stipules. Fls in whorls at base of lvs, open, with five notched petals; calyx in two rings, the outer smaller; stamens numerous, prominently bunched together. Fr a flat disc.

1* MUSK MALLOW Malva moschata. Medium/tall perennial, stem hairs often purple-based. Lvs deeply and narrowly cut. Fls 40-60 mm, solitary, rose-pink; hairs on stalks unbranched, outer calyx-ring 3-lobed, hairless or almost so; July-Aug. Fr hairy. Grassy places, scrub. T. **1a** M. alcea is rather larger, with branched hairs on fl stalks and outer calyx downy. F, G, S.

2* DWARF MALLOW Malva neglecta. More or less prostrate annual. Fls 3-6 together, pale lilac or whitish, 10-15 mm; petals twice as long as sepals; June-Sept. Waste places. T. **2a* Chinese Mallow** M. verticillata is a short/medium biennial, with smaller fls in denser whorls, the calyx enlarging in fr. (B, F, G), from E Asia.

3* LEAST MALLOW Malva parviflora. Short/medium annual, sometimes hairless. Fls 2-4 together, pale lilac-blue, 4-5 mm; petals slightly longer than short-haired or hairless sepals, which greatly enlarge in fr; June-Sept. Fr segments net-veined, winged on angles. Waste places. (B), F. **3a* Small Mallow** M. pusilla is usually prostrate, with up to ten pinkish fls in each cluster, sepals long-haired, not enlarging; angles of fr segments sharp but not winged. T, northern, but (B).

4* COMMON MALLOW Malva sylvestris. Variable, often sprawling, medium/tall annual/perennial. Lvs often with a small dark spot. Fls 2 or more together, pale to dark pink-purple, 25-40 mm; outer sepal-ring not joined at base; June-Oct. Fr segments with sharp angles. Waste places, waysides. T. **4a** Lavatera thuringiaca has larger solitary fls and outer sepal-ring joined at base. G, (S).

5* TREE MALLOW Lavatera arborea. Tall perennial, to 3 m, softly downy, woody at base. Fls pink-purple, veined purple; outer sepal-ring 3-lobed, cup-shaped; Apr-Sept. Rocks and bare ground by sea. B, F.

6* SMALLER TREE MALLOW Lavatera cretica. Medium/tall annual/biennial, differing from 4 especially in its less deeply lobed lvs, paler and pinker fls, outer sepal lobes broader and joined at base, and fr segments with rounded angles; Apr-July. Bare places. B (rare), F.

7* MARSH MALLOW Althaea officinalis. Tall perennial, covered with velvety down, whose roots are used to make the sweetmeat. Fls pink, 25-30 mm; outer sepal-ring 6-9-lobed; Aug-Sept. Damp saline places, especially near sea. B, F, G. **7a* Hollyhock** Alcea rosea is very much larger, with fls up to 100 mm, often white or mauve. Waste places. (B, F, G), of unknown origin.

8* ROUGH MALLOW Althaea hirsuta. Short/medium, sometimes semi-prostrate annual. Lvs becoming more deeply cut up the stems. Fls pale pink, becoming bluer; outer sepal-ring with 6-9 narrow lobes; May-July. Fr rough, hairless. Cf. 1. Grassy and bare places, weed of cultivation; likes lime. (B), F, G.

On p. 151
8* WILD PANSY Viola tricolor (see p. 151). Variable low/short annual/biennial, hairless or downy. Lvs oval to broad lanceolate; stipules lfy, pinnately lobed, the end lobe longer. Fls 10-25 mm, violet, yellow or both, petals longer than sepals, spur longer than appendages; Apr-Nov. Grassy and bare places, dunes. T. **8a* Mountain Pansy** V. lutea ☐ is a creeping perennial with longer fl stalks and 15-30 mm fls, the end segment of the stipules not larger than the others; May-Aug. Mountain grassland. **8b** V. hispida has hairy stems. F, north-western, rare.

9* FIELD PANSY Viola arvensis (see p. 151). Variable low/short annual, shortly hairy. Lvs broad to narrow; stipules lfy, semi-pinnate, end segment lanceolate. Fls 10-15 mm, cream-coloured, sometimes tinged yellow or violet, petals usually shorter than sepals, spur equalling appendages; Apr-Nov. Weed of cultivation. T. **9a* Dwarf Pansy** V. kitaibeliana ☐ is smaller, and downy, with more rounded lower lvs and stipule lobes, and 4-8 mm fls never violet-tinged; Apr-July. Dry bare ground. B (rare), F, G, southern.

St John's Wort Family Guttiferae

Hairless, except 4, 9. Lvs opposite, untoothed, with translucent veins. Fls in branched, usually lfy, clusters (except 2), open, yellow, with five petals and sepals and many stamens. Fr a dry capsule, except 1.

1* TUTSAN *Hypericum androsaemum.* Medium/tall half-evergreen undershrub; stems 2-winged, often reddish. Lvs broad oval, faintly aromatic when crushed. Fls 20 mm; sepals oval, unequal; June-Aug. Fr a fleshy *berry*, green, then red, finally purplish-black. Woods, shady places, cliff ledges. B, F. **1a* Stinking Tutsan** *H. hircinum* has stems usually 4-angled, lvs narrower and foetid when crushed, larger (30 mm) fls with narrower sepals, and dry fr. (B, F), from the Mediterranean. **1b* Tall Tutsan** *H. inodorum,* intermediate between 1 and 1a, is taller than both, to 2 m, with no smell when lvs crushed, and fr fleshy and red at first, then dry. (B, F), from Madeira.

2* ROSE OF SHARON *Hypericum calycinum.* Creeping medium evergreen undershrub. Lvs elliptical. Fls solitary, *70-80 mm,* anthers red; June-Oct. Banks, shrubberies. (B, F), from S E Europe.

3* PERFORATE ST JOHN'S WORT *Hypericum perforatum* agg. Medium perennial, with two raised lines down stem. Lvs oval to linear, with *translucent dots.* Fls 20 mm, petals black-dotted, especially on margins; pointed sepals sometimes also black-dotted; July-Sept. Grassy and bushy places. T. **3a* Imperforate St John's Wort** *H. maculatum* has square but not winged stems, normally no translucent dots on lvs, black dots usually on surface rather than margins of petals, and blunt sepals. Cf. 5. Hybridises with 3. Shadier places.

4* HAIRY ST JOHN'S WORT *Hypericum hirsutum.* Medium/tall *downy* perennial. Lvs elliptical, with translucent dots. Fls pale yellow, sometimes red-veined, 15-20 mm; sepals edged with black dots; July-Sept. Woods, scrub, hedge-banks. T. **4a* Pale St John's Wort** *H. montanum* is stiffer, less branched and less downy, with no translucent dots in the lvs but a row of black dots on the margin beneath, and fewer paler fls. Usually on lime.

5* SQUARE-STALKED ST JOHN'S WORT *Hypericum tetrapterum.* Medium/tall perennial; stems *4-winged.* Lvs with translucent dots, almost clasping the stem. Fls pale yellow, 10 mm; sepals pointed, usually not black-dotted; July-Sept. Cf. 3a. Damp places, often by streams. T, southern. **5a* Wavy St John's Wort** *H. undulatum* has slenderer and less conspicuously winged stems, with wavy-edged lvs, larger, usually red-tinged fls, and black-dotted sepals. B, F.

6* SLENDER ST JOHN'S WORT *Hypericum pulchrum.* Slender short/medium perennial. Lvs with translucent dots and often inrolled margins. Fls rich yellow, tinged *red* beneath, 15 mm; petals and sepals with black dots, petals also red-dotted; July-Sept. Heaths, scrub, open woods; not on lime. T, western.

7* TRAILING ST JOHN'S WORT *Hypericum humifusum.* Slender, usually prostrate perennial. Lvs with translucent dots, often also with black dots on margins beneath. Fls 10 mm, sepals sometimes black-dotted at margins; July-Oct. Open woods, heathy places; not on lime. T, southern. **7a* Flax-leaved St John's Wort** *H. linarifolium* □ is short/medium, often erect, and has narrower lvs, usually with neither translucent nor black dots, the margins often inrolled and larger fls, often red beneath. Hybridises with 7. Rocky places. B (rare), F.

8* IRISH ST JOHN'S WORT *Hypericum canadense.* Low/short annual/perennial; stems with four raised lines. Lvs narrow, 1-3-veined. Fls with no black dots, 6 mm; July-Sept. Damp places. B (Ireland), G. **8a** *H. mutilum* is taller with broader 3-5-veined lvs. F, G. **8b** *H. majus* is taller with broader 5-7-veined lvs and larger fls. F, G.

9* MARSH ST JOHN'S WORT *Hypericum elodes.* Creeping low/short *greyish* perennial, conspicuously hairy. Lvs rounded. Fls less open than other St John's Worts, 15 mm; sepals red-dotted at margins; June-Sept. Shallow fresh water, damp mud. B, F, G, western.

7a* Flax-leaved St John's Wort

1

2

3

4

5

6

7

8

9

Violet Family   Violaceae

VIOLETS AND PANSIES. *Viola*. Lvs alternate, toothed, with stipules. Fls solitary, open, with five unequal petals, the lowest spurred; sepals with short appendages. Fr an ovoid capsule. Hybrids frequent. Violets (1-7) are perennials, with more or less heart-shaped lvs and inconspicuous stipules. Pansies (8-9) may be annual and have more deeply toothed, more or less broad lanceolate lvs, with conspicuous lflike stipules, and flatter fls.

1* SWEET VIOLET *Viola odorata*. Low creeping perennial, downy, with rooting runners. Lvs in a tuft, rounded, enlarging in summer. Fls 15 mm, *fragrant,* blue-violet or white, rarely lilac, pink or yellow; sepals *blunt;* Mar-May, Aug-Sept. Woods, scrub, hedgebanks. T. **1a* Hairy Violet** *V. hirta* has no runners, narrower lvs with longer hairs on the stalks, and unscented fls less often white. Grassland, open scrub, often on lime. **1b White Violet** *V. alba* is hairless and has non-rooting runners, more oval lvs persisting through the winter, narrower stipules and larger (15-20 mm) fls. F, G, southern. **1c** *V. collina* has no runners, narrower lvs and stipules, and pale blue fls with a whitish spur. On lime. F, G, S, southern. **1d** *V. suavis* has shorter runners, very large (to 200 mm) long-stalked summer lvs and fls violet with white throat. F, (G).

2* COMMON DOG VIOLET *Viola riviniana*. Low perennial, almost hairless. Lvs in a tuft, heart-shaped, stipules usually toothed. Fls 15-25 mm, blue-violet, *unscented;* spur stout, curved, pale, often creamy, notched at tip; sepals *pointed,* appendages enlarging in fr; Mar-May, July-Sept. Woods, grassy places, mountains. T. **2a* Teesdale Violet** *V. rupestris* is smaller and minutely downy, with rounder lvs, fls 10-15 mm with spur pale violet, and fr downy. Hybridises with 2. Open ground, on lime. T, rare in B. **2b* Early Dog Violet** *V. reichenbachiana* □ has narrower lvs, paler 12-18 mm fls with narrower petals and an unnotched violet spur and sepal appendages less prominent in fr. Shady places. Southern. **2c** *V. mirabilis* □ is larger, with a line of hairs on stem, untoothed stipules turning brown, and pale violet fls with whitish spur. On lime. F, G, S. 2a-2c occasionally fl in autumn.

3* HEATH DOG VIOLET *Viola canina*. Low/short perennial, hairless or slightly downy. Lvs not in a tuft, broad to narrow. Fls 15-25 mm, bluer than 2, sometimes white, with spur yellowish or whitish. Sepals *pointed,* appendages not enlarging in fr; Apr-June. Heaths, open woods, fens. T. **3a* Pale Dog Violet** *V. lactea* □ is erecter, with narrower lvs, not heart-shaped, and pale bluish-white fls with a short greenish spur; May-June. Heaths. B, F.

4 MEADOW VIOLET *Viola pumila*. Low/short perennial, hairless. Lvs *narrow.* Fls 15 mm, roundish, *pale blue,* spur greenish, short; May-June. Grassland. F, G, S, eastern. **4a* Fen Violet** *V. persicifolia* has heart-shaped and smaller whitish fls, sometimes tinged blue, with violet veins. Fens, marshes. T, rare in B. **4b** *V. elatior* is taller, with much longer stipules, at least equalling lf-stalks, heart-shaped lvs and 20-25 mm fls. Damp woods and grassland. F, G, S.

5 NORTHERN VIOLET *Viola selkirkii*. Low perennial, almost hairless. Lvs tufted, heart-shaped. Fls 15 mm, pale violet, long-spurred, sepals pointed; Apr-May. *Coniferous* woods, damp places. S.

6* MARSH VIOLET *Viola palustris*. Low perennial, usually hairless. Lvs kidney-shaped. Fls 10-15 mm, pale lilac veined purple, short spurred, bracts at or below middle of stalks; Apr-July. Bogs, marshes. T. **6a** *V. epipsila* ⊔ is larger, with lvs in pairs, some hairs beneath lvs, 15-20 mm fls and bracts above middle of stalk. G, S. **6b** *V. uliginosa* has heart-shaped lvs and larger violet fls. G, S, northern, rare.

7 YELLOW WOOD VIOLET *Viola biflora*. Low creeping perennial, sparsely hairy. Lvs in a tuft, kidney-shaped. Fls 15 mm, bright yellow; June-Aug. Damp, shady places, mainly in mountains. F, G, S.

8* WILD PANSY *Viola tricolor*: see p. 146.

9* FIELD PANSY *Viola arvensis*: see p. 146.

151

Rock-Rose Family   Cistaceae

Downy. Lvs untoothed. Fls 5-petalled, open, sepals unequal, stamens numerous.

1* COMMON ROCK-ROSE *Helianthemum nummularium.* Variable, more or less prostrate undershrub. Lvs lanceolate to roundish, 1-veined, usually downy *white* beneath, opposite, with lflike stipules. Fls 20-25 mm, yellow, rarely white, cream, orange or pink; May-Sept. Grassy and rocky places on lime. T, southern. **1a* White Rock-rose** *H. apenninum* □ has narrow lvs with markedly inrolled margins and smaller stipules, and fls white with yellow centre. T, rare in B. **1b** *H. oelandicum* has narrow lvs, not white beneath. Mainly in mountains. F, G, S, eastern. **1c* Hoary Rock-rose** *H. canum* □ is smaller, with narrower lvs, sometimes downy grey above, and 10-15 mm fls; May-June.

2* SPOTTED ROCK-ROSE *Tuberaria guttata.* Low annual. Lvs elliptical, 3-veined, opposite. Fls 10-20 mm, yellow, often with a *red spot*, petals dropping by noon, sepals black-dotted; May-Aug. Dry, bare places, often on lime. B (rare). F, G.

3 COMMON FUMANA *Fumana procumbens.* Spreading undershrub to 40 cm. Lvs alternate, linear, no stipules. Fls yellow, usually with a deeper yellow spot; May. Dry, bare, often rocky places. F, G, S, southern.

4* SEA-HEATH *Frankenia laevis* (Frankeniaceae). Prostrate mat-forming heather-like undershrub, variably downy. Lvs opposite, margins inrolled, sometimes with a white crust. Fls 5-petalled, open, pink, purplish or whitish; July-Aug. Fr a capsule. Sands, shingle and salt-marshes by sea. B, F.

5* WHITE BRYONY *Bryonia cretica* (Cucurbitaceae). Climbing perennial to 4m, hairy. Lvs *ivy-shaped,* not glossy, with unbranched tendrils opposite the stalks. Fls 10-18 mm, 5-petalled, greenish-white with darker veins; stamens and styles on different plants, stigma downy; May-Sept. Fr a *red* berry. Hedges, scrub. B, F, G. **5a** *B. alba* has stamens and styles in different fls on same plant, stigma hairless and berry black. G, (F, S). **5b Prickly Cucumber** *Echinocystis lobata* has branched tendrils, more sharply palmate lvs, male fls smaller, in stalked spikes, female fls solitary, on same plant, and fr prickly. (G) from N America.

Willowherb Family   Onagraceae

6* ENCHANTER'S NIGHTSHADE *Circaea lutetiana.* Short/medium patch-forming downy perennial. Lvs pointed oval, sometimes slightly heart-shaped, opposite. Fls 4-8 mm, in lfless spikes elongating before fls fall; 2-petalled, white; June-Sept. Fr covered with hooked bristles. Woods, shady places, a garden weed. T. **6a* Alpine Enchanter's Nightshade** *C. alpina* is smaller and much less downy, with markedly heart-shaped, more deeply toothed lvs and fl spike not elongating until after fls fall. Mountain woods. Northern. **6 × 6a* Upland Enchanter's Nightshade** *C. × intermedia* is sterile and often found away from the parents.

EVENING PRIMROSES *Oenothera.* Medium/tall hairy biennials, with alternate lanceolate lvs. Fls yellow, 4-petalled; June-Sept. Fr an elongated capsule. Bare and waste places, dunes. (B, F, G), of garden origin.

7* LARGE-FLOWERED EVENING PRIMROSE *Oenothera erythrosepala.* Has conspicuous red blotches and numerous red-based hairs on stem, lvs crinkled and with white midrib, sepal-tube red-striped, petals 30-50 mm long, styles longer than stamens, and fr tapering to tip. Five other fairly frequent species. **7a* Common Evening Primrose** *O. biennis*, **7b*** *O. fallax* (of hybrid origin from 7 and 7a), **7c* Small-flowered Evening Primrose** *O. cambrica*, **7d** *O. rubricaulis* and **7e** *O. salicifolia* differ as follows: 7d, 7e have no red blotches; 7a has no red-based hairs; 7a, 7c have lvs flat; all, except sometimes 7b, 7e, have lvs with red midrib; 7a, 7b, 7c have petals 20-30 mm; 7d, 7e have petals 10-20 mm; 7a, 7c, 7d have sepals all green; all have short styles.

8* FRAGRANT EVENING PRIMROSE *Oenothera stricta.* Shorter than 7-7e, with no red-based hairs, narrow lvs, buds sometimes red, petals 15-35 mm, reddening when dried, short styles and fr thickened at tip.

Willowherb Family *(contd.)*

WILLOWHERBS *Epilobium*. More or less hairy perennials. Lvs lanceolate (except 6), toothed. Fls in stalked spikes (except 6), with four notched petals and eight stamens; June-Aug. Fr a long narrow 4-sided pod, splitting when ripe to reveal the silky plume of hairs attached to each seed. Hybrids are very frequent.

1* ROSEBAY WILLOWHERB *Epilobium angustifolium*. Tall, patch-forming, almost hairless. Lvs alternate. Fls 20-30 mm, bright pink-purple, petals slightly *unequal*. Open woods, heaths, mountains, waste ground. T. **1a River Beauty** *E. latifolium* has shorter lvs and fewer fls in the spike. Iceland. **1b** *E. dodonaei* has much narrower lvs. Bare dry places. F, G.

2* GREAT WILLOWHERB *Epilobium hirsutum*. Tall, softly hairy. Lvs mostly opposite, the upper half-clasping stem. Fls 15-25 mm, purplish-pink, stigma 4lobed. Damp places. T. **2a* Hoary Willowherb** *E. parviflorum* □ is smaller, with woolly hairs on lower stem, upper lvs alternate and not clasping stem, and much smaller (6-9 mm) fls.

3* BROAD-LEAVED WILLOWHERB *Epilobium montanum* agg. Short/medium perennial, stem rounded. Lvs broad lanceolate, opposite, very shortly stalked. Fls 6-9 mm, pinkish-purple, petals well notched, stigma 4-lobed. Shady and waste places. T. **3a** *E. collinum* has smaller fls with blunt not pointed buds. F, G, S. **3b* Spear-leaved Willowherb** *E. lanceolatum* has longer-stalked and mostly alternate lvs, stem slightly 4-angled and fls white becoming pink. B, F, G. **3c* Square-stemmed Willowherb** *E. tetragonum* □ has unstalked strap-shaped lvs and less notched petals. **3d* Short-fruited Willowherb** *E. obscurum* has unstalked lvs and prefers damper places. **3e* Pale Willowherb** *E. roseum* □ has mostly unstalked long-stalked lvs and fls white becoming pink, the latest to flower, Aug-Sept. **3f* American Willowherb** *E. adenocaulon* □ has mostly alternate lvs and smaller, paler fls. (T), from N America. **3g** *E. glandulosum* is like 3f with larger purplish-pink fls. (S) from N America. **3h** *E. alpestre* has lvs whorled. Mountains. G. All have somewhat narrower lvs. 3c-3h have four raised lines on stem and club-shaped stigma.

4* MARSH WILLOWHERB *Epilobium palustre*. Slender short/medium perennial, with thread-like runners. Lvs unstalked, narrow, not or scarcely toothed. Fls 4-6 mm, pink to white, stigma club-shaped. Wet places. T. **4a** *E. davuricum* has no runners, narrower lvs and white fls. S. **4b** *E. nutans* has lvs broader and fls pale violet. Mountain moors. G, southern.

5* ALPINE WILLOWHERB *Epilobium anagallidifolium*. Low creeping perennial, almost hairless, with runners above ground. Lvs little toothed. Fls 4-5 mm, reddish-pink, sepals red. Pods red when ripe. Wet places on mountains. T. **5a** *E. hornemanni* has runners below ground and larger, pale violet fls. S. **5b** *E. lactiflorum* is like 5a but fls white. **5c* Chickweed Willowherb** *E. alsinifolium* is larger with more pointed and toothed lvs, and larger, more purplish fls.

6* NEW ZEALAND WILLOWHERB *Epilobium brunnescens*. Prostrate creeping perennial, almost hairless. Lvs stalked, rounded, weakly toothed, often bronzy. Fls 4 mm, long-stalked, very pale pink, petals deeply notched. Bare damp places. (B, F), from New Zealand. **6a/6b** *E. komarovianum* and *E. pedunculare*, with elliptical lvs and sharply toothed lvs respectively, are two closely related and locally established New Zealand plants which may spread.

Loosestrife Family Lythraceae

7* PURPLE LOOSESTRIFE *Lythrum salicaria*. Tall perennial, variably downy; stems with four raised ˜nes. Lvs opposite or in whorls of 3, untoothed, unstalked, lanceolate. Fls 10-15 mm, in whorled spikes, *6-petalled*, bright red-purple; June-Aug. Fr egg-shaped. Damp places, especially by rivers. T.

8* GRASS POLY *Lythrum hyssopifolia*. Low/short hairless annual. Lvs alternate, linear to oblong. Fls tiny, 5-6-petalled, pale pink, unstalked at base of lvs; July-Sept. Bare damp ground, especially winter-wet hollows. T, southern.

155

Key to Umbellifers

(Pages 158-171)

The great majority of umbellifers, members of the Carrot Family, Umbelliferae, are easily distinguished by the characteristic umbrella-shape of their flower-heads, which are usually white, sometimes tinged pink, but yellow in a few species. A small number, however, notably Sanicle and Sea Holly (p. 168) and the *Bupleurum* species (p. 168), are quite untypical, while some other plants, such as Danewort and Yarrow (p. 242) have superficially umbellifer-like flower-heads. These are all shown in the Main Key on p. 13. The fruits of umbellifers are important in identification.

Plant

aromatic: Sweet Cicely, *Chaerophyllum bulbosum* 160; Spignel 162; Hemlock Water Dropwort, Cambridge Milk-parsley 164; Wild Celery, Corn Parsley 166; Scots Lovage 168; Wild Parsnip, Alexanders, Lovage 170

foetid: Coriander 160; Hemlock 164; Stone Parsley 166

growing in water: Marsh Pennywort 158; Cowbane, Greater Water-parsnip 164; Lesser Water-parsnip, Fool's Watercress, Water Dropworts 166

Stems

purplish: Rough and Bur Chervils 160; Angelica, Hemlock 164; Stone Parsley 166; Milk Parsley, Cnidium, Scots Lovage 168; Thorow-wax 170

ridged: Burnet Saxifrage, Moon Carrot, Fool's Parsley 162; Hemlock Water Dropwort, Greater Water-parsnip, Pleurospermum, Cambridge Milk-parsley 164

angled: Honeywort 162; Biennial Alexanders 170

rooting: Fool's Watercress 166

Leaves

rounded: Marsh Pennywort, Sea Holly 158; Thorow-wax 170

linear/lanceolate: Hare's ears 168 and 170

palmate: Sanicle, Astrantia 158

1-pinnate: Bur Parsleys, Coriander 160; Burnet Saxifrage, Moon Carrot 162; Hogweed, Greater Water-parsnip 164; p. 166; Hartwort 168, Wild Parsnip, Rock Samphire 170

1–3-trefoil: Ground Elder, Bladderseed 162; Pleurospermum, Longleaf 164; Hog's Fennel, Masterwort, Scots Lovage 168, Alexanders 170

thread-like: Whorled Caraway, *Ammi* 160; Spignel 162; Fine-leaved Water Dropwort 166; Fennel 170

spiny: Sea Holly 158

sheathing: Hogweed, Angelica 164; Masterwort, Scots Lovage 168

Flowers

yellow: Spignel 162; Hog's Fennel, Slender Hare's-ear 168, all on p. 170

blue: Sea Holly 158

enclosed by bracts: Thorow-wax, Small Hare's-ear 170

in simple umbels: all on p. 158; Shepherd's Needle 162

with unequal petals: Upright Hedge Parsley, Wild Carrot 160

with lower bracts always present: all on p. 158; Upright Hedge and Greater Bur Parsleys, Whorled Caraway, Wild Carrot 160; Great Pignut, Moon Carrot, Spignel, Bladderseed 162; Hemlock, Hemlock Water Dropwort, Greater Water-parsnip, Pleurospermum, Longleaf 164; Lesser Water-parsnip, Parsley Water Dropwort, Corn and Stone Parsleys 166; *Peucedanum alsaticum*, *P. oreoselinum*, Milk Parsley, Sermountain, Scots Lovage 168; Rock Samphire, Alexanders, Lovage, Small Hare's-ear 170

with petal-like bracts: Astrantia 158

with divided bracts: Wild Carrot 160; Lesser Water-parsnip 166; Milk Parsley 168

with spiny bracts: Sea Holly 158

Fruits

bristly/spiny: Sanicle 158; Bur Chervil, Bur Parsleys, Wild Carrot, 160; Hartwort 168

aromatic: Coriander, Caraway 160

Dogwood Family *Cornaceae*

Lvs opposite, untoothed, pointed oval. Fls 4-petalled, in an umbel. Fr a berry.

1* DOGWOOD *Cornus sanguinea.* Deciduous shrub to 4 m, with *red* twigs, conspicuous in winter. Lvs with 3-5 pairs of prominent veins, reddening in autumn. Fls *white;* May-July. Fr black. Scrub, hedgerows, on lime. T. **1a*** *C. alba* with larger 5-7-veined lvs, yellowish-white fls and white fr, is widely planted and locally naturalised. (B, S), from N Russia. **1b* Red Osier** *C. sericea* differs from 1a in its numerous rooting runners and more sharply pointed lvs. (B), from N America.

2* CORNELIAN CHERRY *Cornus mas.* Deciduous shrub or small tree to 8 m, with greenish-yellow twigs. Umbels unstalked with four yellow-green sepal-like bracts, fls *yellow;* Feb-Mar, *before lvs.* Fr red. Woods, scrub, widely planted. (B), F, G.

3* DWARF CORNEL *Cornus suecica.* Short creeping perennial. Lvs unstalked, rounded to elliptical. Umbels with four conspicuous *white petal-like* bracts, fls purplish-black; June-Aug, shy flowerer. Fr red. Heaths, moors, mountains; not on lime. B, G, S.

Ivy Family Araliaceae

4* IVY *Hedera helix.* Evergreen woody climber, carpeting the ground or ascending by means of tiny roots. Lvs glossy, often bronzy, 3-5-lobed (ivy-shaped) on non-flg shoots, pointed oval on flg shoots. Fls in an *umbel,* green with yellow anthers; Sept-Nov. Berries black. Woods, hedgerows, rocks, walls. T.

Carrot Family Umbelliferae

pp. 158-70. Lvs alternate, no stipules (except Marsh Pennywort, p. 158). Fls small, 5-petalled, in umbels, usually in turn arranged in an umbel. The whole flhead thus looks like a flat-topped umbrella, whose spokes (rays) are the stalks of the secondary umbels. Fls of white-fld species often tinged pink. The strap-like upper and lower bracts at the base of the primary and secondary umbels respectively, and the dry, usually ridged fr, are both important in identification.

Umbellifers on this page are untypical, all having their fls in simple umbels.

5* MARSH PENNYWORT *Hydrocotyle vulgaris.* Prostrate creeping perennial. Lvs almost *circular,* shallowly lobed, erect on long stalks. Fls tiny, pinkish-green, in one or more whorls from base of lvs, often hard to detect among vegetation; June-Aug. Fr rounded. Damp, often grassy places, shallow fresh and brackish water. T.

6* SANICLE *Sanicula europaea.* Short/medium hairless perennial. Lvs long-stalked, shiny, *palmately* 3-5-lobed, toothed. Fls pale pink or greenish-white, in clusters of tight umbels; May-July. Fr roundish with hooked spines. Woods, often on lime. T.

7* ASTRANTIA *Astrantia major.* Medium/tall perennial. Lvs long-stalked, palmately lobed, toothed. Fls white, scented, in umbels with pinkish *petal-like* bracts; June-Sept. Woods, mountain meadows. (B, rare), F, G.

8* SEA HOLLY *Eryngium maritimum.* Short/medium hairless perennial. Lvs leathery, rounded, spiny, blue-green with whitish veins and edges. Fls powder-blue, in tight umbels with spiny lflike bracts; June-Sept. Fr egg-shaped. Sand and shingle by sea. T. **8a* Field Eryngo** *E. campestre* is much branched and greener, with smaller heads of usually greenish-white fls. Bare, dry places. B (rare), F, G. **8b** *E. planum* is smaller with shorter bracts. Dry places inland. G (rare).

Carrot Family   Umbelliferae (see p. 158)

1* COW PARSLEY *Anthriscus sylvestris* agg. Medium/tall perennial, slightly hairy, stems often purple. Lvs 2-3-pinnate. Fls white, no lower bracts; Apr-June. Fr oblong, beaked, black, *smooth*. Hedge-banks, shady places. T. **1a* Garden Chervil** *A. cerefolium* is a shorter annual, with umbels opposite lvs and narrower long-beaked fr. Escape from cultivation. (B, F, G), from E Europe. **1b* Sweet Cicely** *Myrrhis odorata* □ is aromatic when crushed, with lvs flecked whitish and much longer ribbed fr. Waysides. (T), from S Europe.

2* ROUGH CHERVIL *Chaerophyllum temulentum*. Medium/tall hairy biennial; stem *purple* or purple-spotted. Lvs 2-3-pinnate, turning purple. Fls white, no lower bracts; May-July. Fr long, ridged, narrowed to tip. Hedge-banks, shady places. T, southern. **2a** *C. hirsutum* agg. has green stems and hairy petals. F, G. **2b** *C. bulbosum* has a tuberous rootstock, and lower stem and upper bracts both hairless. F, G, S. **2c* Golden Chervil** *C. aureum* is taller and less hairy, with yellow-green aromatic lvs. (B), F, G. **2d** *C. aromaticum* has undivided lf lobes. G (rare).

3* BUR CHERVIL *Anthriscus caucalis*. Medium annual, slightly hairy; stem often purplish below. Lvs 2-3-pinnate, finely cut. Fls white, umbels *opposite* lvs, no lower bracts; May-June. Fr with hooked bristles. Bare, often sandy places. T, southern.

4* UPRIGHT HEDGE PARSLEY *Torilis japonica*. Medium/tall hairy annual; stems *stiff*. Lvs 1-3-pinnate, narrowly triangular. Fls white or pink, petals slightly unequal, lower bracts 4-6; July-Sept. Fr egg-shaped, with purple *hooked* bristles. Hedge-banks, shady places. T. **4a* Spreading Bur Parsley** *T. arvensis* is shorter, with 0-1 lower bracts and fr bristles not hooked. A weed of cultivation. B, F, G.

5* KNOTTED BUR PARSLEY *Torilis nodosa*. More or less prostrate annual, roughly hairy. Lvs 1-2-pinnate. Fls white or pink, umbels few-fld, *opposite* lvs; no lower bracts, upper bracts longer than fls; May-July. Fr with warts and *straight* bristles. Dry, sparsely grassy places. B, F, G.

6* CORIANDER *Coriandrum sativum*. Short hairless, *foetid* annual. Lvs 1-2-pinnate, the lower with broader segments. Fls white or pink, outer petals longer, no lower bracts; June-Aug. Fr globular, ridged, red-brown, *aromatic* when ripe. Bare places. (B, F, G), from W Asia. **6a** *Bifora radians* is not foetid and has lf-lobes linear, fls larger and fr 2-lobed. (F, G).

7* GREATER BUR PARSLEY *Turgenia latifolia*. Medium annual, roughly hairy. Lvs *1-pinnate*. Fls pink, red-purple or white, with lower bracts 2-3, upper bracts broad with pale edges; May-Aug. Fr ridged, with 2-3 rows of *hooked* bristles on each ridge. A weed of cultivation. (B, rare), F, G. **7a* Small Bur Parsley** *Caucalis platycarpos* is shorter and pale green, with 2-3-pinnate lvs, outer petals much longer, and fr with bristles in single rows. **7b** *Orlaya grandiflora* has 2-3-pinnate lvs, outer petals much longer and fr with bristles in 1-2 rows. Grassy places. F, G.

8* CARAWAY *Carum carvi*. Medium hairless biennial. Lvs 2-3-pinnate. Fls white, usually no bracts; June-July. Fr oblong, ridged, *aromatic*, the familiar caraway seed. Grassy places. T, but (B).

9* WHORLED CARAWAY *Carum verticillatum*. Medium/tall hairless perennial. Lvs oblong, with numerous *whorled* threadlike segments, recalling Yarrow (p. 238). Fls white, with both upper and lower bracts; July-Aug. Fr egg-shaped, ridged. Damp grassland, marshes. B, F, G, western.

10* WILD CARROT *Daucus carota*. Medium hairy biennial. Lvs 3-pinnate. Fls white, the centre one of the umbel often red, petals often unequal, *lower bracts conspicuously pinnate or 3-forked;* June-Sept. Fr flattened, with often hooked bristles; fr umbels become hollowly concave. Grassy places; near the sea often fleshy and more luxuriant. T. **10a*** *Ammi visnaga* and **10b*** *A. majus* have tangled linear lf-lobes, resembling Hog's Fennel (p. 164). Casual (T). 10a has upright thick rays.

1

1b

2

3

4

5

6

7

8

9

10

Carrot Family *(contd.)*

1* PIGNUT *Conopodium majus.* Short/medium perennial, almost hairless; root tuberous, edible. Lvs 2-3-pinnate, the upper with almost *thread-like* lobes. Fls white, lower bracts 0-2; May-July. Fr oblong, obscurely ridged, with erect styles. Open woods, shady grassland. B, F, S. **1a* Great Pignut** *Bunium bulbocastanum* is taller and stouter, with 5-10 lower bracts and fr prominently ridged, the styles turned down. Especially on lime. B(rare), F, G.

2* BURNET SAXIFRAGE *Pimpinella saxifraga.* Variable medium perennial, downy; stems slightly ridged. Lvs 1-2-pinnate, the lower usually *1-pinnate* with broader lflets, the upper usually 2-pinnate with narrow lflets. Fls white, no bracts; May-Sept. Fr egg-shaped. Dry grassland, especially on lime. T. **2a* Greater Burnet Saxifrage** *P. major* □ is taller and hairless, with much larger lvs all 1-pinnate, fls often pink and fr prominently ridged. Damper, shadier grassland. **2b** *Ptychotis saxifraga* has lower lf lobes deeply cut, upper thread-like; and small upper bracts. F.

3* SHEPHERD'S NEEDLE *Scandix pecten-veneris.* Short annual, almost hairless. Lvs 2-3-pinnate, lobes linear. Fls white, umbels often simple but sometimes with 2-3 rays, *opposite* lvs; no lower bracts, upper bracts broad, deeply toothed; May-Aug. Fr conspicuously *long* (15-80 mm). Weed of cultivation. T, southern.

4* GROUND ELDER *Aegopodium podagraria.* Creeping, *patch-forming* short/medium perennial. Lvs 1-2-trefoil. Fls white, usually no bracts; June-Aug. Fr egg-shaped, ridged. Shady and waste places, and a tenacious garden weed. T.

5* MOON CARROT *Seseli libanotis.* Variable medium/tall perennial, downy or hairless; stems ridged. Lvs 1-3-pinnate, lobes narrow. Fls white or pink, umbels many-rayed, bracts numerous; July-Sept. Fr egg-shaped. Dry grassland, especially on lime. T, but rare in B. **5a** *S. montanum* is shorter and always hairless, with fewer rays and no lower bracts. F, G, southern. **5b** *S. annuum* is shorter with lvs oval in outline and 0-1 lower bracts. F, G, southern. **5c** *S. hippomarathrum* is always hairless, with lvs 2-3-pinnate and no lower bracts. G(rare), southern.

6* FOOL'S PARSLEY *Aethusa cynapium.* Low/medium annual, hairless; stems ribbed. Lvs 2-3-pinnate, dark green. Fls white, with long upper bracts making umbels look *bearded;* June-Oct. Fr egg-shaped, ridged. Weed of cultivation. T. Subsp. *cynapioides* is taller; woods, eastern.

7* SPIGNEL *Meum athamanticum.* Short/medium perennial, hairless, *aromatic.* Lvs 3-4-*pinnate, lobes thread-like.* Fls white, often tinged yellow or purple, lower bracts 0-2; June-Aug. Mountain grassland. B, F, G. **7a** *Athamanta cretensis* is downy and not aromatic, with lvs 3-5-pinnate, lflets broader, fls always white and 0-5 lower bracts, sometimes pinnately cut. Mountain rocks. G, southern.

8* BLADDERSEED *Physospermum cornubiense.* Medium/tall perennial, almost hairless. Lvs 2-trefoil, *dark green.* Fls white, bracts lanceolate; July-Aug. Fr globular, ridged. Woods. B(rare), F.

9* HONEWORT *Trinia glauca.* Low *greyish* hairless perennial; stems angled, much branched. Lvs 2-3-pinnate, lobes narrow. Fls white, stamens and styles on different plants, no bracts; May-June. Fr egg-shaped, ridged. Grassland, especially on lime. B(rare), F, G.

2a* Greater Burnet Saxifrage

163

Carrot Family (contd.)

1* HOGWEED *Heracleum sphondylium.* Very variable tall *stout* biennial/perennial, to 3 m, hairless or roughly hairy. Lvs 1-3-pinnate, with very broad toothed lflets and stalks expanded to a sheath. Fls white or pink, umbels up to 20 cm across, petals of outer fls *larger,* usually no lower bracts; Apr-Nov. Fr flattened, elliptical to rounded. Grassy places, open woods. T. **1a* Giant Hogweed** *H. mantegazzianum* □ is much stouter and taller, to 5 m, with stems often blotched purple, umbels up to 50 cm across, petals up to 12 mm, and fr narrower; June-July. Especially on damp ground, forming large patches. (T), from S W Asia.

2* ANGELICA *Angelica sylvestris.* Tall perennial, almost hairless; stems often purplish. Lvs 2-3-pinnate, with broad, toothed lflets and *inflated* sheathing stalks. Fls white or pink, sepals untoothed, often no lower bracts, upper bracts thread-like; July-Sept. Fr egg-shaped, flattened, winged. Damp grassy places and woods, cliffs. T. **2a* Garden Angelica** *A. archangelica* has stems usually green, lf-stalks aromatic (the source of the sweetmeat), fls greenish or yellowish, and fr with corky wings. Often by the sea. G, S, (B, F). **2b** *A. palustris* has white sepal-teeth. Wetter places. G (rare). **2c** *Conioselinum tataricum* is always hairless and has fls always white, with no sepals. Stony places. S, Arctic.

3* HEMLOCK *Conium maculatum.* Tall hairless, foetid, very poisonous biennial; stems *purple-spotted.* Lvs 2-4-pinnate, finely cut. Fls white, upper bracts only on *outer edge* of umbel; June-Aug. Fr globular, with wavy ridges. Damp, sparsely grassy places. T.

4* HEMLOCK WATER DROPWORT *Oenanthe crocata.* Tall hairless parsley-scented, very poisonous perennial; stems grooved. Lvs 3-4-pinnate, with broad *wedge-shaped* toothed lflets. Fls white, sometimes no lower bracts; June-Aug. Fr cylindrical, long-styled. Damp grassy places. B, F.

5* GREATER WATER-PARSNIP *Sium latifolium.* Tall hairless perennial; stems ridged. Lvs *1-pinnate* with about five pairs of lflets; also submerged 2-3-pinnate lvs. Fls white, lower bracts large, *lfy;* June-Sept. Fr egg-shaped, ridged, short-styled. In and by fresh water, fens, marshes. T.

6 PLEUROSPERMUM *Pleurospermum austriacum.* Tall biennial/perennial, slightly hairy; stems ridged. Lvs 2-3-trefoil, *triangular,* lobes wedge-shaped, toothed. Fls white, upper bracts unequal, down-turned; June-Sept. Fr egg-shaped, ridged. Mountain grassland. F, G, S, eastern.

7* COWBANE *Cicuta virosa.* Tall, hairless, very poisonous perennial; stems sometimes half-floating. Lvs 2-3-pinnate, lobes toothed. Fls white, no lower bracts; July-Aug. Fr globular, ridged. In and by fresh water. T.

8* LONGLEAF *Falcaria vulgaris.* Medium/tall patch-forming annual/perennial; stems much branched. Lvs 1-2-trefoil with long *strap-like,* finely toothed lflets. Fls white, umbels rather open, bracts linear; July-Sept. Fr oblong, ridged. Grassy and waste places. T but (B, S).

9* CAMBRIDGE MILK-PARSLEY *Selinum carvifolia.* Tall perennial, almost hairless, parsley-scented; stems ridged. Lvs 2-3-pinnate, with winged ridges. Damp grassland and open woods. T, eastern, but rare in B. **9a** *S. pyrenaeum* has unwinged, almost lfless stems and few-rayed umbels with yellowish fls. F, mountains, rare.

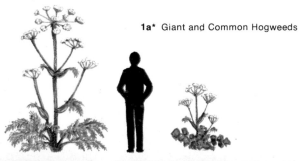

1a* Giant and Common Hogweeds

Carrot Family *(contd.)*

1* LESSER WATER-PARSNIP *Berula erecta*. Medium, often sprawling hairless perennial. Lvs yellow-green, 1-pinnate with 7-10 pairs of toothed lflets; also 3-4 pinnate submerged lvs with linear lobes. Fls white, umbels *opposite* lvs, bracts usually lflike, often *3-cleft* or more or less pinnate; July-Sept. Fr globular, almost divided in two, faintly ridged, long-styled. Wet places. T.

2* FOOL'S WATERCRESS *Apium nodiflorum*. More or less prostrate perennial; stems rooting at lower lf-junctions. Lvs 1-pinnate, lflets oval to lanceolate, toothed. Fls white, umbels not or very shortly stalked, *opposite* lvs; *no* lower bracts; June-Sept. Fr egg-shaped, ridged. Wet places. B,F,G. **2a* Creeping Marshwort** *A. repens* is smaller, with stem rooting at every lf-junction, broader lflets and stalked umbels. T, rare in B. **2b* Lesser Marshwort** *A. inundatum* is smaller and usually largely submerged, with linear or thread-like lflets, fewer-rayed and longer-stalked umbels, and narrower fr. Hybridises with 2.

3* WILD CELERY *Apium graveolens*. Medium/tall yellow-green biennial, hairless, *celery-scented*. Lvs 1-2-pinnate, shiny, with large toothed lflets. Fls white. umbels shortly or not stalked, often opposite lvs; no bracts; June-Aug. Fr egg-shaped. Damp grassy places near sea, saline meadows inland. T.

4* TUBULAR WATER DROPWORT *Oenanthe fistulosa*. Medium greyish hairless perennial, little branched. Lvs 1-2-pinnate, the upper with linear lobes and *inflated* stalks. Fls white, umbels few-rayed, becoming globular in fr; usually no lower bracts; June-Sept. Fr cylindrical, long-styled. Wet places, shallow fresh water. T, southern.

5* PARSLEY WATER DROPWORT *Oenanthe lachenalii*. Medium/tall perennial. Lvs 2-pinnate, the upper with narrower pointed lobes and sometimes 1-pinnate. Fls white, umbel rays not thickened in fr, sometimes no lower bracts; June-Sept. Fr egg-shaped, ridged, short-styled. Damp grassland, fens. T, western. **5a* Corky-fruited Water Dropwort** *Oe. pimpinelloides* has rays thickened in fr and fr cylindrical, shorter-styled. B,F,G. **5b* Narrow-leaved Water Dropwort** *Oe. silaifolia* has lower lvs 2-4-pinnate with narrower lobes, rays thickened in fr; cylindrical, long-styled fr. B,F,G. **5c** *Oe. peucedanifolia* is like 5b but short-styled fr. F,G.

6* FINE-LEAVED WATER DROPWORT *Oenanthe aquatica*. Medium/tall, pale green hairless perennial; much branched. Lvs 3-4-pinnate, upper with pointed lobes, lower with lobes linear to thread-like. Fls white, umbels *opposite* lvs as well as terminal, rays not thickened in fr; usually no lower bracts; June-Sept. Fr oblong, often curved, short-styled. Slow or still fresh water. T. **6a* River Water Dropwort** *Oe. fluviatilis* usually floats on the water and has blunter-lobed 2-pinnate upper lvs and still shorter-styled fr; a shy flowerer. B,F,G, western.

7* CORN PARSLEY *Petroselinum segetum*. Medium greyish hairless biennial, *parsley-scented*. Lvs 1-pinnate, lflets broad, toothed. Fls white, umbels few-rayed and few-fld, bracts thread-like; Aug-Oct. Fr egg-shaped. Dry, sparsely grassy places. B,F,G, western. **7a* Wild Parsley** *P. crispum* □ has shiny 3-pinnate lvs, much crisped in cultivated forms, and more numerous, yellowish fls; June-Sept. (T).

8* STONE PARSLEY *Sison amomum*. Medium/tall perennial, hairless, *foetid*, pale green turning purple; much branched. Lvs and fls like 7. Fr globular. Hedge-banks, grassy places. B,F.

7a* Wild Parsley

1

2

3

4

5

6

7

8

Carrot Family *(contd.)*

1* HOG'S FENNEL *Peucedanum officinale.* Tall dark green hairless perennial. Lvs 2-6-trefoil, the repeatedly divided linear lobes making a *tangled mass.* Fls *yellow,* umbels many-rayed, often no lower bracts; July-Sept. Fr egg-shaped, flattened, ridged. Grassy places. B (rare), F, G. **1a** *P. carvifolia* ☐ has 1-2-pinnate lvs with broader lobes, no bracts and fls sometimes greenish. F, G. **1b** *P. alsaticum* has 2-4-pinnate lvs and always some lower bracts. F, G. **1c** *P. gallicum* is like 1a but has lower lvs 3-pinnate and fls white or pinkish. F. **1d** *P. lancifolium* has shorter pinnate lvs, not tangled; bracts 4-7. F (rare).

2* MILK PARSLEY *Peucedanum palustre.* Tall perennial, almost hairless; stems often purplish. Lvs 2-4-pinnate, lobes blunt. Fls white, bracts unequal, sometimes *forked,* down-turned; July-Sept. Fr egg-shaped, flattened, ridged. Marshes, damp grassland. T. **2a** *P. cervaria* has very sharply toothed lvs, lower bracts often pinnately cut, and fr sometimes rounded. Grassland, woods. F, G.

3* MASTERWORT *Peucedanum ostruthium.* Medium/tall perennial, almost hairless. Lvs 2-trefoil, with broad, toothed lflets and *inflated* sheathing stalks. Fls white or pink, no lower bracts; June-July. Fr rounded, flattened, ridged. Grassy places in hills and mountains. F, G, (B, S). **3a** *P. oreoselinum* is often tinged red and has solid stems, 2-3-pinnate, rather leathery lvs, numerous lower bracts, all bracts down-turned, and more egg-shaped fr; Aug-Sept. F, G, S.

4* HARTWORT *Tordylium maximum.* Medium/tall, coarsely hairy annual/biennial; stems with *down-turned* hairs. Lvs 1-pinnate, root lvs with broader lflets. Fls white, outer petals *larger;* June-Aug. Fr rounded, flattened, ridged, bristly. Grassy and waste places. (B, rare), F, G.

5 SERMOUNTAIN *Laserpitium latifolium.* Tall greyish perennial, almost hairless. Lvs 2-pinnate, stalks of upper lvs *inflated,* lobes heart-shaped. Fls white, lower bracts numerous, *down-turned;* June-Aug. Fr egg-shaped with wavy ridges. Mountain woods and rocks. F, G, S. **5a** *L. prutenicum* is a downy green biennial with lf-lobes narrow, lf-stalks not inflated, fls sometimes tinged yellow, upper bracts more numerous, and fr ridges straight. Grassy places, woods. F, G. **5b** *L. siler* has lvs 2-4 pinnate and lflets lanceolate, untoothed. Mountains, on lime. G, south-eastern. **5c** *Laser trilobum* is hairless and green, with 2-3-trefoil lvs, the stalks not inflated, few bracts and broader umbels. On lime. F (rare), G.

6* SLENDER HARE'S-EAR *Bupleurum tenuissimum.* A most atypical umbellifer, hard to detect among long grass. Medium greyish hairless annual. Lvs undivided, more or less *linear.* Fls *yellow,* very small, in head-like umbels with 1-3 rays; no lower bracts; Aug-Sept. Fr globular. Grassy places by the sea, saline meadows inland. T, southern. **6a** *B. gerardi* is green, with lvs partly clasping stem and 5-7 rays. F (rare).

7 CNIDIUM *Cnidium dubium.* Medium/tall perennial, almost hairless. Lvs 2-3-pinnate, lobes with somewhat *recurved* edges and pointed whitish tip; stalks often purplish, sheathing the stem. Fls white, umbels with numerous narrowly winged rays, few or no lower bracts; July-Oct. Fr globular, ridged. Damp woods, disturbed ground, often near sea. G, S.

8* SCOTS LOVAGE *Ligusticum scoticum.* Medium/tall hairless perennial, *celery-scented;* stems often purple. Lvs 2-trefoil, glossy, with broad, toothed lobes and *inflated* stalks sheathing the stem. Fls white, bracts narrow; June-July. Fr egg-shaped, ridged. Cliffs, rocks and shingle by sea. B, G, S. **8a** *L. mutellinum* has pinnate lvs with narrower lobes. G, rare, southern.

1a *Peucedanum carvifolia*

169

Carrot Family (contd.)

1* WILD PARSNIP *Pastinaca sativa.* Medium/tall biennial, roughly hairy, pungent. Lvs 1-pinnate with broad, toothed lflets. Fls yellow, few or no bracts; June-Sept. Fr egg-shaped, winged. Grassy and bare places. T.

2* ALEXANDERS *Smyrnium olusatrum.* Tall hairless pungent biennial. Lvs 3-trefoil, dark green, *glossy,* with broad, toothed lflets. Fls yellow, few or no bracts; Apr-June. Fr globular, black. Hedge-banks and waste ground near sea. (B),F,G. **2a* Biennial Alexanders** *S. perfoliatum* has yellow-green lvs, the upper clasping the angled stems, and no bracts. Woods, rocky places. (B,rare),F,G, southern.

3* ROCK SAMPHIRE *Crithmum maritimum.* Short/medium greyish hairless perennial. Lvs 1-2-pinnate, with narrow *fleshy* lobes. Fls yellow, bracts eventually down-turned; July-Oct. Fr egg-shaped. Cliffs, rocks and sands by sea. B,F,G, southern.

4* FENNEL *Foeniculum vulgare.* Tall greyish strong-smelling biennial/perennial. Lvs 3-4-pinnate, feathery, with long *thread-like* lobes. Fls yellow, no bracts; July-Sept. Fr egg-shaped. Bare and waste ground, often by the sea. (B),F,G. **4a* Dill** *Anethum graveolens* is smaller and slenderer, with flattened, winged, caraway-tasting fr. Waste ground. (B,F,G), from S Asia.

5* PEPPER SAXIFRAGE *Silaum silaus.* Medium/tall hairless perennial. Lvs 2-4-pinnate, lobes linear, pointed. Fls yellow, few or no lower bracts; June-Sept. Fr egg-shaped, short-styled. Grassland. T, southern.

6 LOVAGE *Levisticum officinale.* Tall pungent perennial. Lvs 2-3-pinnate, *glossy,* with broad, toothed lobes. Fls greenish-yellow, lower bracts numerous, upper bracts joined at base; June-Aug. Fr egg-shaped, yellow brown. Grassy places in mountains. (F,G,S), from S W Asia.

7* THOROW-WAX *Bupleurum rotundifolium.* Short/medium, greyish or purplish annual, looking quite unlike most umbellifers. Lvs undivided, roundish, the upper narrower and *joining* round the stem, the lower short- or unstalked. Fls yellow, umbels with 5-10 rays, secondary umbels *cupped* by enlarged upper bracts, no lower bracts; June-Aug. Fr oblong, smooth, blackish. Weed of cultivation, bare ground. (B),F,G. **7a* Narrow Thorow-wax** *B. lancifolium* □ with narrower lvs, 2-3 rays and warty fr is now commoner than 7 in some areas, especially where bird-food is thrown down. (B),F. **7b** *B. longifolium* is much taller with narrower lvs, the upper clasping, the lower long-stalked, and fr strongly ridged. Grassy and rocky places. F,G.

8* SMALL HARE'S-EAR *Bupleurum baldense.* Low hairless greyish annual; another very untypical umbellifer, often hard to detect in short turf. Lvs undivided, *narrower.* Fls yellow, in tiny 3-9-rayed head-like umbels, *enfolded* by the enlarged greyish, yellowish or brownish *bracts;* June-July. Fr egg-shaped, smooth. Dry bare and grassy places; likes lime. B(rare),F.

9* SICKLE HARE'S-EAR *Bupleurum falcatum.* Medium/tall hairless perennial. Lvs *undivided,* variable, often *curved.* Fls yellow, umbels with thread-like rays, lower bracts unequal; July-Oct. Fr egg-shaped. Grassy and bare places. (B,rare),F,G.

7a* Narrow Thorow-wax

Wintergreen Family   Pyrolaceae

Evergreen creeping hairless perennials with 5-petalled fls and fr a dry capsule.

1* ONE-FLOWERED WINTERGREEN *Moneses uniflora.* Low perennial. Lvs rounded, toothed, pale green, opposite, tapering to the stalk. Fls white, *solitary,* open, 15 mm; style long, straight; May-July. Woods, especially coniferous, in hill and mountain districts. T, but rare in B.

2* TOOTHED WINTERGREEN *Orthilia secunda.* Low perennial. Lvs in rosettes, pointed oval, toothed, rather pale green. Fls greenish-white, globular, 5 mm, in a nodding *one-sided* spike; style straight, protruding conspicuously from fl; June-Aug. Woods and rock ledges in hill and mountain districts. T.

3* COMMON WINTERGREEN *Pyrola minor.* Low/short perennial. Lvs in a rosette, rounded, toothed. Fls white or pale pink, globular, 6 mm, in a stalked spike; style *straight,* shorter than stamens, *not* protruding from fl; June-Aug. Woods, scrub, marshes, moors, mountains. T. **3a* Intermediate Wintergreen** *P. media*□ has fls 10 mm, in longer spikes, with style protruding from fl and slightly curved.

4* ROUND-LEAVED WINTERGREEN *Pyrola rotundifolia.* Short perennial. Lvs like 3 but more rounded. Fls pure white, larger (12 mm) and more *open* than 3, with a much longer and *S-shaped* style; June-Sept. Woods, short turf, rock ledges, dune slacks. T. **4a** *P. norvegica*□ is shorter, with even more rounded lvs. Bare places in mountains. S.

5 YELLOW WINTERGREEN *Pyrola chlorantha.* Short perennial. Lvs in a rosette, roundish, toothed. Fls greenish- or yellowish-white, in a stalked spike, with style protruding, curved; June-July. Woods, rocks. F, G, S, eastern, but only on mountains in the south.

6 UMBELLATE WINTERGREEN *Chimaphila umbellata.* Short perennial. Lvs in *whorls,* narrowly oval, toothed, dark green, leathery. Fls pink, open, in *umbel-like* head, style not protruding; June-July. Woods, rocks. F, G, S.

7* DIAPENSIA *Diapensia lapponica* (Diapensia Family, *Diapensiaceae*). Low evergreen hairless undershrub, forming *cushions.* Lvs oval, untoothed, shiny, leathery. Fls white, 5-petalled, stigma 3-lobed, sepals reddening in fr; May-June. Fr a dry capsule. Bare mountains and tundra. B (rare), S.

8* YELLOW BIRDSNEST *Monotropa hypopitys* agg. Low perennial with no green colouring matter but stems and scale-like lvs *yellow,* turning brown; a saprophyte, feeding on decaying organic matter. Fls yellow, tubular, 4-5-petalled, in nodding spikes, erect in fr; June-Aug. Fr egg-shaped. Cf. Birdsnest Orchid (p. 286). Woods, in Britain especially of beech and pine, and dunes. T.

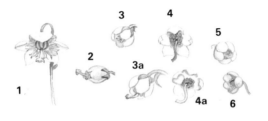

1* One-flowered Wintergreen
2* Toothed Wintergreen
3* Common Wintergreen
3a* Intermediate Wintergreen
4* Round-leaved Wintergreen
4a *Pyrola norvegica*
5 Yellow Wintergreen
6 Umbellate Wintergreen

Heath Family   Ericaceae

pp. 174-76. Undershrubs or small shrubs with lvs undivided, usually alternate, untoothed and evergreen, and with margins inrolled downwards in many species, including all *Ericas*. Fls globular, bell-shaped or flask-shaped (except Cranberry, p. 176), the petals usually joined with 4-5 lobes. Fr usually a dry capsule, but sometimes a berry. The commonly naturalised rhododendron *Rhododendron ponticum* with conspicuous heads of purple fls in June, belongs here, also the less frequent yellow-flowered *R. luteum*.

1* HEATHER *Calluna vulgaris.* Short/medium carpeting undershrub, sometimes downy. Lvs in opposite rows, linear. Fls *pale* purple, in lfy stalked spikes; July-Sept. Heaths, moors, bogs, open woods, fixed dunes, often dominating extensive tracts. T.

2* BELL HEATHER *Erica cinerea.* Short hairless undershrub. Lvs in whorls of three, linear, dark green, often bronzy. Fls *red-purple,* in stalked spikes; May-Sept. Drier heaths and moors, open woods. T, western.

3* CROSS-LEAVED HEATH *Erica tetralix.* Short greyish downy undershrub. Lvs in whorls of four, linear. Fls *pink,* globular, 6-7 mm, in compact heads; June-Oct. Fr downy. Wet heaths and moors, bogs. T, eastern. **3a* Dorset Heath** *E. ciliaris* □ may be taller and more straggly, with lvs in whorls of three and broader; fls larger (8-10 mm), more elongated and in spikes; and fr hairless; Apr-Oct; hybridises with 3; B(rare), F. **3b* Mackay's Heath** *E. mackaiana* has shorter, broader, darker green lvs, hairless above, and deeper pink fls; July-Sept; hybridises with 3. Ireland(rare). **3c** *E. carnea* has bright red fls in one-sided spikes. Mountain woods. G, southern.

4* CORNISH HEATH *Erica vagans.* Short/medium hairless undershrub. Lvs in whorls of 4-5, linear. Fls pink, lilac or white, in lfy stalked spikes, *chocolate-brown* anthers protruding; July-Sept. Drier heaths. B(rare), F.

5* IRISH HEATH *Erica erigena.* Medium/tall hairless shrub *to 2 m.* Lvs in whorls of four, linear. Fls pale pink-purple, in lfy stalked spikes, *reddish* anthers protruding; March-May and occasionally in winter. Wet moors, bogs. Ireland(rare), F.

6* ST DABEOC'S HEATH *Daboecia cantabrica.* Short hairy undershrub. Lvs narrowly elliptical, *whitish* beneath, the margins inrolled downwards. Fls pink-purple, in stalked spikes, calyx-teeth 4; May-Oct. Heaths, moors, open woods. B(Ireland), F.

7* MOUNTAIN HEATH *Phyllodoce caerulea.* Low/short undershrub. Lvs linear, rough-edged, flat, *green* beneath. Fls purple, in head-like clusters of 1-6, calyx-teeth 5; June-July. Moors. B(rare), S.

8* BOG ROSEMARY *Andromeda polifolia.* Low/short hairless undershrub. Lvs narrowly elliptical, greyish and shiny above, whitish beneath. Fls pink or white, in a terminal cluster; May-June. Bogs, wet heaths. T.

9 ARCTIC RHODODENDRON *Rhododendron lapponicum.* Low shrub. Lvs oval, leathery, dark green, reddish hairy beneath. Fls tubular, violet; May-June. Mountain heaths. S.

R. luteum

R. ponticum

9 Arctic Rhododendron

Heath Family (contd.)

1* BILBERRY *Vaccinium myrtillus.* Short/medium erect *deciduous* hairless undershrub; stems angled, green. Lvs oval, bright green, slightly toothed. Fls pink or greenish-pink, solitary or paired; Apr-July. Fr an edible *black* berry with characteristic purplish bloom. Hybridises with 2. Heaths, moors, open woods, not on lime. T. **1a* Northern Bilberry** *V. uliginosum* □ has rounded brownish stems, bluish-green untoothed lvs, clusters of 1-4 pale pink fls, and bluer berries. Mainly higher moors.

2* COWBERRY *Vaccinium vitis-idaea.* More or less prostrate creeping *evergreen* undershrub; twigs rounded, slightly downy when young. Lvs oval, broadest in the *middle,* untoothed, rather leathery, glossy, margins inrolled downwards. Fls white or pink, open-mouthed, in clusters; May-July. Fr an edible *red* berry. Hybridises with 1. Moors, heaths, mountains, open woods. T.

3* CRANBERRY *Vaccinium oxycoccos.* Low slender creeping evergreen undershrub. Lvs dark green, oval, whitish beneath. Fls pink, with four spreading or *down-turned* petals and prominent stamens, slightly downy, 1-4 in stalked spikes; June-Aug. Fr an edible round or pear-shaped red berry, spotted white or brown. Bogs. T. **3a* Small Cranberry** *V. microcarpum* has more triangular lvs, fls pinker, fls and stalks hairless, fr sometimes lemon-shaped. B, G, S.

4* ALPINE BEARBERRY *Arctostaphylos uva-ursi.* Prostrate *mat-forming* evergreen undershrub to 2 m, almost hairless. Lvs dark green, untoothed, broadest at *tip,* rather leathery, margins flat. Fls pink, in clusters; Apr-July. Fr a shiny edible *red* berry. Moors, mountains, rocks, open woods, scrub. T. **4a* Black Bearberry** *A. alpina* has brighter green, finely toothed, deciduous lvs; white fls and fr black when ripe. B, S.

5* LABRADOR TEA *Ledum palustre.* Evergreen bush to 1 m, with *rust-coloured* down on stems and under the oblong leathery dark green lvs. Fls creamy white, in umbels, with five free petals; May-July. Fr dry. Bogs. (B), G, S.

6* WILD AZALEA *Loiseleuria procumbens.* Prostrate mat-forming undershrub, hairless. Lvs *opposite,* oblong. Fls small, pink, in umbels; May-July. Mountain tops, Arctic heaths. T, northern.

7 CASSIOPE *Cassiope hypnoides.* Prostrate mat-forming und⌐ shrub. Lvs scale-like, unstalked, *overlapping* up the stems. Fls bell-shaped, wɪ ɪ white and pink petals and *deeper pink* sepals, solitary, long-stalked, drooping; June-Aug. Mountain tops, Arctic heaths. S. **7a** *C. tetragona* is larger and more erect, with fl-stalks downy. On lime. Northern.

8 LEATHERLEAF *Chamaedaphne calyculata.* Short shrub. Lvs leathery, pointed oval. Fls white, rim yellow, bell-shaped, in a stalkeᴄ spike; June. Damp places. S, northern.

9* STRAWBERRY TREE *Arbutus unedo.* Small evergreen *tree* or shrub, to 12 m, with roughish red-brown bark. Lvs elliptical, leathery, dark green, shiny, slightly toothed, alternate. Fls creamy-white, bell-shaped, in drooping clusters; Aug-Dec. Fr □ a warty berry, 2 cm across, reddening the following autumn, and supposedly strawberry-like. Rocks, scrub. Ireland (rare), F, western.

Crowberry Family   Empetraceae

10* CROWBERRY *Empetrum nigrum.* Prostrate mat-forming or erecter tufted evergreen undershrub, resembling the Heath Family. Lvs oblong, dark green, shiny, margins *inrolled* downwards, alternate. Fls pink, 6-petalled, very small, at base of lvs; stamens and styles usually on separate plants; Apr-June. Fr a *berry,* green, then pink, then purple, finally *black.* Moors, bogs. T.

9* Strawberry Tree

Primrose Family Primulaceae

pp. 178-80. Lvs undivided. Fls 5-petalled (except Chickweed Wintergreen, p. 180). Fr a capsule. *Primula* (1-5) has elongated oval lvs, all from the roots, and open fls in an umbel (except 1) on lfless stalks.

1* PRIMROSE *Primula vulgaris.* Low hairy perennial. Lvs *tapering* to stalk. Fls pale yellow, 20-30 mm, *solitary* on long hairy stalks; rarely the common stalk rises just above ground to make a long-stalked umbel; Mar-May and sporadically in autumn and winter. Hybridises with 2 and 3; see 2a. Woods, scrub, grassy banks, sea cliffs, mountains, often in quantity. T, southern.

2* OXLIP *Primula elatior.* Low hairy perennial. Lvs *abruptly* narrowed at base. Fls pale yellow, 15-20 mm, not fragrant, 1-20 in a nodding one-sided cluster; Apr-May. Hybridises with 1 and 3. Woods, scrub, grassland, usually in quantity. T, eastern. **2a* False Oxlip** *P. veris × vulgaris* □ is the not uncommon hybrid between 1 and 3, with leaves more gradually tapered to base, fls deeper yellow and umbels not one-sided; grows singly, usually near one or both parents.

3* COWSLIP *Primula veris.* Low/short hairy perennial. Lvs *abruptly* narrowed at base. Fls deeper yellow than 1 and 2, with orange spots in centre, 10-15 mm, *fragrant,* 1-30 in an often nodding one-sided cluster; Apr-May. Hybridises with 1 and 2; see 2a. Grassland, open scrub and woods, often in quantity, on lime. T. **3a Auricula** *P. auricula* □ has smaller, fleshier lvs and fls twice as large with a white throat and no orange spots. Mountains, G, southern.

4* BIRDSEYE PRIMROSE *Primula farinosa.* Low perennial, mealy white on young stems and beneath lvs. Lvs broadest near *tip, toothed,* in a basal rosette. Fls lilac-pink with a yellow eye; sepal-teeth pointed; May-July. Grassy places in hills and mountains. T.

5* SCOTTISH PRIMROSE *Primula scotica.* Low biennial, mealy white on stems and lvs. Lvs broadest in the *middle, untoothed,* in a basal rosette. Fls purple with a yellow eye, sepal-teeth blunt; May-June and again in July-Aug. Short coastal turf and dunes. N Scotland. **5a** *P. scandinavica* is very similar; meadows, by streams, on lime. S. **5b** *P. stricta* is taller and has no meal on stems and little on lvs; fls lilac or violet. Damp mountain meadows. S, northern. **5c** *P. nutans* is like 5b with lvs rounded not tapering at base. S, northern.

6* YELLOW LOOSESTRIFE *Lysimachia vulgaris.* Medium/tall downy perennial. Lvs broad lanceolate, often black-dotted, very short-stalked, in whorls of 2-4. Fls yellow, *15-20 mm,* in lfy branched clusters, sepal-teeth with reddish margins; July-Aug. Wet places and by fresh water. T. **6a* Dotted Loosestrife** *L. punctata* has longer-stalked, never black-dotted lvs with hairy margins, larger (35 mm) fls and narrower all-green sepal-teeth. (T), from SE Europe.

7* TUFTED LOOSESTRIFE *Lysimachia thyrsiflora.* Medium perennial, almost hairless. Lvs lanceolate, unstalked, opposite. Fls yellow, 5 mm, with prominent stamens, in *globular clusters* at base of middle lvs, top of stem having lvs only; June-July, a shy flowerer. Fens, marshes, by fresh water. T.

8* CREEPING JENNY *Lysimachia nummularia.* Prostrate creeping hairless perennial. Lvs rounded, opposite. Fls yellow, *bell-like,* 15-25 mm, with broad sepal-teeth; June-Aug. Fr uncommon in B. Damp shady and grassy places. T. **8a* Yellow Pimpernel** *L. nemorum* is smaller, with narrower pointed lvs and longer-stalked open 12 mm fls with narrow sepal-teeth, and fr frequent; May-Aug.

9 ALPINE SNOWBELL *Soldanella alpina.* Low perennial, slightly hairy. Lvs thick, kidney-shaped, long-stalked. Fls blue or white, bell-like, 2-3 together on lfless stems, the petals deeply cut to form a fringe. Mountains. G, southern.

3a Auricula   **9** Alpine Snowbell

Primrose Family (contd.)

1* WATER VIOLET *Hottonia palustris.* Pale green *aquatic* perennial, almost hairless, with short erect lfless stems above surface. Lvs *pinnate*, submerged. Fls pale lilac with yellow eye, in a whorled cluster; May-July. Still fresh water. T, southern.

2 CYCLAMEN *Cyclamen purpurascens.* Low perennial, almost hairless. Lvs long-stalked, all arising direct from a corm, rounded, heart-shaped, not angled or lobed, sometimes finely toothed, pale-blotched above, reddish-purple beneath. Fls carmine-pink, petals *turned back*, throat rounded; June-Oct, with the lvs. Woods, scrub. F, G, eastern. **2a* Sowbread** *C. hederifolium* □ has sharply angled or lobed lvs, and longer-stalked paler pink or white fls with 5-sided throat; Aug-Nov, before the lvs. (B), F.

3* CHICKWEED WINTERGREEN *Trientalis europaea.* Low/short hairless perennial. Lvs lanceolate, in a *single whorl* near the top of the slender stem. Fls white, star-like, 5-9-petalled, usually solitary; May-Aug. Conifer woods, moors, heaths. T.

4* BROOKWEED *Samolus valerandi.* Low/short hairless perennial, often unbranched. Lvs oval, in a basal rosette, and alternate up the stems. Fls white, in a stalked spike, petals *joined half-way;* June-Aug. Fr globular. Damp places, especially near sea. T, southern.

5* SCARLET PIMPERNEL *Anagallis arvensis.* Prostrate annual, almost hairless; stems square. Lvs pointed oval, black-dotted beneath, unstalked, paired or whorled. Fls usually pale scarlet, but sometimes pink, lilac or blue, *star-like,* solitary at base of lvs, opening in sunshine; petals blunt, often finely toothed at tip, with hairs along margins; May-Oct. Weed of cultivation, dunes. T. **5a* Blue Pimpernel** *A. foemina* □ has fls always blue with narrower pointed hairless petals.

6* BOG PIMPERNEL *Anagallis tenella.* Delicate, often mat-forming hairless perennial. Lvs oval or rounded, short-stalked, opposite. Fls pink, *bell-like,* opening in sunshine; May-Sept. Boggy, peaty and marshy places, fens. B, F, G, western.

7* CHAFFWEED *Anagallis minima.* One of the tiniest European plants, rarely over 5 cm high, often *under 2 cm;* a hairless annual. Lvs oval, with a thin *black line* round the underside, mostly alternate. Fls pink, hidden at base of lvs; June-Aug. Fr pink. Bare damp sandy places on heaths and in open woods. T.

8 NORTHERN ANDROSACE *Androsace septentrionalis.* Low/short downy annual/biennial. Lvs lanceolate, toothed, all in a basal rosette. Fls white or pink, 5 mm, long-stalked, in an *umbel;* May-July. Dry grassy places, mountains. F, G, S, northern. **8a** *A. elongata* □ has lvs almost or quite hairless, and smaller (2-3 mm) white fls with a yellow throat; May. G. **8b** *A. maxima* □ has broader lvs, smaller (3 mm) white fls, and umbels with conspicuous hairy bracts; Apr-May. Lowland grassland. F, G, S, eastern. **8c** *A. lactea* □ has lvs unstalked, untoothed, and fls white, 10 mm. G, southern.

9* SEA MILKWORT *Glaux maritima.* More or less prostrate creeping hairless perennial. Lvs elliptical, almost unstalked, opposite. Fls pale pink, petal-less, unstalked, solitary at base of lvs; May-Sept. *Saltmarshes*, on saline soils inland. T.

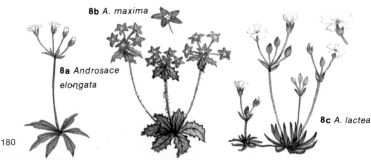

8b *A. maxima*

8a *Androsace elongata*

8c *A. lactea*

Sea-Lavender Family   Plumbaginaceae

Perennials. Lvs undivided. Fls 5-petalled, with *papery* bracts. Often in saline habitats. *Limoniums* are hairless and have lvs all basal and fls in one-sided spikes in a branched cluster.

1* THRIFT *Armeria maritima.* Low/short downy cushion-forming perennial. Lvs linear, 1-veined, all basal. Fls dark to pale pink, in *roundish* heads, with a brown bract; sepal-teeth with short bristles; Apr-Aug. Coastal cliffs, rocks and saltmarshes, also inland on mountains; ssp. *elongata* in sandy grassland. T. **1* Jersey Thrift** *A. alliacea* is taller and stouter, with broader 3-5-veined lvs, hairless fl-stalks, fls usually deep pink, sepal-teeth with long bristles, and outer bract enfolding the head. F, G.

2* COMMON SEA-LAVENDER *Limonium vulgare.* Low/short carpeting perennial; stems rounded. Lvs elliptic to broad lanceolate, long-stalked, with *pinnate* veins. Fls lilac-lavender, tightly packed in flat-topped clusters, at the end of the stalks, which usually start branching well above the middle; July-Sept. Coastal saltmarshes. T, western. **2a* Lax-flowered Sea-lavender** *L. humile* has stems branched below the middle and often angled, lvs narrower and fls widely scattered along whole length of branches.

3* MATTED SEA-LAVENDER *Limonium bellidifolium.* Low perennial, with stems branched almost from the base into numerous *zigzag* branches, rough to the touch. Lvs oblong. Fls pale-lavender, with whitish bracts, only on the uppermost branches; June-Aug. Coastal saltmarshes. B (rare), F.

4* ROCK SEA-LAVENDER *Limonium binervosum* agg. Variable low/short perennial; stems branched from near base. Lvs oval to lanceolate, not pinnately veined, with a winged 3-veined stalk. Fls lilac-lavender, on all but the lowest branches, spikes curved; July-Sept. Coastal cliffs and rocks, less often shingle and sand. B, F. **4a** *L. auriculae-ursifolium* is taller and stouter, with broader greyish, often sticky lvs, lf-stalks 5-9-veined and inner bracts red-edged. F.

Olive Family   Oleaceae

5* WILD PRIVET *Ligustrum vulgare.* Half-evergreen shrub to 4 m. Lvs lanceolate, untoothed, opposite, rather leathery, often bronzing. Fls white, strong-smelling, in short spikes, with four joined petals; May-June. Fr a shiny *black* berry. Scrub, open woods. T. **5a* Garden Privet** *L. ovalifolium* has much broader lvs and is commonly planted. (B, F, G), from China.

Bogbean Family   Menyanthaceae

6* BOGBEAN *Menyanthes trifoliata.* Short hairless aquatic or creeping perennial. Lvs *trefoil,* held conspicuously above surface. Fls pink and white, in spikes, petal-tube with five lobes, *fringed* with long white hairs; Apr-June. Fr globular. Shallow fresh water, marshes, fens, bogs. T.

7* FRINGED WATER-LILY *Nymphoides peltata.* Hairless aquatic perennial. Lvs rounded, shallowly toothed, floating, purple beneath and sometimes purple-spotted above; much smaller than true water-lilies (p. 66). Fls yellow, petal-lobes *fringed*; June-Sept. Fr egg-shaped. Still and slow-moving fresh water. B, F, G.

Periwinkle Family   Apocynaceae

8* LESSER PERIWINKLE *Vinca minor.* More or less prostrate evergreen undershrub, almost hairless; stems rooting. Lvs elliptic, opposite, short-stalked, rather leathery. Fls *blue-violet,* 25-30 mm, solitary, with five joined petals; Feb-May. Fr elongated, forked. Woods, hedge-banks, rocks. (B), F, G. **8a* Greater Periwinkle** *V. major* □ is quite hairless and has stems rooting only at tip, much broader longer-stalked lvs, and larger (40-50 mm) fls with fringed sepals.

Milkweed Family   Asclepiadaceae

9 VINCETOXICUM *Vincetoxicum hirundinaria.* Variable medium/tall perennial, slightly hairy. Lvs *heart-shaped,* sharply pointed, opposite. Fls greenish- or yellowish-white, in clusters at base of upper lvs, with five joined petals; June-Sept. Fr elongated, forked. Woods, rocks, bare ground; on lime. F, G, S.

183

Gentian Family   Gentianaceae

pp. 184-86. Hairless. Lvs undivided, untoothed, opposite, usually unstalked; no stipules. Fls in a branched cluster, with joined petals, often opening only in sunshine. Fr a dry capsule.

1*   COMMON CENTAURY *Centaurium erythraea.* Variable low/short annual. Lvs elliptical, 3-7-veined, mostly in a basal *rosette* but a few narrower stem lvs. Fls pink, 5-petalled, almost *unstalked* in tightly packed stalked clusters, but occasionally solitary; June-Sept. Grassy places, dunes. T. **1a* Seaside Centaury** *C. littorale*□ is often shorter, with much narrower, rather leathery, 3-veined lvs, and larger paler fls. Dunes, sandy ground near sea. Western.

2*   LESSER CENTAURY *Centaurium pulchellum.* Slender low/short annual, varying from unbranched to well branched. Lvs elliptical, with *no* basal rosette. Fls reddish-pink, with five (rarely four) rather narrow petals, *stalked,* in loose, widely spreading clusters; June-Sept. Damp grassy places, drier salt-marshes. T.   **2a* Slender Centaury** *C. tenuiflorum*□ has broader lvs and erecter tighter flat-topped clusters of fls. B(rare),F.

3*   PERENNIAL CENTAURY *Centaurium scilloides.* Low/short perennial; stems numerous, semi-prostrate, some *not* flg. Lvs roundish and stalked on the non-flg stems. Fls pink, larger than 1, in heads of 1-6, 5-petalled; July-Aug. Grassy sea cliffs. B(rare),F.

4*   YELLOW-WORT *Blackstonia perfoliata.* Short/medium greyish annual. Lvs of basal rosette oblong, stem lvs more or less triangular and *joined* round stem. Fls yellow, 6-8-petalled, petals longer than sepals, in a loose branched cluster; June-Nov. Grassy places, dunes; on lime. B,F,G. **4a** ssp. *serotina* □ has lvs not completely joined, fls paler and sepals equalling petals. G.

5*   YELLOW CENTAURY *Cicendia filiformis.* Low annual, sometimes only 2 cm high, often little branched. Lvs linear. Fls yellow, solitary, 4-petalled; June-Oct. Fr egg-shaped. Bare damp places, often near sea. B,F,G, western.

6   GUERNSEY CENTAURY *Exaculum pusillum.* A minute low annual, sometimes only 1 cm high, well branched, very hard to detect in short turf. Lvs linear, greyish. Fls pink, pale yellow or white, 4-petalled, minute; sepal-teeth long, narrow; June-Oct. Bare damp places. F.

7   GREAT YELLOW GENTIAN *Gentiana lutea.* Medium/tall perennial. Lvs pointed oval, up to 30 cm, bluish-green, the upper clasping the stem, the lower stalked. Fls yellow, 5-9-petalled, in tight whorls at base of upper lvs; June-Aug. Grassy places in mountains. F,G.

8   NORTHERN GENTIAN *Gentianella aurea.* Short biennial, branched from the base. Lvs oval, the upper unstalked. Fls in a lfy spike, the lvs almost concealing the dull blue-violet fls, dull yellow at the base; sepals 4-5; July-Aug. Near coast. S, northern.

1a* Seaside Centaury   **2a*** Slender Centaury   **4a** *Blackstonia* ssp. *serotina*

185

Gentian Family *(contd.)*

1* SPRING GENTIAN *Gentiana verna.* Low tufted perennial. Lvs oval to elliptical, mostly in a basal rosette. Fls bright blue, *solitary,* petal-tube 5-lobed, 15-25 mm long; Apr-June. Short, often stony turf in hills and mountains. B, F, G **1a* Alpine Gentian** *G. nivalis* □ is a slenderer annual, often unbranched, with fls half the size (10-15 mm). Also on Arctic heaths. T, but rare in B. **1b Bladder Gentian** *G. utriculosa* is an annual with fls intermediate between 1 and 1a and a slightly inflated calyx. Damp places. G, southern. **1c** *Gentianella detonsa* may be taller and more branched, with no marked rosette and larger fls. Often near the sea. S, northern.

2* MARSH GENTIAN *Gentiana pneumonanthe.* Low/short perennial. Lvs linear. Fls bright blue, striped *green* outside, trumpet-shaped, 25-40 mm long, often solitary, petal-tube 5-lobed; July-Sept. Wet heaths, bogs. T. **2a Willow Gentian** *G. asclepiadea* □ is much taller, with arching stems, numerous lanceolate lvs, and even larger fls with pale but not green stripes outside. F, G.

3 CROSS GENTIAN *Gentiana cruciata.* Medium perennial. Lvs oval to broad lanceolate, rather leathery, the upper clasping the stem, the lower stalked. Fls *dull* blue, oblong, in tight clusters up the stem, petal-tube 4-lobed; June-Sept. Dry grassy places, open woods. F, G, eastern.

4 PURPLE GENTIAN *Gentiana purpurea.* Short/medium perennial. Lvs elliptical, greyish, the lower stalked. Fls bright *red-purple,* yellow at base and with green stripes and spots, bell-shaped, petal-tube split to base on one side; July-Aug. Grassy and rocky places in mountains. F, G, S. **4a Spotted Gentian** *G. punctata* □ is taller and has fls yellow with purple spots, and sepal-tube not split. G.

5* FIELD GENTIAN *Gentianella campestris.* Short annual/biennial, variably branched. Lvs oval and blunt at base, narrower and pointed on stem. Fls bluish-mauve, bell-shaped, 15-30 mm long, in small clusters up the stem, *petal-tube 4-lobed;* sepal-tube with two outer lobes much larger and *overlapping* the inner ones; June-Oct. Grassy places, dunes. T.

6* AUTUMN GENTIAN *Gentianella amarella.* Low/short biennial. Lvs elliptical to lanceolate. Fls dull purple, sometimes whitish, bell-shaped, 12-22 mm long, in small clusters up the stem, petal-lobes four or five; sepal-teeth equal; July-Sept. Hybridises with 6a. Grassy places, dunes; often on lime. T. **6a* Chiltern Gentian** *G. germanica* □ is larger with broader lvs and longer (25-35 mm) fls, wider at the mouth. Always on lime. B (rare), F, G. **6b* Early Gentian** *G. anglica* differs chiefly in flg May-June. Always on lime. **6c* Dune Gentian** *G. uliginosa* is annual and much smaller (rarely over 7 cm), and has unequal sepal-teeth; Aug-Nov. T, rare in B. **6d** *G. aspera* is hairier than 6a. G, southern, rare.

7 SLENDER GENTIAN *Gentianella tenella.* Slender low annual, often unbranched. Lvs oblong, the lower *broader* at the tip. Fls dull blue-violet, 8-12 mm long, *long-stalked,* with usually four petal-lobes; July-Aug. Grassy places in mountains. F, G, S.

8* FRINGED GENTIAN *Gentianella ciliata.* Slender low/short perennial. Lvs linear, not in a rosette. Fls blue, in small clusters at base of upper lvs, the four spreading oval petal-lobes 40 mm across and *fringed* with hairs; Aug-Oct. Dry grassy, rocky places. B (rare), F.

9 MARSH FELWORT *Swertia perennis.* Medium perennial, unbranched. Lvs lanceolate, yellowish-green, the upper clasping the stem, the lower long-stalked. Fls purple, 5-petalled, *star-like,* 25-35 mm across; July-Sept. Wet grassy places in mountains. F, G. **9a** *Lomatogonium rotatum* □ is shorter with much broader petals; fls long-stalked, blue. S, northern.

9a *Lomatogonium rotatum*

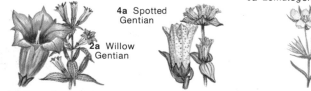

4a Spotted Gentian

2a Willow Gentian

GENTIANS

187

Bindweed Family   Convolvulaceae

Often climbers, twining anticlockwise. Lvs alternate, no stipules. Fr a capsule.

1* HEDGE BINDWEED Calystegia sepium. Creeping or climbing perennial to 3 m, more or less hairless. Lvs arrow-shaped. Fls white, rarely pink, unscented, 30-35 mm across; two large but not overlapping sepal-like bracts enclosing the five narrower sepals; June-Sept. Fr a capsule. Bushy and waste places, woods, fens. T. **1a* Great Bindweed** C. silvatica☐ has larger (60-75 mm) fls, occasionally pink or pink-striped, and larger, almost inflated overlapping bracts, hiding the sepals. (T) from S Europe, often commoner in and around human settlements. **1b* Hairy Bindweed** C. pulchra differs from 1a chiefly in being sparsely hairy and having fls always pink; widely naturalised, of garden origin.

2* FIELD BINDWEED Convolvulus arvensis. Creeping or climbing perennial to 2 m, slightly downy when young. Lvs arrow-shaped. Fls pink and/or white, 15-30 mm across, faintly scented; June-Sept. Weed of cultivation, waste places. T.

3* SEA BINDWEED Calystegia soldanella. Prostrate creeping hairless perennial. Lvs kidney-shaped, fleshy. Fls pink with white stripes, 40-50 mm across, bracts shorter than sepals; June-Sept. Coastal sand or shingle. B, F, G, western.

4* COMMON DODDER Cuscuta epithymum. Slender climbing annual, parasitic on Gorse (p. 122), Heather (p. 174) and many smaller plants; stems red. Lvs reduced to scales. Fls pale pink, scented, bell-shaped with five pointed petal-lobes, stamens and styles slightly protruding, sepals pointed, in tight heads; June-Oct. Heaths, grassy places. T. **4a* Greater Dodder** C. europaea has stems sometimes yellowish-green, larger heads of larger fls with petal-lobes and sepals blunt and stamens and styles not protruding; parasitic on Nettles (p. 38) and other large coarse plants. **4b Flax Dodder** C. epilinum has yellowish fls with curved petal-lobes; on Common Flax (p. 136). G, S, rare.

Phlox Family   Polemoniaceae

5* JACOB'S LADDER Polemonium caeruleum. Medium/tall perennial, slightly downy. Lvs pinnate, alternate. Fls in a cluster, purplish-blue, open, with five petal-lobes; June-Aug. Fr a capsule. Grassy places and open woods in hill and mountain districts, also as a garden escape. B (rare), F, G. **5a** P. acutiflorum has bell-shaped fls. On lime. S, northern.

Bedstraw Family   Rubiaceae

pp. 188-90. Stems square, weak, often scrambling over other vegetation. Lvs usually elliptical or lanceolate, unstalked, in whorls, with often lflike stipules. Fls in clusters or loose heads, small, open, with usually four petal-lobes. Only Field Madder has sepals. Fr a 2-lobed nutlet, a berry in Wild Madder (p. 190).

6* FIELD MADDER Sherardia arvensis. More or less prostrate annual, hairy. Lvs in whorls of 4-6, elliptical. Fls pale purple, in heads surrounded by lflike bracts; May-Sept. Fr globular, surrounded by enlarged sepal-teeth. Bare and cultivated ground. T.

7* SQUINANCYWORT Asperula cynanchica. Slender, more or less prostrate hairless perennial. Lvs linear, sometimes unequal, in whorls of four. Fls pale pink outside, white inside, in terminal clusters, petal-lobes 4, bracts lanceolate; June-Sept. Dry grassland, dunes; on lime. B, F, G. **7a** A. occidentalis has sessile fls with a shorter tube and slightly fleshy lvs. Sand dunes; S Wales, W Ireland only. **7b Dyer's Woodruff** A. tinctoria has lower lvs usually in whorls of 6, whiter fls with 3 petal-lobes, and oval bracts. F, G, S.

8* WOODRUFF Galium odoratum. Short carpeting perennial, almost hairless; unbranched. Lvs elliptical, in whorls of 6-9, edged with tiny forward-pointing prickles. Fls white, in loose heads; Apr-June. Fr covered with hooked bristles. Woods. T. **8a** G. glaucum is taller, more branched and greyish, with whitish stems, swollen at the joints, narrower lvs with inrolled margins and minute points, and fr rough but not bristly; hybridises with Lady's Bedstraw (p. 190); May-July. Sandy, grassy places. F, G.

9* BLUE WOODRUFF Asperula arvensis. Slender short annual, hairless. Lvs linear, blunt, in whorls of 6-9. Fls bright blue; in heads; Apr-June. Weed of cultivation. (B, rare), G.

Bedstraw Family (contd.)

1* WILD MADDER *Rubia peregrina*. Scrambling evergreen perennial to 2 m; stems rough on angles with down-turned prickles. Lvs lanceolate, dark green, shiny, *prickly*, rather leathery, in whorls of 4-6. Fls yellowish-green, petal-lobes 5, in small stalked clusters at base of lvs; June-Aug. Fr a black *berry*. Woods, scrub, rocks; in Britain near the sea. B, F.

2* CROSSWORT *Cruciata laevipes*. Short perennial, *softly* hairy. Lvs elliptical, yellowish-green, in whorls of four. Fls *pale* yellow, scented, with bracts on stalks, in whorls at base of lvs; Apr-June. Fr blackish. Grassy places, open woods; on lime. B, F, G. **2a** *C. glabra* is slenderer and hairless, except for young lvs, and has no bracts on fl stalks. F, G.

3* HEDGE BEDSTRAW *Galium mollugo* agg. Variable medium/tall, often scrambling perennial, sometimes downy; stems *smooth*, square. Lvs elliptical, *1-veined*, ending in a point, in whorls of 6-8. Fls white, petal lobes pointed, in loose branched clusters; June-Sept. Fr black. Hybridises with 7. Grassy places, hedge-banks. T. **3a** *G. sylvaticum* has rounded stems, lvs greyish beneath and broader and blunter, and smaller fls. Eastern.

4* NORTHERN BEDSTRAW *Galium boreale*. Short, rather stiff perennial, sometimes downy. Lvs dark green, rough-edged, *3-veined,* in whorls of four. Fls white, in clusters; June-Aug. Fr covered with hooked bristles. Grassy, bushy and rocky places. T. **4a** *G. rotundifolium* has weak stems, smaller, more rounded lvs, and fr sometimes greenish. Mountains. F, G, S.

5* HEATH BEDSTRAW *Galium saxatile*. Low/short, loosely tufted, hairless perennial, *blackening* when dried. Lvs narrow lanceolate, in whorls of 6-8, edged with *forwardly directed* prickles. Fls white, with a sickly fragrance, in stalked clusters; June-Aug. Fruit with *pointed* warts. Dry grassy and heathy places; *not on lime.* T. **5a Slender Bedstraw** *G. pumilum* agg. is slenderer, not blackening when dried, with lvs in whorls of 8-9, sometimes curved and more sharply pointed, the prickles backwardly directed; fr sometimes with rounded warts. **5b Limestone Bedstraw** *G. sterneri* agg. has backwardly directed prickles and shortly pointed warts on fr. B, G, S. **5c* Cheddar Bedstraw** *G. fleurotii* agg. is often more tufted, sometimes greenish when dry, and has more linear lvs and fr with rounded warts. B, F. 5a-5c sometimes have stems red at base and usually grow on limy soils.

6* MARSH BEDSTRAW *Galium palustre* agg. Variable, rather straggling, short/medium hairless perennial; stems *rough* at angles. Lvs elliptical, blunt, in whorls of 4-5. Fls white, anthers red, in loose stalked clusters; June-Aug. Fr black. Wet places. T. **6a* Slender Marsh Bedstraw** *G. debile* is smaller and slenderer, with shorter, narrower, more pointed rough lvs, and pinkish fls. B (rare), F. **6b* Fen Bedstraw** *G. uliginosum* has rougher stems with down-turned prickles on angles, and narrower lvs, ending in a minute point, 6-8 in a whorl, and yellow anthers. **6c** *G. trifidum* is much shorter and has only three petal-lobes. S. **6d** *G. triflorum* is like 6b with much broader lvs. Rocky woods. S.

7* LADY'S BEDSTRAW *Galium verum*. Low/short sprawling perennial, almost hairless. Lvs linear, dark green, shiny, with margins inrolled beneath, in whorls of 8-12. Fls *bright* yellow, in lfy clusters; June-Sept. Fr black. Hybridises with 3 and *Galium glaucum* (p. 188). Dry grassy places. T.

8* COMMON CLEAVERS *Galium aparine*. Straggling medium/tall annual, *clinging* to animal fur and human clothing by numerous tiny down-turned prickles on stems, lvs and fr. Lvs lanceolate, in whorls of 6-8. Fls *dull* white, in small stalked clusters at base of and longer than lvs; May-Sept. Fr green or purplish, bristles with white swollen bases, on straight spreading stalks. Hedge-banks, fens, disturbed and waste ground, shingle. T. **8a* Corn Cleavers** *G. tricornutum* has creamier fls on stalks shorter than the narrower lvs, and fr rough but with no prickles, on down-curved stalks; cornfields. **8b* False Cleavers** *G. spurium* has smaller green fls and smaller blackish fr whose bristles have unswollen bases; July. A weed of cultivation. T, but (B, rare). **8c* Wall Bedstraw** *G. parisiense* has lvs downturned, tiny fls green inside, reddish outside, and blackish, sometimes smooth fr; June-July. Sandy places, old walls. B, F, G.

Borage Family Boraginaceae

pp. 192-96. Hairy (except Oyster Plant, p. 196), often roughly hairy. Lvs undivided, alternate. Fls often pink in bud, later turning blue, five joined petals, usually in *one-sided* stalked spikes, which at first are tightly coiled. Fr four nutlets.

1* COMMON COMFREY *Symphytum officinale*. Tall stout perennial, softly hairy; well branched. Lvs broad lanceolate, *running down* on to winged stems, root lvs largest, to 25 cm. Fls creamy white or mauve, bell-like, in forked clusters; sepal-teeth pointed, at least equalling tube; May-June. Fr black, shiny. By fresh water, fens. T. **1a* Russian Comfrey** *S.* × *uplandicum* □, a hybrid between 1 and 1b, is stiffly hairy and has stem wings narrower and not reaching next If-junction, and blue or purple-blue fls. Dry waysides, waste places, a widespread escape from cultivation. (T). **1b* Rough Comfrey** *S. asperum* is roughly hairy and has stems not winged, stalked heart-shaped lvs, larger blue fls, red in bud, and blunter sepal-teeth; a much less frequent escape. (B, F, G), from SW Asia. **1c* Soft Comfrey** *S. orientale* is smaller, paler green and earlier flg (Apr-May), with rounder upper lvs, whiter fls and short blunt sepal-teeth. Hedge-banks, a garden escape. (B), from SW Asia. **1d* Tuberous Comfrey** *S. tuberosum* is smaller and sometimes unbranched, with middle lvs the longest, fls yellowish-white only, and pointed sepal-teeth three times as long as tube. Damp, shady places. B, F, G. **1e* Dwarf Comfrey** *S. ibiricum* is much shorter and creeping, with shorter broader lvs, creamy fls, buds tipped brick-red, and sepal-teeth cut nearly to base: an early (sometimes winter) flowering garden escape, (B), from S W Asia.

2* HOUNDSTONGUE *Cynoglossum officinale*. Medium/tall greyish biennial, softly downy, *smelling of mice*. Lvs lanceolate. Fls *maroon,* open, in clusters; May-Aug. Fr flattened, covered with short hooked spines and so adhering to fur and clothing, with a thickened flange. Dry grassy places, dunes. T. **2a* Green Houndstongue** *C. germanicum* is not downy, does not smell, and has shiny green lvs, almost hairless above, fls half the size, and no thickened flange on fr. Woods, shady places. B(rare), F, G.

3* AMSINCKIA *Amsinckia intermedia*. Short/medium annual, rather bristly. Lvs lanceolate. Fls *orange-yellow,* on upper side of coiled spikes; Apr-Aug. Fr wrinkled. Bare or disturbed ground. (B, G), from America. **3a Heliotrope** *Heliotropium europaeum* □ is greyish and foetid, with fls white or lilac, unstalked and no bracts. (F, G, S), from Mediterranean. **3b Lesser Honeywort** *Cerinthe minor* □ is like 3a but almost hairless, with lvs often white-spotted and fls yellow, in lfy spikes, not coiled. On lime. G, south-eastern.

4* YELLOW ALKANET *Anchusa ochroleuca*. Medium perennial, roughly hairy. Lvs lanceolate. Fls creamy yellow, open, in clusters; *white-edged* blunt sepals cut to less than half-way; July-Aug. Waste places. (B, F, G), from Caucasus.

5 NONEA *Nonea pulla*. Short/medium greyish perennial. Lvs oblong to lanceolate, hairy but not rough, the upper clasping the stem. Fls dark brown, in a loose lfy one-sided spike, sepals enlarging in fr; Apr-Aug. Fr egg-shaped, beaked. Dry grassy, stony places. (F, S), G. **5a** *N. rosea* is a short annual, with oblong blunt lvs and pink fls with a yellow throat. (G, S), from Caucasus.

6* CORN GROMWELL *Buglossoides arvensis*. Short/medium annual; *little* branched. Lvs strap-shaped to lanceolate, without prominent side veins, the lower shortly stalked. Fls creamy white, in clusters; Apr-Sept. Fr grey-brown, warty. A weed of cultivation. T.

7* COMMON GROMWELL *Lithospermum officinale*. Medium/tall perennial, *well* branched. Lvs lanceolate, unstalked, with prominent side veins. Fls creamy white, in lfy clusters; May-July. Fr white, shiny, smooth. Woods, scrub. T.

8* PURPLE GROMWELL *Buglossoides purpuro-caerulea.* Short/medium perennial, downy, unbranched. Lvs narrow lanceolate, dark green. Fls reddish-purple turning *blue,* in lfy terminal clusters; Apr-June. Fr white, shiny. Woods, scrub, on lime. B(rare), F, G.

Borage Family (contd.)

1* LUNGWORT *Pulmonaria officinalis.* Low/short perennial, downy. Lvs lanceolate, *pale spotted,* abruptly narrowed at base, those from roots greatly enlarging after flg, to 40-60 cm. Fls pink and blue, bell-shaped, sepals with short broad teeth, in lfy clusters; Mar-May. Fr egg-shaped, pointed. Shady places. T, but (B). **1a* Narrow-leaved Lungwort** *P. longifolia* has narrower lvs tapering to base, smaller fls with longer narrower sepal-teeth, and flattened fr. B (rare), F, G, western. **1b** *P. angustifolia* has unspotted lvs 8-10 times as long as broad. G, S, south-eastern. **1c** *P. mollis* is stickily hairy with winged lf stalks. F, G, southern.

2* BLUE-EYED MARY *Omphalodes verna.* Low/short creeping perennial, slightly downy. Lvs *pointed oval,* often heart-shaped, long-stalked. Fls bright blue, 10 mm across, in a loose cluster; Mar-May. Fr with a hairy border. Woods. (B, F, G). **2a** *O. scorpioides* has yellow-centred fls, solitary at base of lvs. G.

FORGETMENOTS *Myosotis.* Lvs oblong, usually unstalked. Fls usually pink in bud, later blue, open. Fr shiny, on usually elongated spikes.

3* WOOD FORGETMENOT *Myosotis sylvatica.* Short perennial, softly hairy. Fls *sky-blue,* 6-10 mm, flat, sepal-tube with spreading hooked hairs; Apr-July. Fr dark brown, on spreading stalks longer than sepal-tube. Woods, grassland, mountains. T. **3a* Alpine Forgetmenot** *M. alpestris* is smaller, with deeper blue fragrant fls, sepals with few hooked hairs, and fr black on erect stalks with spike scarcely elongating. Mountains. B (rare), F, G.

4* FIELD FORGETMENOT *Myosotis arvensis.* Low/short, softly hairy annual/biennial. Lower lvs stalked, in a rosette. Fls *grey-blue,* 3-5 mm across, cup-shaped, sepal-tube with numerous hooked hairs; Apr-Oct. Fr dark brown, shiny, on stalks longer than sepal-tube. Dry bare or disturbed ground, shady places, dunes. T. **4a** *M. stricta* has stems with smaller fls almost from base, lvs often with hooked hairs and fr stalks shorter than sepal-tube. Sandy places. F, G, S, southern. **4b* Early Forgetmenot** *M. ramosissima* is still smaller, often only 2-3 cm high, with smaller bluer fls and sepal-tube longer-stalked with teeth spreading in fr. Apr-May. Dry open places. **4c** *Myosotis sparsiflora* is shorter and almost hairless, with broader lvs and fewer fls; June. G, S.

5* CHANGING FORGETMENOT *Myosotis discolor.* Slender low/short annual. Fls creamy or pale *yellow* at first, turning blue, smaller, almost unstalked; May-June. Fr dark brown, shiny, on stalks shorter than sepal-tube, which has *incurved teeth.* Bare places, usually on light soils. T.

6* WATER FORGETMENOT *Myosotis scorpioides.* Low/short creeping perennial; hairs closely *pressed* to stems and sepal-tube. Fls sky-blue, occasionally pink or white, flat, petals notched, 4-10 mm across, sepal-teeth short, cut up to ⅓; June-Sept. Fr black, on stalks 1-2 times as long as sepal-tube. *Wet* places. T. **6a* Creeping Forgetmenot** *M. secunda* □ has spreading hairs on lower stem, fls 4-6 mm, sepal-tube narrowly cut to at least half-way, and fr dark brown on stalks 3-5 times as long as sepal-tube, down-turned. B, G, S. **6b* Pale Forgetmenot** *M. stolonifera* has much shorter bluish-green lvs, fls pale blue and 5 mm, petals not always notched, sepal-teeth cut up to ⅔, and fr olive-brown on sometimes down-turned stalks. B. **6c* Tufted Forgetmenot** *M. laxa* agg. □ may be hairless and has fls 2-4 mm, petals not notched, sepal-teeth cut up to ½ or less; and fr dark brown on stalks 2-3 times as long as sepal-tube. **6d Jersey Forgetmenot** *M. sicula* □ is smaller with fls pale blue and 2-3 mm, petals not notched, blunt sepal-teeth half as long as sepal-tube, and brown fr on stalks 1-3 times as long as sepal-tube. F.

7* MADWORT *Asperugo procumbens.* Prostrate annual, very bristly; stems *angled.* Fls blue, 3 mm across, 1-3 together on short down-turned stalks much shorter than the elliptical lvs; May-Nov. Fr *surrounded* by enlarged sepals. Bare and disturbed ground. T, but (B, rare).

8 BUR FORGETMENOT *Lappula squarrosa.* Short greyish annual/biennial, roughly hairy; well branched. Lvs lanceolate, unstalked. Fls blue, 2-4 mm, in a loose lfy spike; June-July. Fr with *hooked* spines, surrounded by *star-like* sepal-teeth. Dry bare places, dunes. F, G, S. **8a** *L. deflexa* □ has fls 5-7 mm, larger sepal-teeth and fr-stalks turned down. G, S.

Borage Family *(contd.)*

1* VIPER'S BUGLOSS *Echium vulgare.* Short/medium perennial, roughly hairy. Lvs lanceolate, with no prominent side veins, the upper unstalked. Fls pink turning vivid blue, 15-20 mm, in a branched spike, with all stamens *protruding;* May-Sept. Fr hidden by sepal-teeth which give fr spike a mossy appearance. Dry bare and waste places. T. **1a* Purple Viper's Bugloss** *E. plantagineum* □ is more softly hairy, the lvs with prominent side veins, the upper heart-shaped and clasping the stem, and larger (25-30 mm) fls, red-purple turning blue-purple, and only two stamens protruding. Dry sandy places. B (rare), F. **1b** *Onosma arenaria* has a symmetrical, scarcely toothed petal tube. G, rare.

2* GREEN ALKANET *Pentaglottis sempervirens.* Medium perennial, roughly hairy. Lvs pointed oval, the lower stalked. Fls bright blue, *white-eyed,* 10 mm, in long-stalked lfy clusters, sepal-teeth blunt; Apr-July. Woods, hedge-banks. T, western, but (B).

3* BORAGE *Borago officinalis.* Medium annual, roughly hairy, with cucumber-scented juice. Lvs pointed oval with wavy margins, the lower stalked. Fls bright blue, 20-25 mm, in loose lfy clusters, with narrow petal and sepal lobes and a prominent column of purple stamens; May-Sept. Fr rough. Waysides, waste places. (B, F, G), from S Europe. **3a* Abraham, Isaac and Jacob** *Trachystemon orientalis* □ is shorter, with large heart-shaped lvs mainly from the roots, and smaller (15 mm) purplish fls in almost lfless clusters, and petal-lobes down-turned, making stamen column more conspicuous.

4* ALKANET *Anchusa officinalis.* Medium biennial/perennial, *softly* hairy; well branched. Lvs lanceolate, the lower narrowed to a stalk. Fls blue-purple, 10 mm, in an elongating coiled cluster; June-Aug. Fr conical, unstalked. Grassy places. T, but (B). **4a* Large Blue Alkanet** *A. azurea* □ is taller and roughly hairy, its much larger (15-25 mm) fls with a tuft of white hairs in the middle (F, G), from Mediterranean.

5* BUGLOSS *Anchusa arvensis.* Short annual, roughly hairy. Lvs lanceolate, *wavy-edged.* Fls blue, 5-6 mm, in elongating, often forked clusters, the petal-tube bent; Apr-Sept. Fr unstalked. Bare, often sandy or cultivated places. T.

6* OYSTER PLANT *Mertensia maritima.* Prostrate mat-forming greyish *hairless* perennial. Lvs oval, *fleshy,* oyster-tasting. Fls pink turning blue-purple, in lfy clusters; June-Aug. Fr smooth, fleshy. Coastal shingle. B, S, northern.

Verbena Family   Verbenaceae

7* VERVAIN *Verbena officinalis.* Short/medium perennial, roughly hairy; stems stiff, square. Lvs pinnately lobed, toothed, opposite, the upper unstalked. Fls lilac, more or less *2-lipped,* petals 5-lobed, in long slender lfless spikes; June-Oct. Dry bare or sparsely grassy places. T, southern.

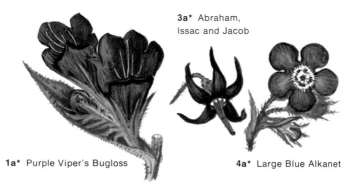

3a* Abraham, Issac and Jacob

1a* Purple Viper's Bugloss

4a* Large Blue Alkanet

Key to Two-lipped Flowers
(Mint and Figwort Families)
(Pages 200-221)

This key does not include all two-lipped flowers, but only those belonging to the Mint and Figwort Families – the Labiatae (pp. 200-208) and Scrophulariaceae (pp. 212-220). Labiates are distinctive for their square stems, and many of them are aromatic, but members of the Figwort Family lack any common distinctive features. Other two-lipped flowers include the Broomrapes (p. 222) and Orchids (p. 278). The aberrant labiates with only one lip, the Bugles and Germanders (p. 200) are excluded. For all these groups, see the Main Key (p. 13).

Plant

white or grey: Cut-leaved Self-Heal, White Horehound, Catmint 202; Downy Woundwort 206; Lesser Calamint 208; Hoary Mullein 212

aromatic: Ground Ivy, White Horehound, Catmints, Winter Savory, Hyssop 202; Red and White Dead-nettles 204; Meadow Clary, Balm, Yellow Woundwort 206; most on p. 208.

foetid: Bastard Balm, Black Horehound 202; Motherwort, Yellow Archangel 204; Hedge Woundwort 206; Common Figwort 212

purplish: Ground Ivy 202; Claries and Wild Sage 206; Red Bartsia 216; Marsh Lousewort 220

woody at base: Ground Ivy 202; Yellow Archangel 204

with winged stems: Water Figwort 212

Leaves

untoothed: Spear-leaved Skullcap 200; Self-heal, *Dracocephalum ruyschiana*, Winter Savory, Hyssop 202; Marjoram 208; most on p. 214; Cow-wheats 220

narrow: *Dracocephalum ruyschiana, D. thymifolium*, Winter Savory, Hyssop 202; Red Hemp-nettle, Yellow Archangel 204; Marsh and Annual Woundworts 206; Spear Mint, Gipsywort 208; most on p. 214; Large Yellow Foxglove, Gratiola, Red Bartsia, Yellow Odontites 216; Yellow Rattle, Cow-wheats 220

pinnate: Cut-leaved Self-heal 202; Wild Clary 206; French Figwort 212; Louseworts and Moor-king 220

heart- or kidney-shaped: Ground Ivy, Catmint 202; Dead-nettles, *Galeopsis pubescens* 204; Wild Sage, Balm, Hedge and Field Woundworts 206; Balm-leaved and Yellow Figworts 212; Cornish Moneywort 220

triangular: Fluellens, Ivy-leaved Toadflax 214

alternate: most on p. 214; Foxgloves and Gratiola 216; Eyebright, Louseworts, Moor-king and Cornish Moneywort 220

Flowers

white: Self-heals, White Horehound, Catmint, Bastard Balm, Winter Savory 202; White Dead-nettle, Common Hemp-nettle 204; Balm 206; Gipsywort 208; *Lindernia pyxidaria* 214; Foxglove, Gratiola 216; Eyebright 220

blue/violet: Skullcap 200; Self-heal, Ground Ivy, Hairless Catmint, Hyssop 202; Claries and Wild Sage 206; Basil Thyme 208; Pale, Purple and Ivy-leaved Toadflaxes 214

yellow: Common and Downy Hemp-nettles, Yellow Archangel 204; Yellow Figwort 212; Snapdragon, Common and Sand Toadflaxes 214; Monkey Flower, Large Yellow Foxglove, Yellow Odontites, Yellow Bartsia 216; Yellow Rattle, Moor-king, Leafy Lousewort, Common Cow-wheat 220

pink/purple: Lesser Skullcap 200; Self-heal, Ground Ivy, Bastard Balm, Black Horehound, Winter Savory 202; Spotted, Red and Henbit Dead-nettles, False Motherwort, Common and Red Hemp-nettles 204; Whorled Clary, Woundworts, Betony 206; all on p. 208; Purple Mullein, French Figwort 212; Snapdragons, Purple and Small Toadflaxes 214; Foxglove, Gratiola, Fairy Foxglove, Alpine and Red Bartsias 216; Lousewort and Marsh Lousewort 220

yellow and purple: Motherwort, Large-flowered Hemp-nettle 204; Yellow Woundwort 206; Small, Pale and Ivy-leaved Toadflaxes, Fluellens 214; Yellow Rattle, Common Cow-wheat, Cornish Moneywort 220

red-brown: Figworts 212

in whorls: Ground Ivy, White Horehound, Catmint, Bastard Balm, Black Horehound, Hyssop 202; most on pp. 206-8

in spikes: Skullcaps 200; Hairless Catmint 202; Betony 206; Spear Mint 208; Snapdragon, Toadflaxes 214; most on pp. 216-18

in heads: Self-heals 202; Water Mint, Wild Thyme 208

in clusters: Winter Savory 202; Common Calamint, Marjoram 208; Figworts 212

solitary: Lesser Snapdragon, Small and Ivy-leaved Toadflaxes, Fluellens 214; Gratiola 216

in pairs: Common Cow-wheat 220

with mouth closed and/or spurred: all on p. 214

bell-shaped: Mints and Gipsywort 208

tubular: Foxgloves and Gratiola 216

with calyx inflated: Yellow Rattle, Louseworts 220

Labiate Family   Labiatae

Often *aromatic.* Stems *square.* Lvs usually undivided, in *opposite* pairs. Fls with both petals and sepals, normally *2-lipped* and open-mouthed (except p. 208, nos. 1-5), though *Ajuga* and *Teucrium* both have upper lip absent or very short, but lower lip 3-lobed and 5-lobed respectively. Fr four nutlets.

1* BUGLE *Ajuga reptans.* Low/short creeping perennial; runners *rooting,* stems hairy on two sides. Lvs oblong, scarcely toothed, hairless, often bronzy, the lower long-stalked. Fls powder-blue, rarely pink or white, in a lfy, often purplish, spike; Apr-June. Damp woods, grassland. T. **1a Blue Bugle** *A. genevensis* has no runners, stems often hairy all round, root lvs larger, fls bright blue, and stamens protruding. Dry grassy, stony places. F, G, (S). **1b* Pyramidal Bugle** *A. pyramidalis* has no runners, stems hairy all round, root lvs hairy, pyramidal bluish spikes of smaller blue-violet fls, and stamens protruding; a shy flowerer; grassy, rocky places in mountains and near sea.

2* GROUND-PINE *Ajuga chamaepitys.* Low greyish annual, faintly *aromatic* of pine resin. Looks somewhat like a bushy pine seedling with its narrowly *linear* 3-lobed lvs. Fls yellow, red-dotted, 1-2 together up the lfy stems; May-Sept. Bare, stony, sparsely grassy places. B, F, G.

3* SKULLCAP *Scutellaria galericulata.* Short creeping perennial, often downy. Lvs lanceolate, bluntly toothed, short-stalked. Fls bright *blue,* in *pairs* up the lfy stem; petal-tube slightly curved; June-Sept. Hybridises with 4. Wet grassy places, by fresh water. T. **3a* Spear-leaved Skullcap** *S. hastifolia* □ is slenderer, with untoothed arrow-shaped lvs, purple beneath, and larger fls in a denser, less lfy spike, with petal-tube strongly curved. Rare in B. **3b*** *S. columnae* has bracts much smaller than lvs and larger fls in a one-sided spike. (B, rare), F.

4* LESSER SKULLCAP *Scutellaria minor.* Low/short creeping perennial, sometimes downy. Lvs lanceolate to oval, scarcely toothed. Fls pale *pinkish-purple,* spotted darker, in *pairs* up the lfy stems; petal-tube almost straight; July-Oct. Damp *heathy* places. T, southern.

5* WOOD SAGE *Teucrium scorodonia.* Short/medium perennial, downy. Lvs heart-shaped, bluntly toothed, *wrinkled,* sage-like. Fls *greenish-yellow,* in pairs in lfless stalked spikes, with prominent maroon stamens; July-Sept. Open woods, scrub, heaths, fixed dunes, screes; not on lime. T, western.

6 MOUNTAIN GERMANDER *Teucrium montanum.* Low/short spreading undershrub. Lvs linear/lanceolate, *white* beneath, margins inrolled. Fls yellowish-white, in rounded clusters; May-Aug. Rocky places in mountains. F, G.

7* WALL GERMANDER *Teucrium chamaedrys.* Short tufted perennial, slightly hairy. Lvs pointed oval, *shiny,* dark green, rather leathery, with rounded teeth. Fls dark pinkish-purple, in short lfy whorled spikes; May-Sept. Dry bare places, including walls. (B, rare), F, G.

8* WATER GERMANDER *Teucrium scordium.* Low/short sprawling perennial, softly hairy. Lvs oblong, toothed, unstalked. Fls pinkish-purple, in whorls up lfy stems; June-Aug. Damp places, by *fresh water,* dune slacks. T, rare in B.

9* CUT-LEAVED GERMANDER *Teucrium botrys.* Low/short downy annual/biennial. Lvs oval, *deeply cut,* almost pinnate. Fls pink or pinkish-purple, in whorls along lfy stems; June-Oct. Bare, stony ground; on lime. B (rare), F, G.

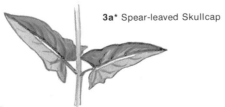

3a* Spear-leaved Skullcap

Labiate Family (contd.)

1* SELF-HEAL *Prunella vulgaris.* Low creeping perennial, downy, not aromatic. Lvs pointed oval, *scarcely toothed,* the lower stalked. Fls *violet,* rarely pink or white, 10-15 mm long, in *oblong* or square purplish heads; June-Nov. Flhds with lvs at base. Hybridises with 1a and 2. Grassy places, bare ground. T.
1a Large Self-heal *P. grandiflora* □ is larger, with fls 20-25 mm long. Flhds with no lvs at base. F, G, S.

2* CUT-LEAVED SELF-HEAL *Prunella laciniata.* Low creeping perennial, downy, not aromatic; stems with white hairs. Lvs pointed oval, the upper *pinnate* or pinnately lobed. Fls creamy *white,* occasionally tinged violet (indicating hybridisation with 1), in oblong or square heads; June-Oct. Dry grassy and bare places. (B), F, G.

3* GROUND IVY *Glechoma hederacea.* Low creeping, often purplish perennial, softly hairy, aromatic; runners long, rooting. Lvs *kidney-shaped,* blunt-toothed, long-stalked. Fls blue-violet, rarely pink, in loose whorls at base of lvs; Mar-June. Woods, hedge-banks, sparsely grassy and bare places, sometimes carpeting. T. **3a** *Dracocephalum ruyschiana* □ has narrow lanceolate untoothed unstalked lvs and dense heads of larger fls; July-Aug. G, S. **3b** *D. thymiflorum* □ is like 3a but with toothed lvs and looser heads of smaller bright blue fls. G.

4* WHITE HOREHOUND *Marrubium vulgare.* Medium perennial, *white* with down, *thyme-scented.* Lvs roundish, bluntly toothed, wrinkled, stalked. Fls white, in dense whorls up lfy stems; sepal-teeth 10, hooked; June-Sept. Dry bare and waste places. T, rare in B.

5* CATMINT *Nepeta cataria.* Short/medium perennial, *grey* with down, *mint-scented.* Lvs heart-shaped, more markedly toothed than 4, stalked. Fls white, red-dotted, in whorls up lfy stems; sepal-teeth 5, straight; June-Sept. Hedge-banks, waysides, rocks. T.

6 HAIRLESS CATMINT *Nepeta nuda.* Medium perennial, almost *hairless.* Lvs oblong, toothed. Fls violet, in branched spikes; June-Sept. Open woods, scrub, grassy and rocky places. F, G.

7* BASTARD BALM *Melittis melissophyllum.* Medium perennial, hairy, *strong smelling.* Lvs pointed oval, toothed, stalked. Fls pink, or white with pink spots, *large* (35-45 mm long), in few-fld whorls at base of lvs; May-July. Woods, hedge-banks, shady rocks. B, F, G.

8* BLACK HOREHOUND *Ballota nigra.* Medium, rather straggly perennial, hairy, *strong smelling.* Lvs pointed oval, toothed, stalked. Fls pinkish-purple, in whorls at base of upper lvs, sepal-teeth *curved* back in fr; June-Sept. Waste places, waysides. T.

9* WINTER SAVORY *Satureja montana.* Short half-evergreen undershrub, almost hairless, pleasantly aromatic; a culinary herb. Lvs narrow lanceolate, *pointed,* untoothed, rather leathery, hairy on margins, *shiny-dotted,* unstalked and joined at base. Fls pale pink or white, in clusters; July-Oct. Old walls. (B, rare), from S Europe; cultivated T.

10* HYSSOP *Hyssopus officinalis.* Medium perennial, almost hairless, pleasantly aromatic; a culinary herb. Lvs linear oblong, *blunt.* Fls violet-blue, in few-fld whorls making a long narrow spike; July-Sept. Old walls, dry banks. (B, F, G, rare), from S Europe.

3a *Dracocephalum ruyschiana*

3b *D. thymiflorum*

203

Labiate Family *(contd.)*

1* WHITE DEAD-NETTLE *Lamium album.* Short creeping perennial, hairy, faintly aromatic; unbranched. Lvs heart-shaped, toothed, stalked, resembling benign Nettle (p. 38) lvs. Fls *white* with greenish streaks on lower lip, 20-25 mm long, in whorls at base of lvs; Mar-Nov. Waysides, waste places. T. **1a* Spotted Dead-nettle** *L. maculatum* ☐ is strong smelling and has lvs often pale-spotted and fls usually pink-purple. T, but (B).

2* RED DEAD-NETTLE *Lamium purpureum.* Low/short, often purplish annual, downy, aromatic. Lvs heart-shaped, bluntly toothed, wrinkled, *all stalked.* Fls pale to dark pinkish-purple, 10-15 mm long; all year. A weed of cultivation. T. **2a* Cut-leaved Dead-nettle** *L. hybridum* ☐ has more deeply toothed lvs and smaller fls.

3* HENBIT DEAD-NETTLE *Lamium amplexicaule* agg. Low/short annual, downy, little branched. Lvs rounded, bluntly toothed, the topmost usually *unstalked* and often half-clasping the stem. Fls pink-purple, 15 mm long, often also smaller and unopened; in whorls at base of upper lvs; Mar-Oct. A weed of cultivation. T.

4* MOTHERWORT *Leonurus cardiaca.* Medium/tall perennial, slightly downy, strong smelling; unbranched. Lvs *palmately* 3-7-lobed, whitely downy beneath, stalked. Fls pinkish-purple or white, the lower lip spotted purple, furry outside, 12 mm long, in whorls up the lfy stems; July-Sept. Waysides, waste places. T, but (B). **4a False Motherwort** *L. marrubiastrum* is biennial, with lower lvs undivided, the upper only toothed, and pale pink, much less hairy fls. F,G,S, south-eastern.

5* COMMON HEMP-NETTLE *Galeopsis tetrahit* agg. Short/medium annual, roughly hairy; well branched, stems *swollen* at lf-junctions. Lvs broad lanceolate, toothed, stalked. Fls pinkish-purple, less often yellow or white, 10-20 mm long, in whorls at base of upper lvs; July-Sept. Probably originated as 5a × 7 hybrid. Open woods, heaths, fens, also a weed of cultivation. T. **5a** *G. pubescens* has lvs square or heart-shaped at base and fls bright purple. G.

6* RED HEMP-NETTLE *Galeopsis angustifolia* agg. Short/medium annual, softly hairy; well branched, stems *scarcely* swollen at lf-junctions. Lvs narrow to broad lanceolate, sparsely toothed, stalked. Fls *deep* pink, white-flecked on lip, 15-25 mm long, in whorls at base of upper lvs; July-Sept. A weed of cultivation, bare stony places. T, southern.

7* LARGE-FLOWERED HEMP-NETTLE *Galeopsis speciosa.* Medium/tall annual, roughly hairy; well branched, stems *swollen* at lf-junctions. Lvs broad lanceolate, toothed, stalked. Fls pale *yellow,* lower lip *purple,* 20-45 mm long, in whorls at base of upper lvs; July-Sept. A weed of cultivation, waste places. T.

8* DOWNY HEMP-NETTLE *Galeopsis segetum.* Short/medium annual, *softly* hairy; well branched, stems *not* swollen at lf-junctions. Lvs lanceolate, toothed, stalked. Fls pale *yellow,* 20-30 mm long, in whorls at base of upper lvs; July-Sept. A weed of cultivation. (B,rare),F,G. Hybridises with 6.

9* YELLOW ARCHANGEL *Lamiastrum galeobdolon.* Short/medium creeping perennial, hairy, strong smelling; *runners* long, lfy, not rooting. Lvs narrow to broad lanceolate, dark green, stalked. Fls *butter* yellow, usually streaked red-brown on lower lip, in whorls at base of upper lvs; May-June. Woods. T.

Nettle (p.38)　　　　　　　　　　　　　　**1*** White Dead-nettle

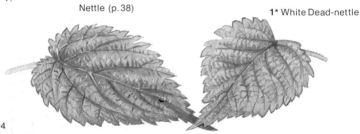

205

Labiate Family *(contd.)*

1* MEADOW CLARY *Salvia pratensis.* Medium perennial, hairy, slightly aromatic. Lvs broad lanceolate, the few on the stem narrower and unstalked, bluntly toothed, wrinkled. Fls *brilliant blue,* in whorled lfless spikes; June-July. Grassland on lime. T. **1a* Whorled Clary** *S. verticillata* is strong smelling, with upper lvs and stems often purplish, all lvs stalked and often with a purplish sheen, smaller pinkish-purple fls, and purple sepals; dry bare and waste places; May-Aug. (T), from S Europe. **1b Jupiter's Distaff** *S. glutinosa* □ has yellow fls. Mountain woods. G, southern.

2* WILD SAGE *Salvia nemorosa.* Short/tall hairy perennial. Lvs heart-shaped, much longer than broad, pointed, the upper and middle ones unstalked. Fls blue, in purplish lfy whorled and branched spikes; June-Aug. Bare and waste places. (B, rare, G), from S Europe.

3* WILD CLARY *Salvia verbenaca.* Variable medium perennial, hairy; little branched, purplish on upper lvs and stems. Lvs oval, *jaggedly* toothed, sometimes almost pinnately lobed, the upper unstalked. Fls blue-violet, often white-spotted on lower lip, 5-15 mm long, in whorled spikes; often not opening and so appearing much smaller than bracts; June-Sept. Dry grassy places. B, F.

4* BALM *Melissa officinalis.* Medium hairy perennial, *lemon-scented* when crushed, a culinary and medicinal herb. Lvs pointed oval, the lowest heart-shaped, toothed, stalked, yellow-green. Fls *white,* in lfy whorled spikes; July-Sept. An escape from cultivation. (B, F, G), from S Europe.

5* HEDGE WOUNDWORT *Stachys sylvatica.* Medium creeping perennial, roughly hairy, *strong smelling.* Lvs heart-shaped, toothed, stalked. Fls dark *beetroot-purple* with whitish blotches, in whorled spikes; June-Oct. Hybridises with 5a. Hedge-banks, shady places. T. **5a* Marsh Woundwort** *S. palustris* □ is only faintly aromatic and has narrower short- or unstalked lvs, and pale pinkish-purple fls. Damp places, by fresh water.

6* DOWNY WOUNDWORT *Stachys germanica.* Medium/tall perennial, thickly covered with a *felt of white hairs.* Lvs pointed oval, toothed, the upper short- or unstalked. Fls pale pinkish-purple, in a lfy whorled spike; July-Sept. Bare ground; on lime. B(rare), F, G. **6a* Limestone Woundwort** *S. alpina* □ is less thickly hairy.

7* YELLOW WOUNDWORT *Stachys recta.* Variable short/medium hairy perennial, pleasantly aromatic. Lvs toothed, oval to oblong, stalked, the upper much narrower and unstalked. Fls *pale yellow* with purple streaks, in lfless whorled spikes, the stamens protruding; June-Sept. Dry rocky or waste places. (B, rare), F, G. **7a* Annual Woundwort** *S. annua* is a shorter annual, with all lvs lanceolate and stalked, and smaller fls; T, but (B, rare). **7b** *Sideritis montana* has stamens enclosed within the petal-tube. (F), G, (S).

8* FIELD WOUNDWORT *Stachys arvensis.* Low/short, often sprawling hairy annual. Lvs heart-shaped, bluntly toothed, the upper unstalked. Fls dull purple, *scarcely* exceeding sepals; Apr-Oct. A weed of cultivation; not on lime. T.

9* BETONY *Stachys officinalis.* Variable short/medium hairy perennial, usually unbranched. Lvs oblong, slightly toothed, all but the topmost stalked. Fls reddish-purple, in tight *oblong* spikes; June-Oct. Grassy and heathy places. T.

1b Jupiter's Distaff

207

Labiate Family *(contd.)*

MINTS *Mentha. Aromatic* perennials, each with a distinctive scent. Lvs stalked, except 3. Fls small, usually lilac or pale purple, bell-shaped, with four more or less equal petal-lobes, in *tight whorls* at base of lvs; sepal-tube 5-toothed, stamens four, protruding except in the numerous hybrids.

1* WATER MINT *Mentha aquatica.* Variable short/medium, often purplish hairy perennial, pleasantly aromatic. Lvs pointed oval, toothed. Fls lilac or pinkish-lilac, with a *round* terminal head; sepal-tube hairy, long-toothed; July-Sept. Hybridises with 2 (q.v.) and with 3, from which the resultant hybrid is the often escaping culinary herb Peppermint *M.* × *piperita,* with narrower, more sharply toothed lvs, reddish-lilac fls, fl stalks and sepal-tube usually hairless and stamens not protruding. Wet places. T.

2* CORN MINT *Mentha arvensis.* Low/medium hairy perennial, rather acridly aromatic. Lvs pointed oval, toothed. Fls lilac, with *no* terminal head; sepal-tube hairy, short-toothed; July-Sept. Hybridises with 1, the resulting *M.* × *verticillata* having no terminal head but longer sepal-teeth and stamens not protruding. Damp places, and as a weed of cultivation. T.

3* SPEAR MINT *Mentha spicata.* Short/medium perennial, hairy or hairless; the most commonly grown garden mint. Lvs lanceolate, almost *unstalked,* either green, shiny and hairless or greyish, wrinkled and downy. Fls lilac, with a *pointed* terminal spike; July-Oct. Hybridises with 1 (q.v.) and 3a. Damp and waste places. T, southern. **3a* Round-leaved Mint** *M. suaveolens* is apple-scented and always thickly downy, with rounder lvs and pinker fls. B,F,G. **3** × **3a*** *M.* × *villosa* var. *alopecuroides* is a large coarse escape from cultivation with sharply toothed rounded lvs and stamens not protruding. (B,F,G).

4* PENNYROYAL *Mentha pulegium.* Prostrate, sometimes erect and short, perennial, downy, strongly aromatic. Lvs oval, scarcely toothed. Fls lilac, with *no* terminal head; sepal-tube hairy, well ribbed, the two lower teeth narrower; July-Oct. Damp places. B,F,G.

5* GIPSYWORT *Lycopus europaeus.* Medium/tall perennial, slightly hairy, *not* aromatic; unbranched. Lvs lanceolate, deeply toothed or lobed, short-stalked. Fls small, white spotted purple, bell-shaped, in tight whorls at base of upper lvs; July-Sept. By fresh water, marshes. T.

6* COMMON CALAMINT *Calamintha sylvatica.* Short/medium hairy perennial, *mint-scented.* Lvs oval, slightly toothed, stalked, dark green. Fls lilac-purple spotted darker, 10-15 mm, in short-stalked opposite clusters at base of lvs; sepal-tube often purple with no hairs protruding from throat after flg; July-Sept. Dry grassy places, on lime. In open woods or scrub fls may be 17-22 mm **(6a*□).** B,F,G. **6b* Lesser Calamint** *C. nepeta* has greyer lvs, paler, sometimes almost white, less spotted fls, and hairs protruding from sepal-tube after flg.

7* WILD BASIL *Clinopodium vulgare.* Short/medium perennial, softly hairy, faintly aromatic; little branched. Lvs pointed oval, slightly toothed, stalked. Fls pinkish-purple, in whorls at base of upper lvs; bristle-like bracts and purplish sepal-tube *white-haired*; July-Oct. Dry grassy places, scrub, on lime. T.

8* MARJORAM *Origanum vulgare.* Medium downy perennial, pleasantly aromatic; a culinary herb. Lvs oval, often slightly toothed, stalked. Fls purple, darker in bud and with *dark purple* bracts, in loose clusters; July-Sept. Dry grassland, scrub, screes; on lime. T.

9* BASIL THYME *Acinos arvensis.* Low sprawling hairy annual, not or slightly aromatic. Lvs oval, slightly toothed, stalked. Fls *violet* with *white* patch on lower lip, in whorls at base of upper lvs; June-Sept. Dry bare places; on lime. T.

10* WILD THYME *Thymus serpyllum* agg. Variable prostrate hairy *mat-forming* undershrub, aromatic; with runners. Lvs oval, sometimes quite woolly, short-stalked. Fls pinkish- or reddish-purple, in *rounded* heads, sometimes also in whorls below; June-Sept. Dry grassy and heathy places, dunes. T. **10a* Large Thyme** *T. pulegioides* is larger, with no runners, and more aromatic, and has lvs always almost hairless and fls in elongated whorled spikes. Often on lime.

Nightshade Family   Solanaceae

Often poisonous. Lvs alternate, normally stalked. Fls 5-petalled, joined at base.

1* BITTERSWEET *Solanum dulcamara.* Clambering, sometimes prostrate downy perennial, to 2 m. Lvs pointed oval, often with two lobes at base. Fls in loose clusters, bright *purple,* petals turning down, yellow anthers in a conspicuous *column;* May-Sept. Fr a poisonous egg-shaped berry, green then yellow then *red.* Woods, scrub, hedges, waste and damp places, shingle. T.

2* BLACK NIGHTSHADE *Solanum nigrum.* Variable low/short annual, hairless or downy; stems often blackish. Lvs pointed oval, often toothed or lobed. Fls in loose clusters, *white,* petals turning down, with prominent column of yellow anthers; July-Oct. Fr a poisonous berry, green then *black.* A weed of cultivation, waste places. T. **2a* Green Nightshade** *S. sarrachoides* agg. is hairier, with stems never blackish and sepals enlarging to exceed the always green berry. **2b* Tomato** *Lycopersicon esculentum* □, a straggling plant with 2-pinnate lvs, fls all yellow and large red fr, is a widespread escape from cultivation.

3 HAIRY NIGHTSHADE *Solanum luteum.* Short hairy annual. Lvs pointed oval, more deeply lobed than 2. Fls like 2 but larger. Fr *orange* or yellowish-brown. Bare and waste places. H.

4* DEADLY NIGHTSHADE *Atropa bella-donna.* Tall stout perennial, sometimes downy; well branched. Lvs pointed oval, up to 20 cm long. Fls dull purple or greenish, *bell-shaped,* solitary, 25-30 mm; June-Sept. Fr a glossy *black* berry, extremely poisonous. Woods, scrub, rocky places; on lime. B,F,G. N.B. Bittersweet is often miscalled Deadly Nightshade.

5* HENBANE *Hyoscyamus niger.* Medium/tall annual/biennial, *stickily* hairy, unpleasantly foetid, very poisonous. Lvs oblong, sometimes with a few teeth, up to 20 cm long, the upper unstalked, clasping the stem. Fls dull creamy yellow with purple veins, bell-shaped, in a lfy cluster; May-Sept. Fr a capsule. Bare and disturbed ground, especially by sea. T.

6* APPLE OF PERU *Nicandra physalodes.* Medium/tall annual, often hairless, foetid, very poisonous. Lvs pointed oval, toothed or lobed. Fls blue or pale *violet* with white throat, bell-shaped, solitary, 30-40 mm across, opening only for a few hours; June-Oct. Fr a brown berry, encased in the net-veined *bladder-like* swollen sepals. Bare and waste places, waysides. (B,F,G), from Peru.

7* THORN-APPLE *Datura stramonium.* Stout medium/tall hairless annual, foetid and poisonous. Lvs pointed oval, jaggedly toothed. Fls white, rarely purple, *trumpet-shaped,* solitary, 60-80 mm across; July-Oct. Fr egg-shaped, *spiny.* Bare, cultivated and waste ground. (T).

8 SMALL TOBACCO PLANT *Nicotiana rustica.* Medium/tall annual, stickily hairy, strong smelling and poisonous. Lvs *heart-shaped*, shiny, with unwinged stalks. Fls greenish-yellow, bell-shaped, in clusters; June-Aug. Fr a capsule. An escape from cultivation. (F,G, southern), from N America. **8a* Tobacco Plant** *N. tabacum* has lf stalks winged or absent and may have fls white, pink or red. (T).

2b* Tomato

Figwort Family Scrophulariaceae

pp. 212-20. Lvs with no stipules. Fls either open and more or less flat with four or five joined petals (Mulleins, below, and Speedwells, p. 218), or two-lipped with the lips either open or closed. Fr a capsule.

MULLEINS *Verbascum.* Biennials, usually tall. Lvs in a basal rosette and alternate up stems. Fls flat, 5-petalled, in spikes; anthers orange. Fr egg-shaped.

1* GREAT MULLEIN *Verbascum thapsus.* Tall, stout, to 2 m, covered with *thick* white woolly down; stem round, usually unbranched. Lvs broad lanceolate, bluntly toothed, running down on to the *winged* stems. Fls yellow, almost flat, 15-30 mm, with three whitely hairy and two hairless stamens; June-Aug. Dry grassy and bare places, open scrub. T. **1a*** *V. densiflorum* □ has yellowish-grey down, more pointed and coarsely toothed lvs, and sometimes branched spikes of larger (30-50 mm), completely flat fls. (B), G. **1b* Orange Mullein** *V. phlomoides* has lvs not or hardly running down stem and larger (30-50 mm), sometimes orange-yellow fls. (B), F, G.

2* DARK MULLEIN *Verbascum nigrum.* Medium/tall, hairy, stem ridged, usually unbranched, often purplish. Lvs heart-shaped, toothed, dark green, stalked. Fls yellow with *purple* hairs on all stamens; June-Sept. Dry grassy places. T.

3* WHITE MULLEIN *Verbascum lychnitis.* Medium/tall, downy; stems angled at top, *branched* like a candelabrum. Lvs lanceolate, bluntly toothed, shortly stalked, dark green and almost hairless above, *white* with down beneath. Fls yellow or white, 15-20 mm, with all stamens whitely hairy; June-Sept. Dry bare and sparsely grassy places. T. **3a* Hoary Mullein** *V. pulverulentum* is thickly covered all over with mealy white down, and has rounded stems, broader lvs and always yellow fls. B (rare), F, G.

4* MOTH MULLEIN *Verbascum blattaria.* Medium/tall, hairless below but with sticky hairs above; stems angled, usually unbranched. Lvs lanceolate, *shiny,* toothed, gradually narrowed to base. Fls yellow, 20-30 mm, solitary, *long-stalked;* stamens with *purple* hairs; June-Aug. Bare and waste places. B, F, G. **4a* Twiggy Mullein** *V. virgatum* is more stickily hairy above, with 1-5 short-stalked fls together. B (rare), F. **4b Purple Mullein** *V. phoeniceum* □ has purple fls. (F), G.

FIGWORTS *Scrophularia.* Square-stemmed perennials. Lvs opposite, stalked, toothed. Fls two-lipped, open-mouthed, with five small blunt lobes; in a branched terminal cluster.

5* COMMON FIGWORT *Scrophularia nodosa.* Medium/tall, almost hairless, foetid; stems not winged. Lvs *pointed oval,* not lobed. Fls red-brown, sepals *all green;* June-Sept. Woods, shady places. T. **5a* Water Figwort** *S. auriculata* has winged stems and lf-stalks, lvs blunter, often with two lobes at base, and sepals white-edged. Wet places. B, F, G. **5b* Green Figwort** *S. umbrosa* is completely hairless and has lf stalks running down on to even more broadly winged stem than 5a, and olive-brown fls.

6* BALM-LEAVED FIGWORT *Scrophularia scorodonia.* Medium/tall, *downy* all over. Lvs *heart-shaped,* double-toothed, wrinkled. Fls red-brown, sepals *white-edged;* June-Sept. Dry shady places near sea. B, F.

7* FRENCH FIGWORT *Scrophularia canina.* Medium/tall, hairless; well branched. Lvs *pinnately lobed,* the lobes toothed. Fls *purplish-black,* upper lip sometimes white, with purple stamens and orange anthers; June-Aug. Bare and waste places. (B, rare), F, G.

8* YELLOW FIGWORT *Scrophularia vernalis.* Medium/tall biennial/perennial, softly hairy. Lvs heart-shaped, wrinkled, well toothed, yellowish-green. Fls pale *greenish-yellow;* Apr-June. Woods, scrub, bare places, old walls. T, but (B).

Figwort Family (contd.)

All plants on this page have 2-lipped fls, with *mouth closed,* except 9, and spurred, except 1 and 2. Some are semi-parasitic on the roots of other plants.

1* **SNAPDRAGON** *Antirrhinum majus.* Short/medium perennial, downy; woody at base; well branched. Lvs narrow lanceolate, untoothed. Fls very variable, most commonly red-purple, less often pale yellow, with a *pouch* instead of a spur, *30-40 mm;* in stalked spikes; May-Oct. Dry bare places, rocks, walls. (B),F,G.

2* **LESSER SNAPDRAGON** *Misopates orontium.* Short annual, usually downy. Lvs linear, untoothed. Fls pink-purple, *10-15 mm, pouched,* scarcely stalked at base of upper lvs; July-Oct. A weed of cultivation, bare sandy places. T.

3* **COMMON TOADFLAX** *Linaria vulgaris.* Short/medium greyish hairless perennial. Lvs linear, untoothed, numerous up stems. Fls *yellow,* with orange spot on lower lip, and long straight spur 15-30 mm, in stalked spikes; June-Oct. Bare and waste places. T. **3a** *L. genistifolia* is more branched with broader lvs and paler fls with a browner orange spot. (F),G. **3b* Prostrate Toadflax** *L. supina* is smaller and more or less prostrate, and has smaller (10-15 mm) fls in a roundish cluster. B,F.

4* **SAND TOADFLAX** *Linaria arenaria.* Low annual, stickily hairy; well branched. Lvs linear, untoothed. Fls yellow with short spur often *violet,* 4-6 mm, in stalked spikes; May-Sept. Coastal dunes. (B,rare),F. **4a** *L. incarnata* is rather taller with purple upper lip and orange and white lower lip. (G).

5* **PALE TOADFLAX** *Linaria repens.* Short greyish hairless perennial. Lvs linear, untoothed, numerous up stems. Fls pale *lilac,* veined violet, with orange spot on lower lip and short curved spur, 7-14 mm, in stalked spikes; June-Sept. Dry bare and sparsely grassy places. T. Hybridises with 3. **5a** *L. arvensis* □ has lower lvs whorled in fours and smaller (4-8 mm) blue-lilac fls with a white spot. F,G, decreasing. **5b Alpine Toadflax** *L. alpina* □ has broader whorled lvs and larger (20 mm) violet fls with an orange spot. Mountain screes and river beds. F,G.

6* **PURPLE TOADFLAX** *Linaria purpurea.* Medium greyish hairless perennial; often unbranched. Lvs linear, untoothed, numerous up stems. Fls bright *violet,* occasionally pink, 8 mm, with long curved spur, in stalked spikes; June-Aug. Walls, bare ground. F,(B,G,S). **6a Jersey Toadflax** *Linaria pelisseriana* □ is shorter and annual, with lower lvs whorled and larger (10-20 mm) bluer fls with a white spot and a straight spur. Bare and disturbed ground. F.

7* **IVY-LEAVED TOADFLAX** *Cymbalaria muralis. Trailing,* often purplish hairless perennial. Lvs palmately lobed, ivy-like, long-stalked. Fls lilac with yellow spot, 8-10 mm, with a short curved spur; solitary on long stalks at base of lvs; Apr-Nov. *Walls,* rocks. (B,F,G), from S Europe.

8* **SHARP-LEAVED FLUELLEN** *Kickxia elatine.* More or less prostrate annual, stickily hairy. Lvs all *pointed,* almost triangular, stalked. Fls with upper lip and throat *purple* and *lower lip yellow,* 8-11 mm, spur straight; solitary on hairless stalks at base of lvs; July-Oct. A weed of cultivation, bare places. T. **8a* Round-leaved Fluellen** *K. spuria* □ has lvs oval or roundish, fls with upper lip maroon, spur curved, stalks hairy. **8** × **8a** has hairy fl-stalks, pointed lvs. F,G.

9* **SMALL TOADFLAX** *Chaenorhinum minus.* Slender low downy annual. Lvs linear, untoothed, greyish, scarcely stalked. Fls pale purple, mouth *slightly open,* 6-8 mm, with a short blunt spur, solitary on long stalks at base of lvs; May-Oct. A weed of cultivation, railway tracks, bare places. T, southern. **9a** *Lindernia procumbens* has unstalked lvs and tubular white fls, petals shorter than sepals. G, southern.

10 DAISY-LEAVED TOADFLAX *Anarrhinum bellidifolium.* Medium/tall hairless biennial/perennial. Lvs in basal rosette oval to narrowly elliptic, toothed, those up stems narrowly *palmately lobed.* Fls blue or violet, 3-5 mm, in long slender lfless spikes, spur slender, curved; Mar-Aug. Dry, rather bare places, rocks, walls, coniferous woods. F,G, southern.

Figwort Family (contd.)

1* MONKEY FLOWER *Mimulus guttatus.* Short creeping perennial, downy above, hairless below. Lvs oblong to roundish, toothed, opposite, the upper clasping the stem. Fls bright yellow with small red *spots,* 2-lipped, open-mouthed, 25-45 mm, in lfy stalked spikes; June-Sept. Hybridises freely with 1a. Wet places. T, from N America. **1a* Blood-drop Emlets** *M. luteus* □ is smaller and almost hairless, and has large red blotches on fls. Less widely naturalised, from Chile. **1b* Musk** *M. moschatus* □ is smaller and stickily hairy all over, and has much smaller (10-20 mm) all yellow fls. Less widely naturalised.

2* FOXGLOVE *Digitalis purpurea.* Tall perennial to 1½ m, downy; unbranched. Lvs broad lanceolate, wrinkled, *soft* to the touch. Fls *pink-purple,* occasionally white, tubular, 2-lipped, in long tapering spikes; June-Sept. Woods, scrub, heaths, mountains. T.

3 LARGE YELLOW FOXGLOVE *Digitalis grandiflora.* Medium/tall hairy perennial; unbranched. Lvs lanceolate, toothed, hairless above. Fls pale *yellow* with purple-brown veins, tubular, 2-lipped, hairy outside, 30-40 mm long, in a long slender spike; June-Sept. Open woods, rocks in mountains. F, G. **3a Small Yellow Foxglove** *D. lutea* □ has lvs hairless on both sides, and smaller (15-20 mm) plain yellow fls hairless outside. Woods, stony hillsides, mainly on lime. **3 × 3a** and **2 × 3a** occasionally occur.

4 GRATIOLA *Gratiola officinalis.* Medium hairless perennial. Lvs lanceolate, opposite, unstalked, toothed near the tip. Fls white or pale pink, tubular, scarcely 2-lipped, long-stalked at base of upper lvs; May-Oct. Wet places. F, G.

5* FAIRY FOXGLOVE *Erinus alpinus.* Low tufted hairy perennial. Lvs oval, broadest at tip, toothed, mostly in rosettes. Fls purple, with five *notched* petal-lobes, in short stalked spikes; May-Oct. Rocks, screes, mainly in mountains. (B), F.

6* RED BARTSIA *Odontites verna.* Low/short semi-parasitic, often purplish downy annual; well branched. Lvs lanceolate, toothed, unstalked, opposite. Fls *pink,* 2-lipped, open-mouthed, lower lip 3-lobed, in lfy one-sided spikes; June-Sept. Bare and disturbed ground. T.

7* YELLOW ODONTITES *Odontites lutea.* Short/medium semi-parasitic annual, almost hairless; *well branched.* Lvs narrow lanceolate, toothed, unstalked, opposite, with margins inrolled. Fls 6 mm, bright yellow, 2-lipped, open-mouthed, lower lip 3-lobed, stamens protruding, in one-sided spikes; July-Sept. Dry bare or sparsely grassy places, often on lime. (B, rare), F, G. **7a** *O. jaubertiana* is downy, with fls 8 mm, sometimes reddish and stamens not protruding; Aug-Oct. F.

8* ALPINE BARTSIA *Bartsia alpina.* Low semi-parasitic downy perennial; unbranched. Lvs oval, toothed, unstalked, opposite. Fls dark *purple,* 2-lipped, open-mouthed, upper lip hooded, lower 3-lobed, in a short spike with *purple* bracts; July-Aug. Rocks, grassland in hills and mountains. T, but rare in B.

9* YELLOW BARTSIA *Parentucellia viscosa.* Short semi-parasitic annual, *stickily* hairy; usually *unbranched.* Lvs lanceolate, toothed, opposite, unstalked. Fls yellow, 2-lipped, open-mouthed, lower lip 3-lobed, upper hooded, in lfy spikes; June-Sept. Damp grassy and sandy places, dune slacks. B, F.

On p. 219

8* WALL SPEEDWELL *Veronica arvensis* (see p. 219). Low *erect* annual, hairy. Lvs pointed oval, toothed, short-stalked. Fls blue, tiny, shorter than sepals, in dense lfy spikes; Mar-Oct. Dry bare places, walls. T. **8a* American Speedwell** *V. peregrina* has narrowly elliptical, often untoothed lvs, and lilac fls in even lfier spikes. A garden weed. (B, F, G), from N America. **8b* Spring Speedwell** *V. verna* is shorter and has lvs pinnately lobed. Rare in B. **8c* Fingered Speedwell** *V. triphyllos* is half-prostrate, with palmately lobed lvs and long-stalked dark blue fls longer than sepals; Mar-May. Rare in B. **8d* Breckland Speedwell** *V. praecox* has more deeply toothed lvs, often purple below, and petals longer than sepals. T but (B, rare). **8e* French Speedwell** *V. acinifolia* is like 8d but has lvs less toothed and fr clearly 2-lobed. (B, rare), F, G. **8f** *V. dillenii* is like 8b but larger, with fleshy lvs and larger bright blue fls. G.

217

Figwort Family *(contd.)*

SPEEDWELLS *Veronica.* Lvs opposite, usually toothed. Fls blue, with four joined petals and four sepals. Fr flattened, 2-lobed or notched.

1* SPIKED SPEEDWELL *Veronica spicata.* Short/medium downy perennial; unbranched. Lvs oval, stalked, often in a rosette, the upper narrower and unstalked. Fls bright blue, 5 mm, with prominent stamens and blunt sepals, in long dense *lfless* spikes; July-Oct. Dry grassland, rocks, often on lime. T, but rare in B. **1a*** *V. longifolia* □ is taller and hairless, with longer, more strongly toothed lvs, and looser, sometimes branched spikes of larger fls with pointed sepals. (B), F, G.

2 LARGE SPEEDWELL *Veronica austriaca.* Short/medium hairy perennial. Lvs oblong, well toothed, un- or short-stalked. Fls bright blue, 10 mm, *5-sepalled*, in stalked spikes at base of upper lvs; June-Aug. Grassy places. F, G. **2a** *V. prostrata* is half-prostrate with lvs less deeply toothed and fls 6-8 mm with hairless sepals. On lime.

3* GERMANDER SPEEDWELL *Veronica chamaedrys.* Low/short sprawling hairy perennial, with two *opposite* lines of hairs on stems. Lvs pointed oval, short- or unstalked. Fls bright blue with a *white eye,* 10 mm, in opposite stalked spikes at base of upper lvs; Apr-June. Grassy places. T. **3a* Wood Speedwell** *V. montana* □ has stems hairy all round, longer-stalked paler green lvs, purplish beneath and smaller (7 mm) paler blue fls. Damp woods. **3b* Rock Speedwell** *V. fruticans* □ is almost hairless and has much less toothed lvs and fls with a red eye; July-Aug. Mountains. **3c** *V. urticifolia* is like 3a with pale pink fls in longer spikes. Mountain woods, on lime. G.

4* COMMON FIELD SPEEDWELL *Veronica persica.* Low sprawling hairy annual. Lvs oval, short-stalked, pale green. Fls sky-blue with darker veins, the lowest petal usually *white,* 8-12 mm, solitary on long stalks at base of upper lvs; all year. Fr with widely diverging lobes. A weed of cultivation. (T), from W Asia. **4a* Green Field Speedwell** *V. agrestis* □ is smaller, with paler 4-8 mm fls on shorter stalks, and fr only notched; Mar-Nov. T. **4b* Grey Field Speedwell** *V. polita* □ is smaller and greyish, with darker and uniformly blue 4-8 mm fls on shorter stalks and fr only notched; Mar-Nov. **4c** *V. opaca* □ is like 4a but has smaller, bluer fls. F, G, S. **4d* Slender Speedwell** *V. filiformis* □ is a matforming perennial with rounder lvs, mauver fls on very long slender stalks, and rounded fr; Apr-June. Increasing on lawns and in grassy places. (B, F, G), from W Asia.

5* LILAC IVY-LEAVED SPEEDWELL *Veronica hederifolia* ssp. *lucorum.* Low sprawling downy annual. Lvs palmately lobed, *ivy-like,* stalked. Fls *purple-lilac* to white, with whitish anthers, rather small, stalked, solitary at base of lvs; Mar-Aug. A weed of cultivation. T. **5a* Blue Ivy-leaved Speedwell** *V. hederifolia* ssp. *hederifolia* has more sharply toothed lvs and bright blue fls with blue anthers.

6* THYME-LEAVED SPEEDWELL *Veronica serpyllifolia.* Low creeping perennial, almost hairless. Lvs oval, shiny, *scarcely* toothed, short- or unstalked. Fls pale blue or white with purple veins, 5-6 mm (larger on mountains, ssp *humifusa* □ **(6a)**), stalked, in *lfy* spikes; Apr-Oct. Bare and sparsely grassy places. T. **6b* Alpine Speedwell** *V. alpina* □ is scarcely creeping and has bluish-green lvs and deep blue fls with a white eye; July-Aug. Mountains, Arctic heaths.

7* HEATH SPEEDWELL *Veronica officinalis.* Low creeping hairy perennial. Lvs oval, short-stalked. Fls *lilac* with darker veins, in stalked spikes at base of lvs; May-Aug. Dry grassy and heathy places. T.

8* WALL SPEEDWELL *Veronica arvensis*: see p. 216.

9* BROOKLIME *Veronica beccabunga*: see p. 220.

10* WATER SPEEDWELL *Veronica anagallis-aquatica.* Short creeping rooting hairless fleshy perennial. Lvs *lanceolate,* slightly toothed, unstalked. Fls pale blue, 5-6 mm, in opposite stalked spikes at base of upper lvs; June-Aug. Fr rounded. Wet places. T. **10a* Pink Water Speedwell** *V. catenata* □. Stems often purplish, fls pink. **10b* Marsh Speedwell** *V. scutellata.* Slenderer, lvs often purplish, whiter fls in alternate spikes, fr deeply notched. **10c** *V. anagalloides* has narrower lvs and fr egg-shaped. F, G, southern.

219

Figwort Family (contd.)

1* EYEBRIGHT *Euphrasia officinalis* agg. Variable low/short, often bronzy green semi-parasitic hairy annual; often well branched. Lvs more or less oval, deeply toothed, the upper sometimes alternate. Fls *white,* often tinged violet or with purple veins or a yellow spot; 2-lipped, open-mouthed, the lower lip 3-lobed; in lfy spikes with broad, toothed bracts; June-Oct. Fr hairy. Grassy places, mountains. T. **1a* Irish Eyebright** *E. salisburgensis* has narrower, less markedly toothed bracts and fr hairless or almost so.

2* YELLOW RATTLE *Rhinanthus minor* agg. Variable low/medium semi-parasitic annual, usually almost hairless. Lvs oblong to linear, toothed, unstalked, opposite. Fls *yellow,* 2-lipped, usually open-mouthed, often with two purple teeth, in loose lfy spikes; May-Sept. Fr *inflated,* the seeds rattling inside when ripe. Grassy places, cornfields. T. **2a** *Pedicularis lapponica* is shorter, with very narrow lvs, longer, paler yellow fls and less inflated sepal-tube. S.

LOUSEWORTS *Pedicularis.* Semi-parasites, especially on grasses. Lvs pinnately cut, toothed, short-stalked, alternate. Fls 2-lipped, open-mouthed, in lfy spikes. Sepal-tube inflated in fr.

3* MARSH LOUSEWORT *Pedicularis palustris.* Short/medium biennial, almost hairless; stem *single,* branched, often purplish. Fls reddish-pink, the lips equal, sepal-tube *downy;* May-Sept. Wet grassy and heathy places. T. **3a** *P. hirsuta* has woolly stems and less deeply toothed lvs. S.

4* LOUSEWORT *Pedicularis sylvatica.* Low perennial, almost hairless; stems *numerous,* unbranched. Fls pink, the upper lip longer, sepal-tube usually *hairless;* Apr-July. Moors, damp heaths, bogs. T.

5 MOOR-KING *Pedicularis sceptrum-carolinum.* Stout medium/tall perennial. Lvs with *oval lobes,* mostly in a basal rosette. Fls yellow, *30-35 mm,* mouth often closed but if not lower lip purplish, in almost lfless spikes. Damp grassy places and scrub. G,S. **5a** *P. oederi* is short, with rounded lflets and fls in a dense head, the upper lip red-brown with two red spots. **5b** *P. flammea* is like 5a, but lip unspotted.

6 LEAFY LOUSEWORT *Pedicularis foliosa.* Stout medium perennial. Lvs *2-pinnate,* with pointed *triangular* lobes. Fls pale yellow, 15-20 mm, in dense lfy spikes; June-Aug. Mountain grassland. F,G.

7* COMMON COW-WHEAT *Melampyrum pratense.* Variable short semi-parasitic annual, slightly downy. Lvs lanceolate, untoothed, unstalked, opposite. Fls yellow, sometimes pinkish-purple, 2-lipped, with mouth *shut,* in pairs facing the same way at base of toothed lfy bracts; May-Sept. Woods, heaths, grassland. T. **7a* Small Cow-wheat** *M. sylvaticum* □ is slenderer with smaller, always deep yellow open-mouthed fls and less toothed bracts; June-Aug. Woods. T, rare in B. **7b** *M. nemorosum* □ has lvs stalked, much broader at base and slightly toothed, rather larger fls with orange lower lip, and upper bracts purple. Heaths, woods. F,G,S.

8* CRESTED COW-WHEAT *Melampyrum cristatum.* Short/medium semi-parasitic annual, slightly downy. Lvs narrow lanceolate, untoothed, unstalked, opposite. Fls yellow and purple, 2-lipped, mouth almost closed, in a short *squarish* spike with conspicuously *purple* short-toothed bracts; June-Sept. Dry grassy, rocky places, wood margins. T. **8a* Field Cow-wheat** *M. arvense* □ has a longer looser fl spike with bracts magenta, broader and much longer-toothed. T, rare in B.

9* CORNISH MONEYWORT *Sibthorpia europaea.* Prostrate *mat-forming* perennial, hairy; stems thread-like. Lvs *kidney-shaped,* palmately lobed with very blunt lobes, long-stalked, alternate. Fls 5-petalled, the two upper pale yellow, the three lower pale pink, tiny, short-stalked at base of lvs; June-Oct. Damp shady places. B,F.

On p. 219

9* BROOKLIME *Veronica beccabunga* (see p. 219). Low creeping rooting hairless fleshy perennial. Lvs *oval,* stalked. Fls blue, 7-8 mm, in opposite stalked spikes at base of upper lvs; May-Sept. Wet places. T.

221

Figwort Family *(contd.)*

1* TOOTHWORT *Lathraea squamaria.* Low/short creamy perennial, slightly downy. Fls pale *pink*, short-stalked, in a drooping one-sided spike; Apr-May. Woods, hedges; hosts: Hazel and various trees. T. **1a* Purple Toothwort** *L. clandestina*□ is much shorter and hairless, with longer-stalked purple fls. (B), F.

Broomrape Family Orobanchaceae

Parasitic on roots of other plants. *No* green pigment. Lvs replaced by *scales,* more or less pointed oval, on stems. Fls usually coloured as rest of plant, 2-lipped, in spikes. Fr a capsule, usually egg-shaped. Broomrapes *Orobanche* (2-5) are a difficult group, with the host often the best clue to identity. Their erect stems persist rigid and brown with the dead fls through the winter. Grassy places.

2* GREATER BROOMRAPE *Orobanche rapum-genistae.* Medium/tall perennial, honey-brown tinged purple. Fls 20-25 mm, upper lip hooded *not lobed,* stamens hairless below and stigma lobes yellow; only one bract; June-July. Hosts: shrubby peaflowers, especially Broom and Gorse (p.118). B,F,G. **2a* Knapweed Broomrape** *O. elatior* is shorter and sometimes reddish, with fls 15-20 mm, upper lip 2-lobed and stamens hairy below; hosts: Knapweeds (p.248) and Globe-thistles (p.248). **2b** *O. flava* is like 2a, but fls browner, and lower lip unequally lobed. On Butterbur (p.242) and Coltsfoot (p.244). F,G. **2c** *O. alsatica* is like 2a but hairless, with fls lilac-veined. On umbellifers. F,G, eastern.

3* COMMON BROOMRAPE *Orobanche minor*: see p.224.

4* PURPLE BROOMRAPE *Orobanche purpurea.* Short/medium annual, bluish-purple; rarely branched. Fls large (18-30 mm), with *three* bracts to each fl, petal lobes pointed; June-July. Hosts: composites (pp.232-60), especially Yarrow (p.242). T, southern, rare in B. **4a* Branched Broomrape** *O. ramosa* is usually branched and sometimes yellowish, with smaller (10-17 mm) fls; on Hemp (p.38), Tomato (p.210), Potato, Tobacco and other plants. B(rare),F,G. **4b** *O. arenaria* has fls 25-35 mm. On Field Wormwood (p.242). F,G, rare.

5* CLOVE-SCENTED BROOMRAPE *Orobanche caryophyllacea.* Short annual, yellowish tinged purple. Fls *fragrant*, large (20-35 mm), much longer than their single bract, stamens hairy below; June-July. Hosts: bedstraws (p.190). B(rare), F,G. **5a** *O. amethystea* is reddish-purple, with smaller, much less hairy fls; on Field Eryngo (p.158). F,G, western.

Moschatel Family Adoxaceae

6* MOSCHATEL *Adoxa moschatellina.* Low carpeting hairless perennial. Lvs 2-trefoil, lobed, long-stalked. Fls green, in tight heads of five, at *right angles* to each other; Mar-May. Fr a green berry. Woods, shady places, also on mountains. T.

Butterwort Family Lentibulariaceae

Butterworts *Pinguicula.* Low perennials, stickily hairy. Lvs oblong, untoothed, all in a starfish-like basal rosette, the margins rolling inwards to trap and digest insects. Fls 2-lipped, spurred, solitary on long lfless stalks.

7* COMMON BUTTERWORT *Pinguicula vulgaris.* Lvs yellow-green. Fls *violet* with a *white* throat-patch, spur pointed, 10-15 mm; May-July. Bogs, fens, wet heaths and moors. T. **7a* Large-flowered Butterwort** *P. grandiflora* □ has much larger (25-30 mm) fls with a stouter spur. B,F.

8* PALE BUTTERWORT *Pinguicula lusitanica.* Lvs olive green with red-brown veins. Fls pale *lilac* with *yellow* throat and down-turned cylindrical spur, 6-7 mm; June-Oct. Bogs, wet heaths. B,F, western. **8a Alpine Butterwort** *P. alpina* □ has larger (8-10 mm) white fls with two yellow spots in throat and a conical spur. Damp places on mountains. S. **8b** *P. villosa* □ is smaller and has brownish lvs and pale violet fls with two yellow spots on lower lip and straight spur. S.

Plantain Family Plantaginaceae

Lvs strongly veined or ribbed, all in a basal rosette (except 4). Fls tiny, 4-petalled, in dense spikes or heads, on long *lfless* stalks (except 4), with prominent *stamens* providing most of the colour. Fr a capsule.

1* GREATER PLANTAIN *Plantago major*. Low/short perennial, hairless or downy. Lvs *broad oval*. Fls pale greenish-yellow and pale brown, in long greenish spikes about equalling the unfurrowed stalks; anthers pale purple then yellowish-brown; June-Oct. Waste and well trodden places, paths, lawns. T. **1a* Hoary Plantain** *P. media* ☐ is always downy and greyish, with longer, often narrower lvs, and shorter long-stalked spikes of fragrant whitish fls with a fuzz of pinkish-lilac stamens; May-Aug. Grassy places; on lime. T.

2* RIBWORT PLANTAIN *Plantago lanceolata*. Low/medium perennial, hairless or downy. Lvs *lanceolate*, slightly toothed. Fls brown, in short blackish spikes on furrowed stalk̲s, anthers pale yellow; Apr-Oct. Grassy and waste places; one of the commonest European plants. T.

3* SEA PLANTAIN *Plantago maritima*. Low/short perennial, usually hairless. Lvs *linear*, fleshy, 3-5-veined, sometimes slightly toothed. Fls brownish-pink with yellow anthers, in greenish spikes on unfurrowed stalks; June-Sept. Coastal saltmarshes, occasionally on mountains inland. T.

4* BRANCHED PLANTAIN *Plantago arenaria*. Short downy annual; much branched. Lvs linear, sometimes obscurely toothed, not in a rosette, the lower with short lfy shoots at their base. Fls pale brown, in long-stalked *egg-shaped* heads up the lfy stems; May-Aug. Waste and sandy places, dunes. T, but (B).

5* BUCKSHORN PLANTAIN *Plantago coronopus*. Low biennial, usually downy. Lvs variable, usually *pinnately lobed*, less often linear and deeply toothed, 1-veined. Fls yellow-brown with yellow anthers, in short greenish spikes on unfurrowed stalks; May-Oct. Dry bare, often sandy places, most frequent on the coast. T.

Arrow-Grass Family Juncaginaceae

A monocotyledonous family. (See p. 10 and p. 262.)

6* MARSH ARROW-GRASS *Triglochin palustris*. Short slender hairless perennial. Lvs linear, deeply furrowed to appear almost *cylindrical*. Fls green, purple-edged, 3-petalled, tiny, short-stalked, in an interrupted spike, the style showing as a white tuft; May-Aug. Fr narrow, opening arrow-shaped when ripe. *Marshes,* fens, damp grassland. T. **6a* Rannoch Rush** *Scheuchzeria palustris* ☐ (Scheuchzeriaceae) has flat lvs with inflated sheathing bases, longer than the small cluster of yellower fls; bog pools. T. Scattered, rare in B.

7* SEA ARROW-GRASS *Triglochin maritima*. Short/medium hairless perennial. Lvs linear, fleshy, *flat,* not veined. Fls green, 3-petalled, tiny, short-stalked, in a long plantain-like spike; May-Sept. Fr egg-shaped. Coastal *saltmarshes*. T.

On p. 223

3* COMMON BROOMRAPE *Orobanche minor* (see p. 223). Variable short/medium annual, purplish, reddish or yellowish. Fls 10-18 mm, shorter than or equalling their single bract, stamens hairless, stigma lobes purple; June-Sept. Hosts: peaflowers (pp. 120-36) and composites (pp. 232-60). B, F, G. **3a* Thyme Broomrape** *O. alba* is usually much shorter and deep purplish-red, with fls longer than bract and stamens hairy below; on labiates (pp. 198-208), especially Wild Thyme (p. 208). **3b* Ox-tongue Broomrape** *O. loricata* is yellowish-white tinged purple, with stamens very hairy below: on ox-tongues (p. 260) and hawksbeards (pp. 258, 260). B (rare), F, G. **3c* Thistle Broomrape** *O. reticulata* is yellowish tinged purple with sparse hairs on fls and stamens sometimes sparsely hairy: on thistles (p. 250). B (rare), F, G. **3d* Ivy Broomrape** *O. hederae* is yellowish tinged purple, with yellow stigma lobes; on Ivy (p. 158). **3e** *O. gracilis* is like 3a, but fls yellow, red inside, anthers yellow; on peaflowers, F, G. **3f** *O. teucrii* is like 3e, but fls bent almost in a right angle; on germanders (p. 200). **3g* Carrot Broomrape** *O. maritima* has stem and bracts always purple; on Wild Carrot (p. 160) by the sea.

225

Valerian Family Valerianaceae

Lvs opposite, no stipules. Fls with five joined petals, small, in clusters.

1* COMMON VALERIAN *Valeriana officinalis* agg. Variable short/tall perennial, hairy mainly below; usually unbranched. Lvs *pinnate,* the lower stalked, lflets lanceolate, toothed. Fls pale pink, pouched at base, in more or less rounded clusters; June-Aug. Woods, grassy places, both damp and dry. T.

2* MARSH VALERIAN *Valeriana dioica.* Short perennial, almost hairless, with creeping *runners.* Root lvs *oval,* untoothed, long-stalked; stem lvs more or less pinnate, unstalked. Fls pale pink, pouched at base, in rounded clusters; stamens and styles on different plants; May-June. Marshes, fens. T. **2a** *V. montana* has no runners and stem lvs oval, scarcely toothed; June-Aug. Mountains. F, G. **2b** *V. tripteris* differs from 2a in its greyish lvs, heart-shaped at roots, 3-lobed on stems, and narrower white-edged bracts.

3* RED VALERIAN *Centranthus ruber.* Medium/tall greyish tufted hairless perennial. Lvs pointed oval or *lanceolate,* the lower stalked and untoothed. Fls red, pink or sometimes white, fragrant, spurred, in rounded clusters; May-Sept. Cliffs, rocks, quarries, walls, steep banks. (B), F, G.

4* CORNSALAD *Valerianella locusta* agg. Low/short annual, almost hairless, well branched. Lvs oblong, unstalked. Fls pale lilac, *tiny,* in flat-topped umbel-like clusters; Apr-Aug. Fr flattened. Bare and cultivated ground, dunes, walls. T. **4a** *V. coronata* has narrower lvs, bluer fls in rounded heads, and fr conspicuously crowned by enlarged sepals. F, G.

Honeysuckle Family Caprifoliaceae

Lvs opposite. Fls with five joined petals.

5* DWARF ELDER *Sambucus ebulus.* Stout tall foetid patch-forming perennial to 1 m. Lvs *pinnate,* lflets lanceolate, sharply toothed, with lfy stipules. Fls white, often tinged pink, in a flat-topped umbel-like cluster; anthers violet; July-Aug. Fr black. Waysides, waste places. T, but (B). **5a* Elder** *S. nigra* is a deciduous shrub to 10 m, with broader lflets, no or tiny stipules, and yellow anthers; May-July. Woods, hedges, waste places. T. **5b* Red-berried Elder** *S. racemosa* has more pointed yellow-green lvs, greenish-white fls in a dense oval cluster, and red berries; Apr-May. Hills, mountains. T, but (B).

6* GUELDER ROSE *Viburnum opulus.* Deciduous shrub or small tree to 4 m, scarcely downy. Lvs palmately 3-5-lobed, *ivy-like.* Fls white, slightly fragrant, in flat umbel-like clusters, outer petals of outer fls *much larger;* May-July. Fr a shiny *red* berry. Damp woods, fens, scrub, hedges. T.

7* WAYFARING TREE *Viburnum lantana.* Downy deciduous shrub to 4 m. Lvs *oval,* minutely toothed, wrinkled. Fls creamy white, in a flat-topped cluster, fragrant; Apr-June. Fr red then *black.* Scrub, hedges; on lime. T.

8* TWINFLOWER *Linnaea borealis.* Delicate prostrate mat-forming downy evergreen undershrub. Lvs oval, toothed, stalked. Fls pink, *bell-shaped,* drooping, fragrant, in pairs; June-Aug. Fr egg-shaped. Coniferous woods. T, northern.

9* HONEYSUCKLE *Lonicera periclymenum.* Deciduous woody *climber* to 6 m, hairless or downy; twining clockwise. Lvs oval, untoothed, short- or unstalked, appearing in midwinter. Fls creamy becoming orange-buff, often reddish outside, tubular, 2-lipped, very fragrant, in a *head;* June-Oct. Fr a red berry. Woods, scrub, hedges. T. **9a* Perfoliate Honeysuckle** *L. caprifolium* is always hairless, with upper lvs and bracts joined at the base, cup-like; May-July. (B), F, G.

10* FLY HONEYSUCKLE *Lonicera xylosteum.* Deciduous shrub to 2 m, downy. Lvs pointed oval, short-stalked, greyish. Fls yellowish, tubular, 2-lipped, *in pairs* at base of upper lvs; May-June. Fr a red berry. Woods, scrub. T, rare in B. **10a Blue Honeysuckle** *L. caerulea* is less hairy, with blunter lvs, bell-shaped fls and black berries fused together. Mountains. G, S. **10b Black-berried Honeysuckle** *L. nigra* is hairless with long-stalked pinkish-white fls and a distinct pair of black berries. F, G. **10c** *L. alpigena* has fls brownish and red berries fused. Mountains. G, southern.

Teasel Family Dipsacaceae

Lvs opposite. Fls small, 4-5-petalled, tightly packed into a composite-like (p. 234) head, but with four stamens projecting from each fl. Each head cupped by sepal-like bracts. Fr small, with one seed.

1* **FIELD SCABIOUS** *Knautia arvensis.* Medium/tall hairy perennial. Lvs *pinnately lobed,* the upper sometimes undivided. Fls bluish-lilac, in rather *flat* heads, 30-40 mm, outer petals of outer fls much larger; bracts in two rows, shorter than fls; anthers pink; June-Oct. Dry grassy places, cornfields. T. **1a Wood Scabious** *K. dipsacifolia* has brighter green lvs, toothed but not lobed, with outer petals only slightly larger, and bracts nearly as long as fls. Shady places in mountains. F, G.

2* **SMALL SCABIOUS** *Scabiosa columbaria.* Variable short/medium hairy perennial; little branched. Lvs *pinnate* with narrow lobes, basal lvs sometimes undivided. Fls like 1, but heads smaller (15-25 mm) with *dark bristles* among the fls, and only one row of bracts; June-Oct. Dry grassland; on lime. T, southern. **2a Yellow Scabious** *S. ochroleuca* □ is annual/biennial, with upper lvs more but lower lvs less deeply cut, and fls yellow. G. **2b** *S. canescens* □ has basal lvs always lanceolate and anthers bright purple. F, G, S, southern.

3* **DEVILSBIT SCABIOUS** *Succisa pratensis.* Medium/tall hairy perennial. Lvs *elliptical,* untoothed, stalked, often blotched purplish, narrower and sometimes toothed up stems. Fls dark blue-purple, occasionally pink, in *rounded* heads, 15-25 mm; June-Oct. Damp grassy places. T.

4* **SMALL TEASEL** *Dipsacus pilosus.* Medium/tall biennial, hairy above, *prickly* on stems and lvs. Lvs oblong, stalked, the upper narrower, all often with a pair of lflets at base. Fls *white,* in woolly spiny *globular* heads, anthers violet; July-Sept. Woods, scrub, streamsides. T, but (S).

5* **TEASEL** *Dipsacus fullonum.* Tall hairless perennial, *prickly* on stems and lvs; dead stems and flhds persist through winter. Lvs lanceolate, covered with white *pimples,* in a basal rosette that withers before flg; narrower up stems, often cupped at base. Fls pale *purple,* in a *conical* spiny head; July-Aug. Bare and sparsely grassy places, often damp. B, F, G. **5a** *D. laciniatus* has deeply cut lvs and paler fls. Waste places. G, southern.

Bellflower Family Campanulaceae: see p. 230

6* **ROUND-HEADED RAMPION** *Phyteuma orbiculare.* Low/short perennial, usually hairless, unbranched. Root lvs variable, toothed, broad lanceolate, sometimes heart-shaped; stem lvs lanceolate to linear, unstalked. Fls *dark* blue, in globular heads, petal lobes narrow; bracts short, triangular to lanceolate; July-Aug. Dry grassland; on lime. B, F, G. **6a** *P. nigrum* has root lvs always heart-shaped, stem lvs broader, sometimes untoothed, and flhds more elongated. F.

7* **SPIKED RAMPION** *Phyteuma spicatum.* Short/medium hairless perennial. Root lvs heart-shaped, toothed, long-stalked; stem lvs narrower, less stalked. Fls yellowish-white, small, in *elongated* heads, petal lobes narrower; May-July. Woods, grassy places. T, rare in B and S.

8* **SHEEPSBIT SCABIOUS** *Jasione montana.* Low/short biennial, slightly hairy; no dead lvs at base. Lvs narrow oblong, untoothed but sometimes *wavy,* the upper unstalked. Fls blue, in globular heads, petal lobes narrow; May-Sept. Dry grassy places, heaths, sea cliffs, shingle; not on lime. T. **8a** *J. laevis* is perennial with runners and has longer lvs persisting at base when dead. F, G. **8b** *Globularia vulgaris* (Globulariaceae) is perennial, with oval lvs, notched at the tip, narrower and unstalked up stems. Fls 2-lipped, the upper lip very short, the lower 3-lobed; Apr-June. Dry grassy or stony places. F, G, S, southern.

1

2a

2b

2

3

4

5

6

7

8

Bellflower Family   Campanulaceae

Lvs undivided, alternate, no stipules. Petals joined, with five lobes. Fr a capsule.
Bellflowers *Campanula* are perennials with more or less bell-shaped fls.

1* HAREBELL *Campanula rotundifolia*. Slender, short/medium, hairless. Root
lvs *roundish*, withering early; stem lvs *linear*, the upper unstalked. Fls blue,
15 mm, on long thin stalks, in loose clusters; petal lobes short; July-Oct. Dry
grassland, heaths. T. **1a** *C. uniflora* ☐ is shorter with lower lvs oblong, and
smaller (5-10 mm) solitary nodding fls. Mountains, arctic heaths. S. **1b** *C.
scheuchzeri* has unstalked lvs. G, S. **1c** *C. cochlearifolia* is smaller than 1 and
white with hairs below. F, G. **1d** *C. rhomboidalis* with oval stem lvs and **1e** *C.
baumgartenii* with lanceolate lvs, both have fl stalks longer than fls. F, G. 1b-1e
are all rare, in mountains.

2* CLUSTERED BELLFLOWER *Campanula glomerata*. Variable, stiff, low/
medium, hairy. Lower lvs oval, sometimes heart-shaped, stalked; upper lvs
narrower, unstalked. Fls deep violet, in tight *heads*; sepal lobes *pointed*; June-
Oct. Grassland; on lime. **2a** *C. cervicaria* has lanceolate lower lvs, smaller
pale blue fls and blunt sepal lobes. F, G, S.

3* NETTLE-LEAVED BELLFLOWER *Campanula trachelium*. Medium/tall, hairy;
stems sharply angled, unbranched. Lvs broad lanceolate, long-pointed, irregularly toothed. Fls violet-blue, 30-40 mm long, 1-3 together in a lfy spike,
opening from the top; petal lobes short, triangular; July-Sept. Woods, scrub,
hedges. T. **3a* Giant Bellflower** *C. latifolia* is stouter and has bluntly angled
stems, more evenly toothed lvs, and larger (40-50 mm) paler fls, with longer,
narrower petal lobes, solitary at base of lvs, the lower ones opening first.
Mainly in hills and mountains. **3b* Creeping Bellflower** *C. rapunculoides*
is smaller, slenderer and creeping, with stems hardly angled, narrower, more
evenly toothed lvs, and smaller (20-30 mm) fls with longer petal lobes. Sometimes
on bare ground. T, but (B). **3c** *C. bononiensis* is like 3b but has lvs white-woolly
beneath and still smaller (10-20 mm) fls in a denser one-sided lfless spike-like
cluster. Mountains. F, G.

4* SPREADING BELLFLOWER *Campanula patula*. Medium/tall biennial/peren-
nial, rough to the touch; stems angled. Lvs oblong, the upper narrower and
unstalked. Fls violet-blue, 15-25 mm, in a spreading cluster, on stalks with
a small bract in the middle; petal-lobes long, sepal-teeth linear; June-July. Woods,
grassy places. T. **4a* Rampion Bellflower** *C. rapunculus* has milky juice, and
bluer fls with shorter petal lobes and small bracts at base of stalks. T, but (B).

5* PEACH-LEAVED BELLFLOWER *Campanula persicifolia*. Medium/tall, *hairless*. Lower lvs oblong, stalked; upper lvs linear/lanceolate, unstalked. Fls
violet-blue, *30-40 mm*, in a stalked spike, petal-lobes broadly triangular; May-
Aug. Woods, scrub, also naturalised from gardens. T.

6 BEARDED BELLFLOWER *Campanula barbata*. Short, hairy; unbranched.
Lvs oblong, wavy, with a basal rosette. Fls pale blue, with long *white hairs*
inside, in one-sided cluster; sepals in two rows, like a Canterbury Bell *(C.
medium)*; June-Aug. Woods, grassy places, in mountains. F, G, S.

7* IVY-LEAVED BELLFLOWER *Wahlenbergia hederacea*. Low creeping, rather
delicate, hairless. Lvs palmately lobed, *ivy-like*, stalked. Fls pale blue, bell-
shaped with short petal lobes, on hairlike stalks at base of lvs; July-Aug. Damp
woods, heaths and moors. B, F, G.

8* VENUS'S LOOKING GLASS *Legousia hybrida*. Low/short hairy annual. Lvs
oblong, wavy, unstalked. Fls purple, in clusters, petal lobes only 8-12 mm across,
shorter than sepal teeth, closing in dull weather and then hard to detect; May-Aug.
A weed of cultivation, bare sandy or stony places. B, F, G. **8a Large Venus's
Looking Glass** *L. speculum-veneris* ☐ has fls violet, opening *wide* to 20 mm;
May-July. F, G.

9* HEATH LOBELIA *Lobelia urens*. Short/medium, more or less hairless;
juice *milky*, acrid. Lvs oblong, toothed. Fls purple-blue, *2-lipped*, upper lip
2-lobed, lower lip 3-lobed, in a stalked spike; Aug-Sept. Damp woods and heaths;
not on lime. B (rare), F.

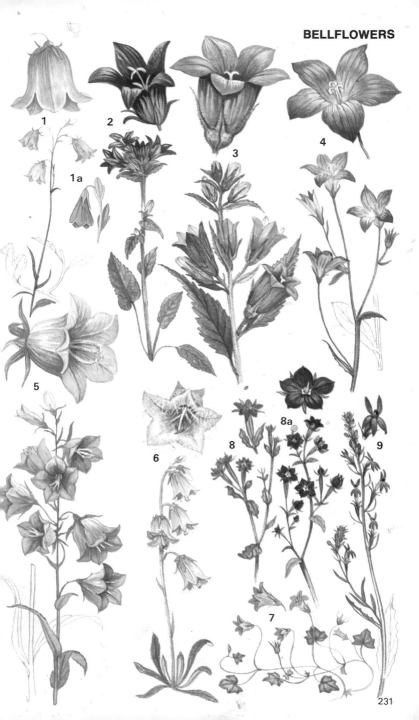

1
1a
2
3
4
5
6
7
8
8a
9

Key to Composites

(Pages 234-261)

Composites are members of the Daisy Family, Compositae; see p. 234, where the nature of their flowers as a tight head of small flowers or florest is explained. There are three main types of composite flower-heads, consisting of disc florets only (thistle-like), ray florets only (dandelion-like) or both (daisy-like).

Plant

aromatic: Stinking Cocklebur 234; Chamomiles, Scented Mayweed 236; Mountain Everlasting 238; Mugworts, Wormwoods, Buttonweed 242; Tansy, Feverfew 244

a shrub: Silver Ragwort 246

spiny: Spiny Cocklebur 234; Star-thistles 252; Prickly, Perennial and Marsh Sow-thistles, Prickly Lettuce 256

Stems

winged: Cotton Thistle 248; Spear, Musk, Slender, Welted and Great Marsh Thistles 250; Yellow Star-thistle 252; Alpine Sow-thistle 254

Leaves

clasping stem: Michaelmas Daisies, Fleabanes 236; Jersey Cudweed, Small Fleabane 238; Common Fleabane, Elecampane, Irish Fleabane, Leopardsbane 240; Garden Marigold 244; Hoary, Marsh and Oxford Ragworts, Marsh Fleawort, Groundsel 246; Carline, Milk and Cabbage Thistles 248; most on pp. 254 and 256; Spotted Catsear 258; Bristly Ox-tongue 260

in rosette: Daisy, Alpine Fleabane 236; Mountain Everlasting, Dwarf Cudweed 238; Arnica, Leopardsbane 240; Carline Thistle 248; Lamb's Succory 252; most on p. 258; Few-leaved and Alpine Hawkweeds, *Crepis praemorsa* 260

fleshy: Golden Samphire 234; Sea Aster 236

opposite: Hemp Agrimony, Ragweed 234; Shaggy Soldier 236; Sunflowers, Arnica 240; Bur Marigolds 242

white-veined: Milk Thistle 248; Goatsbeard, Salsify 254

Heads with ray-florets only (Dandelion-like):

pink/purple: Salsify, Purple Viper's Scarz, Alpine Sow-thistle, Purple Lettuce 254

brown: *Podospermum laciniatum* 254

yellow: Goatsbeard, Viper's Grass 254; Sow-thistles and Lettuces 256; Dandelions and allies 258 and 260

blue: Chicory, Alpine Sow-thistle, Blue Lettuce 254

Heads with disc florets only (Thistle-like):

pink/purple: Hemp Agrimony 234; Mountain Everlasting 238; Butterbur 242; Purple Coltsfoot 244; Burdocks, Cotton and Milk Thistles, Alpine Sawwort 248; all on p. 250 and most on p. 252

red: Mountain Everlasting 238; *Artemisia austriaca* 242

yellow: Ragweed 234; Sea Aster, Pineapple Mayweed 236; all on 238; Bur Marigolds, Buttonweed, Cottonweed 242; Tansy 244; Groundsel 246; Yellow Star-thistle 252

blue: Globo Thistle 248; Cornflowers 252

white: Mountain Everlasting 238

brown: Cockleburs 234; Cudweeds 238; Mugworts, Wormwoods, Bur Marigolds 242; Carline and Cabbage Thistles 248

Heads with discs and rays (Daisy-like):

rays pink/purple: Sea Aster, Michaelmas daisies, Fleabanes 236; Yarrow 242; Purple Coltsfoot 244

rays yellow: most on p. 234; Small Fleabane 238; all on p. 240; Bur Marigolds 242; Coltsfoot, Corn and Garden Marigolds 244; all on p. 246

rays white: Daisy, Mayweeds, Shaggy Soldier 236; Yarrow, Sneezewort 242; Ox-eye Daisy, Feverfew 244

Heads solitary

Golden Samphire 234; Daisy, Mayweeds, Pineapple Mayweed, Alpine Fleabane 236; Small, Dwarf and Heath Cudweeds 238; all on p. 240; Bur Marigolds 242; most on pp. 244, 248 and 252; Goatsbeard, Salsify, Viper's Grasses, Chicory 254; most on p. 258; Alpine Hawkweed 260

Heads more than 30 mm

Golden Samphire 234; Mayweeds 236; all on p. 240; Ox-eye Daisy, Corn Marigold 244; Fen Ragwort 246; Globe, Cotton, Milk and Cabbage Thistles 248; Spear, Woolly, Dwarf, Musk and Welted Thistles 250; Greater and Black Knapweeds 252; Blue Lettuce 254; Perennial and Marsh Sow-thistles 256; Dandelion, Common and Spotted Catsears, Rough Hawkbit 258; Few-leaved and Alpine Hawkweeds 260

Heads in spikes

Goldenrods, Canadian Fleabane, Ragweed 234; Dwarf and Heath Cudweeds 238; Butterbur, Mugworts, Wormwoods 242; Burdocks 248; Chicory, Alpine Sow-thistle 254; Prickly and Least Lettuces 256

Daisy Family   Compositae

The Composites are the largest family of flowering plants. Fls tiny, closely packed into a compound head, surrounded by sepal-like bracts. Petals are joined in a tube, and the ending either in two kinds, with the tube ending either in five short teeth, *disc florets*, or in a conspicuous flat flap, *ray florets*. Composite flhds are thus of three kinds; *rayless*, with disc florets only, like thistles; *rayed*, with disc florets in the centres and ray florets round the edge, like daisies; and *dandelion-like*, with all ray florets. Fr tiny, often surmounted by a feathery pappus on which it floats away in the wind. Sometimes the pappuses form a rounded "clock".

1* HEMP AGRIMONY *Eupatorium cannabinum.* Medium/tall perennial, downy; stems often reddish, usually branched. Lvs *palmate,* the segments lanceolate, toothed, the upper undivided. Flhds whitish pink to reddish mauve, *rayless,* in dense, rather flat-topped clusters; July-Sept. Damp woods, marshes, fens, by fresh water, and on waste ground. T.

2* GOLDEN-ROD *Solidago virgaurea.* Variable short/medium perennial, hairless or downy; little branched. Lvs bluntly oval or lanceolate, toothed, stalked; narrower, untoothed and unstalked up stems. Flhds bright yellow, *shortly* rayed, in branched spikes; June-Sept. Woods, scrub, heaths, grassy and rocky places. T.

3* CANADIAN GOLDEN-ROD *Solidago canadensis.* Tall downy perennial; unbranched. Lvs lanceolate, toothed, 3-veined. Flhds yellow, *very shortly* rayed, in *one-sided* spikes in a branched cluster; July-Sept. Waste places, (T), from N America. **3a*** *S. gigantea* is usually shorter, hairless below and greyish, with more deeply toothed lvs and longer ray florets. (B,F,G), from N America. **3b*** *S. graminifolia* has linear-lanceolate lvs and flhds in flat heads. (B, G).

4* GOLDILOCKS *Aster linosyris.* Short/medium hairless perennial; unbranched. Lvs *linear*, 1-veined, numerous up stems. Flhds yellow, *rayless*, in a flat-topped umbel-like cluster; Sept-Nov. Woods, grassy places, on rocks, often by the sea; on lime. T, southern, rare in B.

5* CANADIAN FLEABANE *Conyza canadensis.* Low/tall annual, slightly hairy; branched above. Lvs narrow lanceolate, sometimes toothed. Flhds with yellow disc florets and *short* whitish rays, in loose branched spikes; June-Oct. Bare and waste places. (T), from N America.

6* PLOUGHMAN'S SPIKENARD *Inula conyza.* Short/tall perennial, downy; stems purplish, branched above. Lvs broad lanceolate, *foxglove-like,* slightly toothed. Flhds dull yellow with no or obscure rays and purplish inner bracts, in a flat-topped umbel-like cluster; July-Sept. Open woods, scrub, dry grassy and rocky places; on lime. B,F,G. **6a* Stink Aster** *Dittrichia graveolens* is foetid and stickily hairy; lvs linear; flhds smaller in a spike. Damp places. (B, rare),F.

7* GOLDEN SAMPHIRE *Inula crithmoides.* Tufted short/medium hairless perennial. Lvs linear, *fleshy,* sometimes 3-toothed at tip. Flhds yellow, *rayed,* in a loose flat-topped cluster; July-Oct. Coastal cliffs, shingle and saltmarshes. B,F.

8* RAGWEED *Ambrosia artemisiifolia.* Medium/tall annual, hairy; stems branched, often reddish, angled. Lvs oval, deeply pinnately cut, *greyish* beneath, mostly opposite. Flhds greenish-yellow, rayless, those with stamens in spikes, those with styles solitary at base of lvs; June-Sept. Fr prickly. Bare and waste ground. (B,F,G), from N America.

9* SPINY COCKLEBUR *Xanthium spinosum.* Stiff short/medium annual, hairless or not. Lvs diamond-shaped, 3-5-lobed, dark shiny green above, downy white beneath, with 1-2 sharp 3-forked orange *spines* at their base. Flhds greenish, rayless, covered with hooked spines, globular male heads with stamens above egg-shaped female heads with styles; July-Oct. Fr covered with hooked spines. A weed of cultivation, waste places. (B, F, G), from N America. **9a* Rough Cocklebur** *X. strumarium* is greyish with broader, roughly hairy lvs. **9b* Stinking Cocklebur** *X. strumarium* ssp. *italicum* is yellowish-green and strong-smelling, the lvs covered with short stiff hair and yellow dots, and male flhds conical. 9a and 9b both have lvs spineless, less lobed, toothed and not white beneath.

Daisy Family (contd.)

1* DAISY *Bellis perennis.* Low hairy perennial. Lvs spoon-shaped, slightly toothed, in a basal rosette. Flhds with disc florets yellow and rays white, often tipped red, *solitary* on lfless stalks; all year. Lawns, short turf. T. **1a* Mexican Fleabane** *Erigeron karvinskianus* is taller and branched with lanceolate lvs; Apr-Oct. Walls. (B,F), from Mexico. **1b** *Aster michelii* is taller and stouter, fls to 40 mm; rays white or pink; Mar-Sept. Mountains. G, southern.

2* SCENTLESS MAYWEED *Matricaria perforata*: Variable short half-prostrate hairless annual/biennial. Lvs 2-3-pinnate, feathery. Fls with yellow disc florets and white rays, 15-40 mm, solitary; sepal-like bracts bordered brown; Apr-Nov. Disturbed ground. T. **2a* Sea Mayweed** *M. maritima* is stouter, slightly fleshy and earlier flg. By the sea. **2b* Corn Chamomile** *Anthemis arvensis* is erecter and grey with down, and has broader lf segments and bracts all green. **2c* Stinking Chamomile** *A. cotula* is erecter and has a sickly smell, slightly hairy stems, broader lf segments and whitish bracts with a green midrib. **2d* Lawn Chamomile** *Chamaemelum nobile* is perennial, spreading, hairy and pleasantly aromatic, and has lf segments with inrolled margins and bracts white-edged; June-Sept. Grassy and heathy places. B,F,G. **2e* Scented Mayweed** *Chamomilla recutita* is erecter and aromatic, with bristle-tipped lf segments, smaller (10-25 mm) fls, rays down-turned and bracts edged greenish-white.

3* PINEAPPLE MAYWEED *Chamomilla suaveolens.* Low/short hairless annual, *pineapple*-scented; well branched. Lvs 2-3-pinnate with thread-like segments. Flhds yellowish-green, *rayless*, solitary; sepal-like bracts pale-edged; May-Nov. Bare and waste, especially well trodden places. (T), from N E Asia.

4* SEA ASTER *Aster tripolium.* Short/medium hairless, *fleshy* perennial; often well branched. Root lvs bluntly oval, untoothed, dark green; stem lvs narrow lanceolate. Fls with yellow disc florets and pale purple or whitish rays, but frequently *rayless*, in loose clusters; sepal-like bracts blunt; July-Oct. Coastal *saltmarshes*, cliffs. T. **4a** *A. amellus* is roughly hairy and not fleshy, and has fls always rayed, solitary on each branch of the cluster, and bracts curved and coloured at the edges. Dry grassy places. F,G, southern.

5* MICHAELMAS DAISY *Aster novi-belgii.* Medium/tall perennial, almost hairless; well branched. Lvs lanceolate, untoothed, the upper unstalked, numerous up stems. Fls with yellow disc florets and purple, violet or whitish rays, in a widely branched cluster; outer sepal-like bracts half as long as the inner, *pointed;* Aug-Nov. Damp and waste places, streamsides. (B,F,G), from N America. **5a*** *A. novae-angliae* is roughly and stickily hairy with rays usually reddish. (B,G). **5b*** *A.* × *salignus* has narrower, sometimes toothed lvs, all unstalked and narrowed at base; rays white turning violet, and bracts purplish. Several other introduced species occur. **5c** *A. alpinus* is much shorter, with shorter, broader lvs and larger (to 50 mm) solitary fls; July-Aug. Mountains. G, southern.

6* BLUE FLEABANE *Erigeron acer.* Low/medium roughly hairy annual/biennial; stems often purple. Lvs lanceolate, untoothed, the upper unstalked. Flhds with yellow disc florets and short *erect* dingy purple rays, 10-15 mm, solitary or in loose clusters; sepal-like bracts dull purple; June-Sept. Dry bare and sparsely grassy places, walls, mountains. T. **6a** *E. annuus* has fls 15-25 mm, with spreading white or purple rays. Lvs toothed, narrower. Waste places. (F,G), from N America.

7* ALPINE FLEABANE *Erigeron alpinus* agg. Low/short perennial, hairy unbranched. Lvs spoon-shaped in basal rosette, narrower and fewer up stems. Flhds with yellow disc florets and spreading pinkish-purple rays, *20-30 mm,* usually *solitary,* bracts *shortly* hairy, grey-violet; July-Sept. Grassy places and rock-ledges in mountains. T, rare in B. **7a** *E. uniflorus* has lvs almost hairless, flhds smaller (10-15 mm) and bracts with long woolly hairs. G,S. **7b** *E. humilis* is shorter, with bracts black or dark violet. S.

8* SHAGGY SOLDIER *Galinsoga quadriradiata.* Low/short annual, whitely hairy. Lvs pointed oval, toothed, dark green, opposite, narrowing up the stem. Flhds with 1-2 green outer sepal-like bracts, yellow disc florets and 4-5 *white* trifid rays to 2.5 mm, in open clusters; June-Oct. Bare and disturbed ground. (T), from S America. **8a* Gallant Soldier** *G. parviflora* is yellower green and much less hairy, with 2-4 brown-edged outer bracts and narrower rays 0-1.5 mm long. Hybridises with 8.

1

2

3

4

5

6

7

8

Daisy Family *(contd.)*

1* MOUNTAIN EVERLASTING *Antennaria dioica.* Low/short creeping perennial, with rooting runners. Lvs whitely woolly *beneath,* those in basal rosette broadest near tip. Flhds rayless, pink or red, with sepal-like bracts pink or white, in umbels; stamens and styles on separate plants; June-July. Heaths, moors, mountain grassland. T. **1a** *A. alpina* is shorter, with rosette lvs grey hairy above and smaller flhds. Mountain rocks on lime. S. **1b** *A. carpatica* is like 1a, but lvs longer, linear and 3-veined. **1c** *A. porsildii* has shorter runners, lvs white or grey woolly on both sides, and bracts greenish or brownish. S, Arctic. **1d* Pearly Everlasting** *Anaphalis margaritacea* is a medium perennial, whitely woolly, with yellow flhds appearing white from the bracts, and stamens and styles not always on separate plants. Waste places. (B,F,G), from N America.

2* COMMON CUDWEED *Filago vulgaris.* Low/short annual, covered with silvery, sometimes yellow, hairs; widely branched. Lvs narrow oblong, *wavy-edged,* blunt or pointed, spirally up stems. Flhds rayless, white tipped red, but appearing yellow from tips of sepal-like bracts; in globular clusters overtopping upper lvs; July-Aug. Heaths, sandy places. T. **2a* Red-tipped Cudweed** *F. lutescens* has hairs always yellowish, and bracts red-tipped. **2b* Broad-leaved Cudweed** *F. pyramidata* has hairs always silvery, flhds well overtopped by 3-5 large lvs and bracts with down-curved points. B,F. 2a and 2b both have lvs broader, scarcely wavy and sharply pointed.

3* SMALL CUDWEED *Logfia minima.* Slender low/short annual, covered with silvery hairs. Lvs linear, spirally up stems. Flhds rayless, white tinged red, but appearing yellow from tips of sepal-like bracts, which spread *star-like* in fr; in narrow conical clusters, overtopping upper lvs; June-Sept. Heaths, sandy places. T. **3a* Narrow-leaved Cudweed** *L. gallica* has longer, even narrower lvs, the topmost exceeding the flhds. (B,rare),F,G. **3b** *L. arvensis* is like 3a but has bracts tipped with white hairs. F,G,S. **3c** *L. neglecta* is like 3a but has longer broader lvs. F,G.

4* MARSH CUDWEED *Filaginella uliginosa.* Low/short silvery grey annual; well branched. Lvs narrow oblong, green above, alternate. Flhds rayless, yellow-brown, in clusters overtopped by upper lvs; July-Oct. Damp bare places. T.

5* JERSEY CUDWEED *Gnaphalium luteoalbum.* Short/medium annual, thickly covered with white woolly hairs; little branched. Lvs narrow lanceolate, margins inrolled, the upper *wavy-edged,* hairy on both sides. Flhds rayless, yellow, with red stigmas, egg-shaped, in umbels; July-Sept. Sandy places. T, but rare in B. **5a Cape Cudweed** *G. undulatum* is taller, bushier and foetid, with broader lvs green above and running down on to stem. (F), from S Africa.

6* HEATH CUDWEED *Omalotheca sylvatica.* Short/medium perennial, grey with wool; unbranched. Lvs linear, 1-veined, green above, white-felted beneath, alternate. Flhds reddish or yellowish, hidden by brown sepal-like bracts, solitary or in small clusters in a *long lfy spike;* July-Sept. Heaths, open woods, dry grassland. T. **6a* Highland Cudweed** *O. norvegica* is shorter, with broader 3-veined lvs downy on both sides, and a much shorter fl spike. Mountains. B (rare),F,S. **6b* Dwarf Cudweed** *O. supina* is smaller and slenderer still with narrower leaves. Mountains.

7 HELICHRYSUM *Helichrysum arenarium.* Short perennial, grey with woolly hairs. Lvs oblong, *flat,* alternate. Flhds *bright yellow,* rayless, in a cluster; July-Sept. Bare ground, often sandy. F,G. **7a** *H. stoechas* has woody stems and is aromatic. F.

8 UPRIGHT CUDWEED *Bombycilaena erecta.* Short annual, covered with woolly hairs. Lvs short, narrow, alternate, numerous up stems. Flhds straw yellow, rayless in a *tight* cluster; June-July. Grassy places. F, southern.

9* SMALL FLEABANE *Pulicaria vulgaris.* Low/short downy annual; much branched. Lvs oblong, *wavy-edged,* alternate. Flhds yellow, *shortly* rayed, rays no longer than sepal-like bracts, solitary in loose clusters; Aug-Sept. Damp bare places, especially where water stands in winter. T, southern, rare in B.

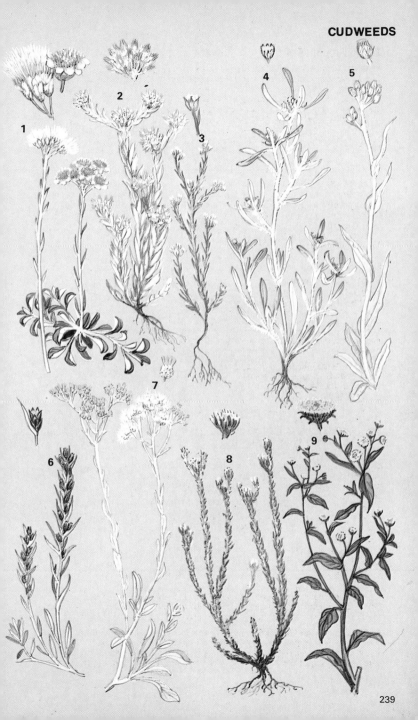

Daisy Family (contd.)

1* COMMON FLEABANE *Pulicaria dysenterica.* Medium hairy perennial. Lvs lanceolate, *wavy-edged,* the upper clasping the stem. Flhds daisy-like, yellow, *15-30 mm,* in a loose flat-topped cluster; July-Sept. Damp grassy places. T, southern.

2* IRISH FLEABANE *Inula salicina.* Medium perennial, more or less hairless; stems brittle. Lvs *narrow* elliptical, the upper unstalked and half-clasping the stem. Fls daisy-like, yellow, 25-40 mm, solitary or in small clusters; sepal-like bracts lanceolate; July-Aug. A shy flowerer in Ireland. Marshes, scrub, rocky slopes. Ireland, F,G,S. **2a** *I. britannica* has lvs lanceolate, bracts linear and both hairy. F,G. **2b** *I. ensifolia* has lvs lanceolate or linear, parallel-veined and hairy on margins, and larger (25-55 mm) fls. G. **2c** *I. hirta* has long hairs on both stems and lvs. Grassy places. F,G. **2d** *I. germanica* has lvs hairy below and flhds 10 mm. G, southern. **2e** *I. helvetica* has lvs hairy below. G (Upper Rhine valley).

3* ELECAMPANE *Inula helenium.* Tall hairy perennial. Lvs elliptical, toothed, basal ones up to *40 cm* long, the rest unstalked, alternate, clasping the stem. Flhds daisy-like, rays *narrow,* yellow, 60-80 mm, in small clusters; June-Sept. Grassy places, woods. T, but (B). **3a Large Yellow Ox-eye** *Telekia speciosa* has more deeply toothed, heart-shaped lower lvs, broader orange-yellow rays and brownish-yellow disc florets. Especially by streams. (F,G), from E Europe.

4 YELLOW OX-EYE *Buphthalmum salicifolium.* Medium/tall hairy perennial; little branched. Lvs oblong-lanceolate, feebly toothed, alternate, clasping the stem. Flhds daisy-like, rays *broad,* yellow, 30-60 mm, *solitary;* July-Aug. Woods, rocks, in hills and mountains. F,G.

5* CONE FLOWER *Rudbeckia laciniata.* Tall perennial, almost hairless. Lvs lanceolate, mostly *lobed,* sometimes toothed, alternate. Flhds daisy-like, rays broad and down-turned, yellow, disc florets blackish-brown, making a conspicuous *cone,* 70-120 mm, solitary or in small clusters; June-Oct. Damp woods, waste places. (B,F,G), from N America. **5a* Black-eyed Susan** *R. hirta* is roughly hairy, with lvs not lobed and fls smaller. (T).

6* PERENNIAL SUNFLOWER *Helianthus rigidus.* Tall patch-forming perennial to 2 m, *rough* to the touch. Lvs broad lanceolate, scarcely toothed, opposite, very shortly stalked. Flhds daisy-like, rays long and broad, yellow, disc florets turning dark *purple,* 60-100 mm, long-stalked; Aug-Oct. Waste places. (B,F,G), from N America. **6a* Jerusalem Artichoke** *H. tuberosus* has coarsely toothed lvs with winged stalks, and smaller all-yellow fls rarely seen; Sept-Nov. **6b* Sunflower** *H. annuus* is annual and can reach nearly 3 m, with broader, often heart-shaped lvs and much larger (100-300 mm) fls. From S America. **6c*** *Guizotia abyssinica* is stickily hairy above with unstalked lvs and flhds 20-30 mm. (T, casual), from Africa.

7 ARNICA *Arnica montana.* Short/medium downy *aromatic* perennial. Lvs lanceolate, mainly in a basal *rosette,* a few opposite on stems. Flhds daisy-like, rays long, yellow, 40-80 mm, solitary; June-Aug. Grassy places in *hills and mountains.* F,G,S. **7a** *A. angustifolia* is shorter, with narrower lvs and smaller (30 mm) flhds. S.

8* YELLOW CHAMOMILE *Anthemis tinctoria.* Short/medium perennial, *grey* with woolly down. Lvs *2-pinnate,* downy beneath. Flhds daisy-like or sometimes rayless, yellow, 25-40 mm, solitary, long-stalked; July-Aug. Dry bare and waste places. T, but (B). **8a** *A. austriaca* has white fls. G, southern.

9* LEOPARDSBANE *Doronicum pardalianches.* Medium patch-forming hairy-perennial. Lvs broad *heart-shaped,* toothed, the upper unstalked. Flhds daisy-like, rays long, yellow, 40-60 mm, in small clusters; May-July. Woods, shady places. (B),F,G. **9a* Green Leopardsbane** *D. plantagineum* has narrower lvs, not heart-shaped, and larger (50-80 mm) solitary flhds.

Daisy Family (contd.)

1* BUTTERBUR *Petasites hybridus.* Low/medium patch-forming perennial, downy. Lvs *very large,* to 1 m across, heart-shaped, toothed, downy grey beneath, long-stalked, all from roots, appearing after fls. Flhds rayless, brush-like, lilac-pink, unscented, in spikes; stalks with strap-like bracts; stamens and styles *on separate plants;* Mar-May. Damp places, roadsides, streamsides. T. **1a* Winter Heliotrope** *P. fragrans* □ is shorter, and its much smaller lvs appear with the fewer fragrant fls; Nov-Mar. (B),F. **1b* White Butterbur** *P. albus* has fragrant white fls and smaller lvs whitely woolly beneath. T, but (B). **1c** *P. frigidus* is like 1b but shorter and with smaller lvs. S. **1d** *P. spurius* has stems pale, lvs triangular and flhd much branched. F,G, southern. **1e*** *P. japonicus* has smaller lvs and creamy flhds whose large pale green bracts make the whole spike resemble a small cauliflower. (B), from E Asia.

2* YARROW *Achillea millefolium.* Short/medium perennial, downy, *aromatic.* Lvs 2-2-pinnate, *feathery,* dark green. Flhds shortly rayed, rays white or pink, disc florest creamy, in flat umbel-like clusters; June-Nov. Grassy places. T. **2a** *A. nobilis* has lvs flat, not feathery, and pale green with winged stalks, and rays very pale yellow; July-Aug. Bare ground. F,G. **2b** *A. collina* has narrower, leathery lvs. **2c** *A. pannonica* is woolly. **2d** *A. setacea* has weak stems and crowded, pointed lflets. 2b-2d are all G, south-eastern.

3* SNEEZEWORT *Achillea ptarmica.* Medium hairy perennial, *not aromatic.* Lvs *linear,* minutely saw-toothed, dark green, half-clasping stem. Flhds shortly rayed, white, with creamy disc florets, in a loose umbel-like cluster; July-Sept. Damp grassy places, on acid soils. T.

4* MUGWORT *Artemisia vulgaris :* see p. 244.

5* SEA WORMWOOD *Artemisia maritima.* Short greyish perennial, downy, *pungently* aromatic. Lvs 2-pinnate. Flhds rayless, egg-shaped, yellow or orange-yellow, in branched spikes; Aug-Oct. *Saltmarshes,* occasionally inland. T.

6* TRIFID BUR MARIGOLD *Bidens tripartita.* Short/medium annual, almost hairless. Lvs lanceolate, usually with two smaller *lobes at base,* toothed, opposite, on winged stalks. Flhds normally unrayed, button-like, yellow, *erect,* often solitary; July-Oct. Fr flattened, adhering to clothing with two barbed bristles. Damp places, by fresh water. T. **6a* Nodding Bur Marigold** *B. cernua* □ is hairier, with unstalked unlobed lvs and larger nodding flhds. **6b* Beggar Ticks** *B. frondosa* □ is usually taller, with lvs pinnate and smaller flhds with outer bracts shorter than inner. (B,G), from America. **6c** *B. connata* has unlobed lvs clasping the stem. (F,G, spreading), from N America. **6d** *B. radiata* has yellowish lvs with teeth bent inwards. (G).

7* COTTONWEED *Otanthus maritimus.* Short creeping perennial, covered with *thick* white down. Lvs oblong, toothed. Flhds rayless, button-like, yellow, in small clusters; Aug-Nov. Coastal sand and shingle. Ireland (rare), F.

8* BUTTONWEED *Cotula coronopifolia.* Low/short yellow-green hairless fleshy *aromatic* annual. Lvs variable, linear to pinnate, alternate, sheathing the stem. Flhds very shortly rayed, appearing rayless and button-like, yellow, long-stalked, *solitary,* always facing the sun. July-Oct. Sandy places, especially on *coast.* (B,F,S), from S Africa. **8a* Scottish Wormwood** *Artemisia norvegica* is shorter, tufted, not fleshy, and grey with silky hairs, and has 1-2-pinnate lvs and larger rayless fls not always solitary; June-Sept. Mountains. B(rare),S.

6a* Nodding Bur Marigold

6b* Beggar Ticks

1

1a

2

3

4

5

6

7

8

243

Daisy Family (contd.)

1* OX-EYE DAISY *Leucanthemum vulgare.* Medium perennial, slightly hairy; little branched. Lvs dark green, variable, the lower rounded or spoon-shaped, toothed or lobed, the upper narrower, clasping the stem. Flhds daisy-like, rays white, disc florets yellow, 25-50 mm, *solitary;* May-Sept. Grassy places. T.
1a* Shasta Daisy *L. maximum,* with much larger (50-80 mm) flhds, is a frequent patch-forming garden escape; July-Sept. (B,G), from the Pyrenees.

2* FEVERFEW *Tanacetum parthenium.* Short/medium perennial, downy, highly *aromatic.* Lvs yellowish, 1-2-pinnate. Flhds daisy-like, rays white (occasionally absent), disc florets yellow (a "double" garden form often escapes), 10-25 mm, in an umbel-like *cluster;* June-Sept. Walls, waste places. (T), from S E Europe.
2a *T. corymbosum* is not aromatic and has lobed lanceolate lvs and larger (25-40 mm) fls. F,G.

3* TANSY *Tanacetum vulgare* Medium/tall perennial, almost hairless, strongly *aromatic;* unbranched. Lvs pinnate, toothed. Fls rayless, *button-like,* yellow, in large umbel-like clusters; July-Oct. Grassy and waste places. T.

4* CORN MARIGOLD *Chrysanthemum segetum.* Short/medium greyish hairless perennial. Lvs fleshy, oblong, toothed or lobed, the upper clasping the stem. Fls daisy-like, bright *yellow,* 35-55 mm, solitary; June-Oct. A weed of cultivation. T.

5* COLTSFOOT *Tussilago farfara.* Low/short creeping downy perennial; unbranched. Lvs heart-shaped with pointed teeth, from roots only, appearing after fls, much smaller than Butterbur (p. 242). Fls yellow, with narrow rays, solitary on stems covered with purplish scales; *Feb-*Apr. Fr a white "clock". Bare and waste ground. T.

6* PURPLE COLTSFOOT *Homogyne alpina.* Low/short creeping hairy perennial; unbranched. Lvs kidney-shaped, dark green, shiny, purplish beneath, all from roots, much smaller than 5. Fls purple, *rayless,* solitary, on stems with narrow scales; June-Sept. Fr a white "clock". Damp grassy places in mountains. B (rare), S.

7* GARDEN MARIGOLD *Calendula officinalis.* Short/medium perennial, roughly hairy. Lvs oblong, often broadest at tip. Fls daisy-like, *orange,* 40-50 mm, solitary; May-Oct. Waste ground, a frequent garden escape. (B,F,G), from Mediterranean.
7a* Marigold *C. arvensis* has lvs narrower and toothed and flhds 10-20 mm. Arable weed, especially in vineyards. F,G.

On p. 243

4* MUGWORT *Artemisia vulgaris* (see p. 243). Medium/tall downy perennial, *slightly* aromatic. Lvs 1-2-pinnate, the upper unstalked, dark green and almost hairless above, silvery downy beneath. Flhds rayless, egg-shaped, yellowish- or purplish-brown, numerous in branched spikes; July-Sept. Waste places, roadsides. T. **4a* Chinese Mugwort** *A. verlotiorum* is shorter and more aromatic, has darker lflets less downy beneath, and is much later flg (Oct-Dec). (B), from China. **4b* Wormwood** *A. absinthium* is strongly aromatic and has lvs whitely downy above and larger yellower flhds. **4c* Field Wormwood** *A. campestris* agg. is shorter and not aromatic, with almost hairless stems, 2-3-pinnate lvs downy above at first, and yellower fls; Aug-Sept. Bare, often sandy places. T, rare in B. **4d** *A. pontica* is taller and shrubby, with lvs downy white on both sides and pale yellow flhds. G. **4e Tarragon** *A. dracunculus,* the culinary herb, is very aromatic and has green linear-lanceolate lvs, the lower 3-lobed at the tip, and greenish flhds. (F,G), from Asia. **4f** *A. rupestris* is shorter and woody at base, with hairless green lvs. **4g** *A. austriaca* is like 4d with lvs silky below and fls sometimes red. G, eastern. **4h** *A. laciniata* is like 4c, but stem hairier and often purplish, and segments of lower lvs blunter. S (Oland).

Daisy Family (contd.)

RAGWORTS and GROUNDSELS *Senecio*. Flhds normally rayed, daisy-like, all yellow, in branched clusters. Fr pappus with simple hairs, not forming a clock. Hybrids not infrequent.

1* RAGWORT *Senecio jacobaea*. Medium/tall biennial, often hairless; little branched. Lvs pinnately lobed, end lobe *small*, blunt. Flhds 15-25 mm, in dense flat-topped clusters; outer sepal-like bracts few, much shorter than *dark-tipped* inner bracts; June-Nov. Dry grassy places. T. **1a* Hoary Ragwort** *S. erucifolius* □ is grey with down and has lvs more narrowly lobed, the end lobe more pointed and longer outer bracts; July-Sept. Southern. **1b* Marsh Ragwort** *S. aquaticus* is more widely branched and has lower lvs varying from oval to pinnately lobed, the end lobe larger and blunt, and looser clusters of larger (25-30 mm) flhds. Damp grassland. **1c** *S. erraticus* is very like 1b, but has lvs always pinnately lobed, the lobes at right angles to axis instead of directed forwards. F,G, eastern.

2* OXFORD RAGWORT *Senecio squalidus* agg. Medium perennial, almost hairless; well branched. Lvs lanceolate to pinnately lobed, the end lobe *sharply* pointed, the upper unstalked. Flhds 15-20 mm, sepal-like bracts black-tipped, the outer row much shorter; Apr-Nov. Walls, bare and waste ground. (B,F), from S Italy. **2a** *S. vernalis* is annual with matted hairs and larger (20-30 mm) flhds. F,G,S.

3 MARSH FLEAWORT *Senecio congestus*. Medium/tall annual/perennial, covered with *white* woolly hairs. Lvs *lanceolate*, toothed or not. Flhds pale yellow, 20-30 mm; June-July. Marshes. T, but extinct in B.

4* FEN RAGWORT *Senecio paludosus*. Medium/tall hairy perennial. Lvs lanceolate, *coarsely* saw-toothed, unstalked, shiny above, cottony *white* beneath. Flhds 30-40 mm, in a flat-topped cluster; July-Aug. Wet places. T, but rare in B. **4a* Broad-leaved Ragwort** *S. fluviatilis* is taller, to 1½ m, and stouter, with less deeply toothed lvs hairless beneath and smaller (30 mm) flhds. Damp places. (B),F,G, southern.

5 ALPINE RAGWORT *Senecio nemorensis*. Medium/tall perennial, almost hairless. Lvs lanceolate, *finely* saw-toothed, short-stalked. Flhds with *few* rays, 15-25 mm, in a flat-topped cluster; July-Sept. Woods and rocks, mainly in hills and mountains. T,but(B).

6* GROUNDSEL *Senecio vulgaris*. Low/short annual, often downy, not foetid. Lvs pinnately lobed, hairless above. Flhds usually *rayless*, brush-like, but occasionally rayed □, yellow, in loose clusters; sepal-like bracts *black-tipped*, the outer very short; all year. A weed of cultivation, disturbed ground. T. **6a* Sticky Groundsel** *S. viscosus* is taller, foetid and greyish with sticky hairs and has larger paler yellow flhds with very short rolled-back rays and bracts not black-tipped, the outer half as long as the inner; July-Oct. Bare and waste places, fens. **6b* Heath Groundsel** *S. sylvaticus* is greyish with sticky hairs (but less so than 6a) and somewhat foetid, with flhds like 6a and bracts purple-tipped, the outer still smaller; July-Oct. Dry sandy and heathy places.

7* SILVER RAGWORT *Senecio bicolor*. Medium undershrub, covered with a dense *silvery* white felt; well branched. Lvs oval to oblong, toothed to pinnately lobed, green above. Flhds 8-12 mm; June-Aug. Cliffs and rocks by sea. (B,F), from Mediterranean.

8* FIELD FLEAWORT *Senecio integrifolius*. Low/short perennial, often whitely downy; *unbranched*. Lvs oval, wrinkled, toothed or not, mainly in a basal rosette, narrower up stems. Flhds *few*, 15-25mm, in a loose cluster; May-June. Short dry turf. T. **8a*** *S. helenitis* is stouter, with lvs more coarsely toothed and larger and more numerous lvs, almost clasping the stem. Grassy places. F,G,S. **8b** *S. rivularis* is like 8a, but lower lvs ace-of-spades shaped, and flhds 25-35 mm; Apr-Aug. Damp places in hills, on lime. G.

1a* Hoary Ragwort

Daisy Family *(contd.)*

1* GLOBE THISTLE *Echinops sphaerocephalus*. Medium/tall thistle-like perennial, to 2 m, stickily hairy and white-woolly. Lvs lanceolate, pinnately lobed, spiny. Flhds rayless, *globular,* pale blue, spiny; June-Sept. Dry bare stony places. T, but (B). **1a*** *E. exaltatus* is less spiny and not stickily hairy, with whiter fls. (B, F, G), from Mediterranean.

2* CARLINE THISTLE *Carlina vulgaris*. Low/short *erect* biennial, spiny. Lvs thistle-like, oblong, pinnately lobed, prickly, the lower cottony, not in a rosette. Flhds rayless, yellow-brown, but the conspicuous spreading *yellow* sepal-like bracts appear like rays, folding up in wet weather; July-Sept. Dead plants survive through winter. Grassland, dunes; on lime. T.

3 STEMLESS CARLINE THISTLE *Carlina acaulis*. Low biennial/perennial, spiny. Lvs all in a *rosette,* pinnately lobed, more or less prostrate. Flhds rayless, whitish, but appearing rayed from the conspicuous spreading *silvery* sepal-like bracts, solitary, unstalked; May-Sept. Grassy, rocky places in mountains. F, G, S.

4* ALPINE SAWWORT *Saussurea alpina*. Short stout hairy perennial. Lvs lanceolate, more or less *toothed*, silvery white beneath, spineless. Flhds rayless, purple, fragrant, in compact clusters; Aug-Sept. Mountain grassland. B, G, S.

5* LESSER BURDOCK *Arctium minus* agg. Medium/tall stout downy spineless biennial; stems arching. Lvs broadly heart-shaped, to 30 cm, longer than broad, with hollow stalks. Flhds 15-30 mm, rayless, egg-shaped, purple, in short-stalked spikes; hooked bracts forming *burs* when dried and lasting through winter; July-Sept. Shady and waste places. T. **5a* Greater Burdock** *A. lappa* has blunter lvs as long as broad with solid stalks and larger (35-40 mm), more globular flhds. **5b** *A. tomentosum* has lf undersides and smaller (20-25 mm) flhds covered with white cottony down. Extensive hybridisation occurs.

6* COTTON THISTLE *Onopordum acanthium*. Tall spiny biennial, to 1½ m, *white* with cottony down; stems *broadly* winged. Lvs oblong, spiny. Flhds rayless, pale purple, solitary; sepal-like bracts cottony at base, ending in yellow spines; July-Sept. Bare and waste ground, often sandy. T.

7* MILK THISTLE *Silybum marianum*. Medium/tall spiny annual/biennial; stems downy, unwinged. Lvs oblong, wavy-edged, dark green with conspicuous *white veins,* spiny. Flhds rayless, purple, solitary; sepal-like bracts ending in sharp yellow spines; June-Aug. Bare and waste places. (B), F, G.

8* CABBAGE THISTLE *Cirsium oleraceum*. Medium/tall perennial, more or less hairless. Lvs lanceolate, pinnately lobed to toothed. Flhds rayless, egg-shaped, *straw-yellow,* with narrow spiny sepal-like bracts, in dense clusters almost hidden by topmost lvs; July-Aug. Damp places. T, but (B, rare). **8a** *C. spinosissimum* is shorter and much spinier, with all lvs pinnately lobed and spiny and flhds whitish. Mountains. F, G. **8b Yellow Melancholy Thistle** *C. erisithales* □ has lemon-yellow flhds not surrounded by upper lvs, usually solitary and nodding. Hill districts. F, G.

8b Yellow Melancholy Thistle

Daisy Family (contd.)

THISTLES: *Carduus* and *Cirsium*. Stems usually winged and spiny. Lvs oblong or lanceolate, and usually pinnately lobed, spiny and wavy-edged. Flhds rayless, brush-like, purple (except 244/8); sepal-like bracts usually spine-tipped. Fr with a pappus, feathery in *Cirsium* but undivided in *Carduus,* not forming a clock.

1* CREEPING THISTLE *Cirsium arvense*. Short/tall *creeping* perennial, usually *hairless;* little branched, stems spineless. Lvs sometimes cottony beneath. Flhds *lilac*, 15-25 mm, fragrant, in clusters; bracts purplish, scarcely spiny; June-Sept. Grassy and waste places, a weed of cultivation. T. **1a** *C. rivulare* is hairier, with lvs scarcely spiny and much larger (25-35 mm) and rounder purple unstalked flhds. G.

2* SPEAR THISTLE *Cirsium vulgare*. Medium/tall biennial, downy, *sharply* spiny. Lvs deeply pinnately lobed. Flhds 20-40 mm across, often *solitary;* bracts often hairless, with yellow-tipped spines; July-Sept. Bare and waste places. T.

3* WOOLLY THISTLE *Cirsium eriophorum*. Tall stout biennial, very spiny, covered with *white wool;* stem unwinged. Lvs almost pinnate. Flhds red-purple, *globular*, 40-70 mm across, usually solitary; bracts covered with cobwebby wool; July-Sept. Bare and grassy places; on lime. B, F, G.

4* MEADOW THISTLE *Cirsium dissectum*. Medium perennial, cottony all over but *scarcely* spiny; little branched, stems unwinged; has creeping runners. Lvs whiter beneath, the upper undivided but clasping the stem. Flhds red-purple, 20-25 mm across, solitary; June-Aug. Damp grassy places, fens. B, F, G. **4a* Melancholy Thistle** *C. helenioides* ☐ is taller and more thickly white-felted, with larger (30-50 mm) flhds. T. **4b* Tuberous Thistle** *C. tuberosum* ☐ is taller and had tuberous roots, no runners, all lvs pinnately lobed and green on both sides, and larger (25-30 mm) flhds. Drier grassland, on lime. B (rare), F, G.

5* DWARF THISTLE *Cirsium acaule*. Low/short, usually prostrate perennial, almost hairless. Lvs all in a basal *rosette,* rather spiny. Flhds red-purple, 20-50 mm across, usually *unstalked* but occasionally stalked to 15 cm, usually solitary; bracts purplish, scarcely spiny; June-Sept. Dry grassland; on lime. T.

6* MUSK THISTLE *Carduus nutans*. Medium/tall biennial, with cottony white hairs. Lvs deeply pinnately lobed. Flhds bright red-purple, 35-50 mm, often solitary, slightly *nodding*, on spineless upper stalks; bracts conspicuous, purple; June-Sept. Bare and grassy places; on lime. T. **6a** *C. defloratus* has lfless unwinged fl-stems and erect bracts. Dry grassy places in hills. G.

7* SLENDER THISTLE *Carduus tenuiflorus*. Short/tall slender annual/biennial, with cottony white hairs. Lvs cottony white beneath. Flhds pinkish, *8 mm* across, in clusters; May-Aug. Grassy and waste places, often by sea. B, F, G.

8 GREAT MARSH THISTLE *Carduus personata*. Tall perennial, to 2 m, with cottony white hairs. Lvs *softly* spiny, the upper undivided, white beneath. Flhds 15-20 mm, unstalked, in tight clusters; bracts *not* spine-tipped, the outer recurved, often blackish; July-Aug. Damp places in mountains. F, G.

9* WELTED THISTLE *Carduus acanthoides*. Tall biennial, to 2 m. Lvs *weakly* spined, cottony white beneath. Flhds red-purple, 10-15 mm across, in a cluster, with stalks spineless *at top;* bracts weakly spined; June-Sept. Grassy places, hedge-banks. T. **9a** *C. crispus* is very similar, but has stronger spines and fl stalks spiny to top. F, G, S. **9b* Marsh Thistle** *Cirsium palustre* ☐ is less branched, and usually purple-tinged, especially on bracts, and has more cottony down and flhds less reddish, occasionally white, on stalks spiny to top. June-Oct. More often in marshes and woods.

9b* Marsh Thistle   **4b* Tuberous Thistle**   **4a* Melancholy Thistle**

Daisy Family *(contd.)*

KNAPWEEDS and STAR THISTLES *Centaurea*. Spineless (except 5, 6). Stems stiff, downy. Lvs spirally arranged. Flhds rayless but with enlarged sterile outer florets often giving appearance of rays, deeply lobed to appear star-like. Sepal-like bracts with toothed or cut appendages. Pappus not forming a clock. Numerous casuals and hybrids also occur.

1* BLACK KNAPWEED *Centaurea nigra*. Short/medium perennial. Lvs *lanceolate*, the lower sparsely toothed. Flhds brush-like, solitary or in clusters, 15-20 mm across; sepal-like bracts blackish-brown with long fine teeth, recurved at tip; June-Sept. Hybridises with 1b. Grassy places. T. **1a Slender Knapweed** *C. debeauxii* ssp. *nemoralis* is slenderer, with flhds usually rayed, and sepal-like bracts paler brown and erect. Prefers limy soils. **1b* Brown Knapweed** *C. jacea* is shorter and slenderer, with smaller (10-20 mm) paler flhds always apparently rayed and browner bracts with a paler irregularly cut margin. T, but (B). **1c** *Jurinea cyanoides* has lvs white-felted beneath and inrolled at edges. G, southern.

2* GREATER KNAPWEED *Centaurea scabiosa*. Medium perennial. Lvs *pinnately lobed*, sparsely bristly. Flhds apparently rayed, solitary, *30-50 mm* across; sepal-like bracts *green*, edges and long slender teeth blackish, June-Oct. Grassy places; on lime. T. **2a* Jersey Knapweed** *C. paniculata* has cottony stems and much smaller fls in a long branched spike. (B, rare), F. **2b** *C. rhenana* agg. has lf-lobes linear, flhds smaller, more numerous, and bracts green. F, G.

3* PERENNIAL CORNFLOWER *Centaurea montana*. Medium creeping downy perennial. Lvs *lanceolate*, unstalked. Flhds apparently rayed, blue, solitary, *60-80 mm* across; May-Aug. Grassy places. (B), F, G.

4* CORNFLOWER *Centaurea cyanus*. Medium annual, grey with down. Lvs *pinnately lobed*, stalked, the upper lanceolate, unstalked. Flhds apparently rayed, bright blue, solitary, *15-30 mm* across; June-Aug. Waste places, cornfields. T.

5* YELLOW STAR-THISTLE *Centaurea solstitialis*. Medium/tall annual, grey-white with down; well branched. Lvs pinnately lobed, the upper linear. Flhds rayless, pale *yellow*, solitary; sepal-like bracts ending in a long stiff yellow *spine*, with smaller spines below; July-Sept. Bare and waste ground. (T, casual), from S Europe. **5a* Cockspur Star-Thistle** *C. melitensis* ☐ has pinnate bracts and sticky hairs on fls. **5b*** *Carthamus lanatus* ☐ has spiny lvs and lf-like spiny bracts. **5c* Safflower** *Carthamus tinctorius* ☐ is like 5b but fls bright reddish-orange. (B, F, G, casual), from Asia. **5d Spanish Oyster Plant** *Scolymus hispanicus* has spiny winged stems. F.

6* RED STAR-THISTLE *Centaurea calcitrapa*. Short/medium perennial, almost hairless; widely branched. Lvs pinnately lobed, *bristle-pointed*. Flhds rayless, thistle-like, red-purple, solitary, 8-10 mm across; sepal-like bracts ending in a long stout yellow *spine*, with shorter spines below; July-Sept. Dry bare places. (B), F, G. **6a* Rough Star-Thistle** *C. aspera* has larger (25 mm) flhds and very short, palmately arranged, spreading or down-turned reddish spines. (B), F.

7* SAWWORT *Serratula tinctoria*. Slender medium perennial, hairless, thistle-like but *spineless;* stems stiff. Lvs pinnately lobed, lobes *finely* saw-toothed. Flhds rayless, purple, in a branched cluster; sepal-like bracts purplish; July-Oct. Pappus yellowish. Damp grassy places. T.

5a* Cockspur Star-thistle

5b* *Carthamus lanatus*   **5c*** Safflower

Daisy Family (contd.)

All plants on this page have dandelion-like flhds, with no disc florets, and all except 5 and 8 have milky juice.

1* GOATSBEARD *Tragopogon pratensis.* Medium hairless annual/perennial; little branched. Lvs *linear,* grass-like. Flhds yellow, solitary, florets either shorter or longer than sepal-like bracts, opening fully only on sunny mornings, whence folk-name of Jack-go-to-bed-at-noon; May-Aug. Pappus makes a large clock. Grassy places. T. **1a** *T. dubius* is taller with stem very swollen beneath flhd. Dry woods. F, G.

2* SALSIFY *Tragopogon porrifolius.* Resembles 1, but has much larger dull purple flhds, whose florets are about *as long as* sepal-like *bracts;* Apr-June. Grassy places, formerly cultivated. (T), from Mediterranean.

3 PURPLE VIPER'S GRASS *Scorzonera purpurea.* Short/medium hairless perennial; little branched. Lvs *linear,* keeled. Flhds *lilac-purple,* solitary, florets much longer than sepal-like bracts; May-June. Grassy and rocky places. F, G.

4* VIPER'S GRASS *Scorzonera humilis.* Low/short perennial, almost hairless; little branched. Lvs narrow *lanceolate,* untoothed, becoming small and scale-like up stems. Flhds pale *yellow,* solitary, florets much longer than sepal-like bracts; May-July. Damp grassy places. T, southern, rare in B. **4a** *S. austriaca* has a ring of scales at base of stem and brighter yellow fls. F, G. **4b** *S. hispanica* has much branched stems. G, southern. **4c** *S. laciniata* □ has lvs with linear lobes, the end lobe broader, and fls brownish. G, southern.

5* CHICORY *Cichorium intybus.* Medium/tall perennial, hairy or not; stems stiff, well branched, *no* milky juice. Lvs pinnately lobed, the upper undivided. Flhds *clear* blue, 25-40 mm, in lfy spikes; June-Sept. Grassy and waste places. T.

6* ALPINE SOW-THISTLE *Cicerbita alpina.* Medium/tall perennial; stems unbranched, with *reddish* hairs. Lvs pinnately lobed, hairless, clasping the stem. Flhds mauvish blue, 20 mm, in lfy spikes; July-Sept. Grassy places in *mountains.* T, rare in B. **6a* Blue Sow-thistle** *C. macrophylla* is taller, stouter and patch-forming, with lvs less lobed but more toothed, and larger (30 mm) flhds. Waysides, waste places. (T), from the Caucasus. **6b*** *C. plumieri* is hairless, with more lf lobes and fls in a cluster. (B), F, G. **6c*** *C. tatarica* is shorter and almost hairless, with lvs more lobed but less toothed, and compacter clusters of darker flhds. (B), G, S. **6d** *C. sibirica* is like 6c with untoothed lanceolate lvs. S.

7 BLUE LETTUCE *Lactuca perennis.* Medium hairless perennial. Lvs pinnately lobed, the upper clasping the stem, greyish. Flhds 30-40 mm, blue-purple or white, long-stalked from *arrow-shaped* bracts, in a loose cluster; May-Aug. Bare and grassy places. F, G.

8 PURPLE LETTUCE *Prenanthes purpurea.* Medium/tall greyish perennial, not unlike Wall Lettuce (p.252), but with purple fls. Lvs oblong, often *waisted* or lobed, clasping the stem. Flhds occasionally white, each with very few florets, 20 mm, in a loose cluster; July-Sept. Pappus white. *Shady places.* F, G.

4c *Scorzonera laciniata*

Daisy Family *(contd.)*

All plants on this page have yellow dandelion-like flhds with no disc florets. All except 7, 8 and 9 have milky juice.

1* SMOOTH SOW-THISTLE *Sonchus oleraceus.* Short/tall greyish annual, hairless. Lvs pinnately lobed, the end lobe the largest, with *softly* spiny margins, clasping the stem with arrow-shaped points. Flhds pale yellow, *20-25 mm,* in a loose cluster; May-Nov. Pappuses making a clock. Bare and waste ground, a weed of cultivation. T. **1a* Prickly Sow-thistle** *S. asper* has lvs less lobed, often undivided, thicker and glossy, with sharper spines and clasping the stem firmly with rounded lobes.

2* PERENNIAL SOW-THISTLE *Sonchus arvensis.* Tall patch-forming perennial, almost hairless. Lvs pinnately lobed, greyish beneath, softly spiny on margins, clasping the stem with rounded lobes. Flhds *40-50 mm,* rich yellow, in clusters; stalks and sepal-like bracts with numerous sticky yellow hairs; July-Sept. Bare and waste places. T. **2a** *S. maritimus* lacks the sticky hairs. F.

3* MARSH SOW-THISTLE *Sonchus palustris. Very tall,* to 3 m, tufted perennial, hairless below. Lvs *deeply* pinnately lobed, greyish, edged with softly spiny teeth, clasping the stem with arrow-shaped lobes. Flhds pale yellow, *30-40 mm,* in tight clusters; stalks and sepal-like bracts with numerous sticky blackish-green hairs; July-Sept. Fresh and brackish marshes, ditches. T, southern, rare in B.

4* WALL LETTUCE *Mycelis muralis.* Slender short/tall hairless perennial, often purple-tinged. Lvs pinnately lobed, toothed, the end lobe large, *triangular* and sharply cut, clasping stem with rounded toothed lobes. Flhds 7-10 mm, with only five florets, in a loose cluster; June-Sept. Shady places, walls, rocks. T.

5* PRICKLY LETTUCE *Lactuca serriola.* Tall foetid biennial, hairless, stems whitish, prickly. Lvs greyish, oblong, sometimes irregularly lobed, alternate, weakly spiny on margins, *prickly* on midrib beneath, clasping the stem with arrow-shaped points; often twisted at an angle to the sun, whence name Compass Plant. Flhds 11-13 mm, numerous, in branched spikes; sepal-like bracts with spreading auricle; July-Sept. Fr pale brown to *grey-green.* Bare and waste ground. T, southern. **5a* Great Lettuce** *L. virosa* is often tinged maroon, holds upper lvs horizontally and has bracts with clasping auricle and fr maroon to blackish. B, F, G.

6* LEAST LETTUCE *Lactuca saligna.* Medium/tall hairless, *spineless* annual/biennial; little branched. Lvs greyish, pinnately lobed, the lobes curved *backwards,* with a white midrib, the upper narrow, undivided and held almost vertically against the stem, which they clasp with arrow-shaped points. Flhds 10 mm, in a slender stalked spike, sepal-like bracts green; July-Aug. Fr beak white. Bare grassy or shingly places, often near the sea. B, F, G. **6a Pliant Lettuce** *L. viminea* has narrower lvs deeply pinnately cut with linear lobes, and unstalked flhds often in threes in a branched spike. F, G. **6b** *L. quercina* has a weak green stem, lvs varying from toothed to pinnate, and fr beak black. G, S (rare).

7* NIPPLEWORT *Lapsana communis.* Short/tall hairy annual; *no* milky juice. Lvs pointed oval, toothed, often pinnately lobed *at base.* Flhds 10-20 mm, in branched clusters, not opening in dull weather; June-Oct. Shady, bare and waste places. T.

8* LAMB'S SUCCORY *Arnoseris minima.* Low/short hairless annual; *no* milky juice. Lvs all from roots, lanceolate, toothed. Flhds very shortly rayed, 7-10 mm, solitary, on lfless stems much *swollen* at top; June-Aug. No pappus. Bare places, especially sandy. T, southern, rare in B.

9 CHONDRILLA *Chondrilla juncea.* Medium/tall biennial, hairy only on lower stem. Lower lvs lanceolate, lobed, withering before flg time; upper lvs narrower, not lobed, sometimes finely toothed. Flhds 10 mm, unstalked, in small clusters up stems, sepal-like bracts *downy;* July-Sept. Dry sandy or stony places. F, G. **9a** *Calycocorsus stipitatus* has stem black hairy, lvs less toothed, mostly in a rosette, and larger flhds. Wet places. G, southern.

Daisy Family (contd.)

Yellow dandelion-like fls, with no disc florets. Lvs all in a basal rosette (except 5a, 8), oblong, broadly toothed or shallowly pinnately lobed (except 9).

DANDELIONS *Taraxacum*. A variable, difficult group divided into eight Sections, each with many microspecies, typical examples of three being shown here. Perennials, mostly hairless, with *milky juice*. Lvs usually well lobed. Flhds solitary, on hollow lfless stalks, often downy at top. Pappus making a clock.

1* DANDELION *Taraxacum* Sect. *Vulgaria*. Low/short. Flhds 35-50 mm, outer florets usually *grey-violet* beneath, sometimes purple, never red; sepal-like bracts broader than 2 but narrower than 3, erect or *recurved;* all year, but especially Apr-June. Fr colour variable, but not purplish-red. Grassy and waste places. T. **1a** *Aposeris foetida* is stouter, with foetid juice, sparser florets and no pappus. G, south-eastern.

2* LESSER DANDELION *Taraxacum* Sect. *Erythrosperma*. Low, delicate. Flhds 15-25 mm, sepal-like bracts narrow, appressed, tips of inner ones appearing *forked;* Apr-June. Fr usually *purplish-red*. Dry grassy and bare places, especially on lime. T.

3 RED-VEINED DANDELION *Taraxacum* Sect. *Spectabilia*. Low/short. Lvs often dark-spotted and with reddish stalk and midrib. Flhds 20-35 mm, outer florets sometimes reddish or purplish beneath, and sepal-like bracts broader and often closely pressed but not recurved; Apr-Aug. Moist and wet places, often in mountains. T. **3a Marsh Dandelion** *T.* Sect. *Palustria*, has lvs narrower, very finely toothed or untoothed, not spotted or reddish; sepal-like bracts broad, closely pressed, with broad pale margin; Apr-June. Marshes, fens, streamsides.

4* COMMON CATSEAR *Hypochaeris radicata*. Short/medium, little branched. Lvs usually roughly hairy, the end lobe blunt. Flhds solitary, on lfless stems, with few *scale-like* bracts; June-Oct. Drier grassland. T. **4a* Autumn Hawkbit** *Leontodon autumnalis* is usually hairless and branched, with shiny, more deeply lobed lvs, end lobe pointed, and smaller (10-35 mm) flhds, outer florets usually reddish beneath and bracts mainly near top of stalk; July-Oct.

5* ROUGH HAWKBIT *Leontodon hispidus*. Short/medium, shaggily hairy; unbranched. Lvs shallowly lobed, the end lobe rounded. Flhds solitary, 25-40 mm, outer florets often *reddish* or orange beneath; stalks lfless but with *0-2* small scale-like bracts; June-Oct. Pappus making a clock. Grassland; on lime. T. **5a** ssp. *hyoseroides* has lvs deeply pinnately lobed, all lobes long and narrow. F, G. **5b** *L. incanus*, grey-hairy, and **5c** *L. pyrenaicus*, with clearly stalked lvs, are both rare in mountains. G, southern.

6* LESSER HAWKBIT *Leontodon taraxacoides*. Low/short, hairy or not; unbranched. Lvs shallowly lobed. Flhds solitary, 12-20 mm, outer florets *grey-violet* beneath; stalks lfless, *bractless* and often hairless; June-Oct. Pappus forming a clock. Dry grassy places, dunes. T, southern. **6a* Smooth Catsear** *Hypochaeris glabra* ☐ is a usually hairless annual, with shiny lvs and smaller (12-15 mm) flhds, opening fully only in bright sunshine, the florets no longer than the sepal-like bracts; stalks with a few scale-like bracts; especially on sandy soils.

7* SPOTTED CATSEAR *Hypochaeris maculata*. Short/medium, hairy. Lvs lanceolate, heavily *spotted* purplish-black. Flhds pale yellow, 40-50 mm, on stalks with a few tiny lvs; June-Aug. Grassland, mainly on lime. T, rare in B.

8* SMOOTH HAWKSBEARD *Crepis capillaris*. Short/medium, mostly hairless; *branched*. Lvs shiny, not all in a basal rosette, the upper narrower and clasping the stem with arrow-shaped points. Flhds 10-15 mm, in loose clusters, outer florets often *reddish* beneath; outer sepal-like bracts *half-spreading;* June-Nov. Grassy and waste places. T.

9* MOUSE-EAR HAWKWEED *Hieracium pilosella* agg. Low/short, creeping shaggy with *white* hairs, unbranched, with lfy runners. Lvs elliptical, untoothed. Flhds lemon yellow, solitary, 20-30 mm, outer florets often *reddish* beneath, on bare stalks; May-Oct. Bare and grassy places. T. **9a*** *H. lactucella* agg. has flhds in clusters, like Orange Hawkweed (p. 260). T, but (B).

Daisy Family (contd.)

HAWKWEEDS *Hieracium.* An exceptionally variable and difficult group, with some hundreds of microspecies. Here only four major groupings can be illustrated. Milky juice, stems unbranched, more or less lfy. Lvs lanceolate, toothed, often sparsely, alternate. Pappus hairs usually pale brownish.

1* FEW-LEAVED HAWKWEED *Hieracium murorum* agg. Short/medium perennials, hairy. Lvs *few,* both on stems and at base. Flhds 20-30 mm, rather *few* in the cluster; June-Aug. Grassy places, walls, rocks. T. **1a* Spotted Hawkweed** *H. maculatum* is the most frequent of a few forms with distinctively purple-blotched lvs. B, F, G.

2* LEAFY HAWKWEED *Hieracium umbellatum* agg. Medium/tall perennials, softly hairy. Lvs *numerous,* all up stems. Flhds 20-30 mm, *numerous,* in clusters; June-Nov. Grassy and heathy places. T.

3* ALPINE HAWKWEED *Hieracium alpinum* agg. Low/short perennials, shaggily hairy. Lvs mostly in a basal *rosette,* few and very small up stems. Flhds usually *solitary,* 25-35 mm; July-Aug. Grassy and rocky places in mountains. T.

4* ORANGE HAWKWEED *Hieracium aurantiacum* agg. Short/medium perennials, covered with *blackish* hairs. Lvs mainly in a basal rosette. Flhds *orange-red* (if yellow, see *H. lactucella,* p. 258), 15 mm, in a tight cluster; June-Aug. Grassy and waste places. T.

5* HAWKWEED OX-TONGUE *Picris hieracioides.* Medium perennial, roughly hairy; *branched.* Lvs lanceolate, *wavy-edged.* Flhds 20-35 mm, in a cluster; sepal-like bracts *narrow,* the outer spreading, covered with blackish hairs; July-Oct. Pappus creamy. Grassy and bare places. T.

6* BRISTLY OX-TONGUE *Picris echioides.* Medium perennial, covered with rough *bristles;* well branched. Lvs oblong, wavy-edged, covered with *whitish pimples.* Flhds pale yellow, 20-25 mm, in a cluster; sepal-like bracts *broadly* triangular; June-Nov. Pappus white. Rough grassy places. B, F, G.

HAWKSBEARDS *Crepis.* Distinguished from other dandelion-like fls by the outer row of sepal-like bracts being *much shorter* and usually spreading. Pappus usually white.

7* BEAKED HAWKSBEARD *Crepis vesicaria.* Medium downy biennial. Lvs *pinnately lobed,* dandelion-like, clasping the stem with pointed lobes. Flhds 15-25 mm, erect in bud, in a cluster, outer florets usually *orange* beneath; May-July. Grassy and waste places. B, F, G. **7a* Rough Hawksbeard** *C. biennis* □ is taller and hairier, with fewer, larger (25-35 mm) richer yellow flhds with outer florets yellow beneath. T. **7b* Bristly Hawksbeard** *C. setosa* has paler yellow flhds with numerous spiny yellow bristles. (B, rare), F, G. **7c* Stinking Hawksbeard** *C. foetida* smells of bitter almonds when bruised and has fewer flhds, drooping in bud. Rare in B. **7d* French Hawksbeard** *C. nicaeensis* has lvs clasping stem with arrow-shaped points and florets often red-tipped. T, but (B, rare, S). **7e** *C. sancta* has stems lfless, lvs more coarsely lobed and no pappus. F, G, range expanding.

8* MARSH HAWKSBEARD *Crepis paludosa.* Medium perennial, almost hairless; stems lfy. Lvs lanceolate, *sharply* toothed, clasping the stem with *pointed* bases. Flhds dull orange-yellow, 15-25 mm, rather few to the cluster; sepal-like bracts downy, with numerous sticky *blackish* hairs; July-Sept. Pappus *brownish white. Damp* grassy places, stream-sides. T. **8a** *C. tectorum* is a shorter, much branched annual with pappus pure white. Dry sandy grassy places. F, G, S. **8b** *C. pulchra* is like 8a but taller, less branched and stickily hairy with hairless bracts. Dry sunny slopes and woods. F, G, southern. **8c** *C. praemorsa* has lvs scarcely toothed, few or none on the stem, and dense yellow flhds. F, G, S, eastern.

9* NORTHERN HAWKSBEARD *Crepis mollis.* Slender medium perennial, hairy or not. Lvs lanceolate, toothed or not, clasping the stem with *rounded* bases. Flhds 20-30 mm, rather few in the cluster; July-Aug. Pappus *pure white.* Woods, streamsides, in hill districts. B (rare), F, G.

261

Water-Plantain Family   Alismataceae

The first of the Monocotyledon families (see p.10). Hairless aquatic plants, with lvs all from roots, 3-petalled fls and fr (nutlets) in a rounded head.

1* COMMON WATER-PLANTAIN *Alisma plantago-aquatica.* Medium/tall perennial, with *broad* lanceolate lvs, rounded or heart-shaped at base. Fls in whorls, pale *lilac,* 8-10 mm; June-Sept. Fr head a close ring of nutlets with long beak arising at or below middle. By fresh water. T. **1a* Narrow-leaved Water-Plantain** *A. lanceolatum* □ has narrower lvs tapering into stalk, pinker fls and nutlets with short beak arising near top. **1b* Ribbon-leaved Water-Plantain** *A. gramineum* has often submerged, ribbon-like lvs, slightly broader at tip, smaller fls and nutlets with coiled beak. T, eastern, rare in B.

2* LESSER WATER-PLANTAIN *Baldellia ranunculoides.* Low/short perennial, with *narrow* lanceolate lvs. Fls pale *pink,* 10-15 mm; in whorls of 10-20; June-Sept. In and by shallow, often peaty, fresh water. T, western. **2a** *B. repens* is creeping, rooting at the lf-junctions, with fls 15-22 mm in whorls of 2-6. F, G, western.

3* FLOATING WATER-PLANTAIN *Luronium natans.* Floating perennial, with elliptical lvs, also narrow tapering submerged ones. Fls *white* with a yellow spot, 12-15 mm; May-Aug. Still and slow fresh water. T, western.

4* STAR-FRUIT *Damasonium alisma.* Low/short annual, with blunt oval lvs, heart-shaped at base, floating or submerged. Fls white, 6 mm; June-Sept. Fr heads with long-beaked nutlets spreading like a 6-pointed *star.* In or by shallow fresh water. B(rare), F.

5 PARNASSUS-LEAVED WATER-PLANTAIN *Caldesia parnassifolia.* Medium/tall perennial, with conspicuously *heart-shaped* long-stalked lvs. Fls white, in whorls; July-Sept. By fresh water. F, G.

6* ARROWHEAD *Sagittaria sagittifolia.* Medium/tall perennial, with lvs *arrow-shaped* and in fast streams also oval or lanceolate (floating) or ribbon-like (submerged). Fls white with a large purple spot, 20 mm; July-Aug. Still and flowing shallow fresh water. T. **6a* Canadian Arrowhead** *S. rigida* is shorter, with pointed oval lvs, all aerial, and fls all white, sometimes with a yellow spot. (B), from N America. **6b** *S. natans* has basal lobes of lvs directed downwards and smaller fls. S(rare).

Flowering Rush Family   Butomaceae

7* FLOWERING RUSH *Butomus umbellatus.* Tall hairless perennial. Lvs long, rush-like, *three-cornered*, all from roots. Fls 3-petalled, bright pink, in an umbel; July-Aug. Fr egg-shaped, purple when ripe. In and by shallow fresh water. T, southern.

Frog-Bit Family   Hydrocharitaceae

8* FROGBIT *Hydrocharis morsus-ranae.* Hairless floating perennial, with bronzy-green, *kidney-shaped* lvs. Fls 3-petalled, white with a yellow spot, 20 mm; July-Aug. Still fresh water. T. **8a Water Chestnut** *Trapa natans* □ (Trapaceae; dicotyledonous) is rooted, with toothed quadrangular lvs and all-white 4-petalled fls; June-July. F, G.

9* WATER SOLDIER *Stratiotes aloides.* Floating perennial, submerged except at flg time, when rosettes of *spine-toothed* lanceolate lvs rise to surface to reveal 3-petalled white 30-40 mm fls, stamens and styles on separate plants; June-Aug. Still fresh water. T.

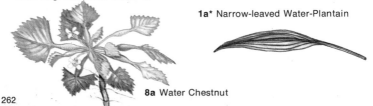

1a* Narrow-leaved Water-Plantain

8a Water Chestnut

Lily Family   Liliaceae

pp. 264-72. Lvs undivided. Fls with three petals and three sepals, often the same colour and so appearing 6-petalled.

1* LILY OF THE VALLEY *Convallaria majalis.* Low/short patch-forming hairless perennial. Lvs *broad* elliptical, two on each stem, nearly opposite. Fls creamy white, *fragrant,* bell-shaped, drooping in a one-sided spike; May-June. Fr a red berry. Drier woodland. T.

2* MAY LILY *Maianthemum bifolium.* Low/short patch-forming hairless perennial; unbranched. Lvs *heart-shaped,* shiny, two on each stem, not opposite. Fls white, very small, fragrant, with prominent stamens, in a spike; May-July. Fr a red berry. Woods. T, rare in B.

3* SNOWDON LILY *Lloydia serotina.* Low slender hairless perennial; unbranched. Lvs grasslike, almost thread-like. Fls white, *purple-veined,* 20 mm, bell-shaped, *solitary;* May. Grassy places, rocks in hills or mountains. B, rare.

4* KERRY LILY *Simethis planifolia.* Low/short hairless perennial. Lvs narrow, grass-like, often *curled,* greyish, all at the base. Fls white, purplish outside, 20 mm, 6-petalled, in a loose head; May-June. Heathy and rocky places. Ireland (rare), F.

5 ST BERNARD'S LILY *Anthericum liliago.* Slender medium hairless perennial, *little branched.* Lvs all at base, long and grasslike. Fls white, starlike, with conspicuous yellow stamens, 30-50 mm, 6-petalled, in a loose *elongated* cluster; bracts long, lanceolate; May-June. Dry grassland, especially in hills. F, G, S.
5a *A. ramosum* □ has smaller (25 mm) fls with shorter bracts in widely branched clusters; June-July. F, G.

6 FALSE HELLEBORINE *Veratrum album.* Medium/tall stout hairy perennial. Lvs broad elliptical, *pleated* lengthwise, unstalked, arranged in *threes.* Fls yellowish-green, often white inside, starlike, 6-petalled, unstalked, in branched spikes; July-Aug. Grassy places in hills and mountains. F, G, S.

7* SCOTTISH ASPHODEL *Tofieldia pusilla.* Low/short, rather slender hairless perennial; unbranched. Lvs flat, sword-shaped, usually *all at base,* occasionally a few small up stem. Fls greenish-yellow, with a *green* 3-lobed bract, in a *short* spike; June-Aug. Wet places on mountains and in tundra. B, F, S.

8 GERMAN ASPHODEL *Tofieldia calyculata.* Short hairless perennial. Lvs flat, sword-shaped, both at base and smaller *up stem.* Fls greenish-yellow, occasionally reddish, with both a green lanceolate and a *papery* 3-lobed bract, in an *elongated* or tight cluster; July-Aug. Damp grassy places, bogs. G, S (Gotland).

Lily Family continued on p. 266.

Pipewort Family   Eriocaulaceae

9* PIPEWORT *Eriocaulon aquaticum.* Short/medium hairless aquatic perennial; unbranched. Lvs *all submerged* at base of stem, narrow, pointed, translucent. Fls white, with *grey* bracts, in flat button-like heads, on lfless stems projecting a few cm out of the water; July-Sept. Shallow still fresh water, bare wet ground. B (W Ireland, Inner Hebrides).

2* May Lily

5a *Anthericum ramosum*

Lily Family (contd.)

1* FRITILLARY *Fritillaria meleagris.* Short hairless perennial. Lvs grass-like, greyish, up the stem. Fls *bell-like,* solitary, nodding, *varying* from dull purple to creamy white, chequered with darker blotches; Apr-May. Damp meadows, forming extensive colonies when these are not ploughed. T, southern, scattered.

2* BOG ASPHODEL *Narthecium ossifragum.* Low/short hairless perennial. Lvs sword-shaped, iris-like, all from roots, in rather flattened tufts, often tinged orange. Fls *orange*-yellow, starlike, 6-petalled, in spikes on lfless stems; July-Aug. Fr deep orange. Bogs, wet heaths. T, western.

3* MARTAGON LILY *Lilium martagon.* Medium/tall hairless perennial; stem red-spotted. Lvs elliptical, dark green, in whorls up stem. Fls in a close cluster, dull pink spotted dark purple, with petals *curled back,* making the pinkish stamens prominent; June-July. Woods, scrub, mountain grassland. T, southern, but (B). **3a* Pyrenean Lily** *L. pyrenaicum* □ is shorter and foetid, with narrower alternate lvs and larger dark-spotted yellow fls. (B), from the Pyrenees. **3b** *L. bulbiferum* has erect, sometimes yellowish fls with petals not curled back. With (ssp. *bulbiferum*) or without (ssp. *croceum*) bulbils at base of lvs. G.

4* YELLOW STAR OF BETHLEHEM *Gagea lutea.* Low slender perennial, almost hairless. Lvs like Bluebell (p. 270), but solitary, yellower green, more hooded at tip, and with 3-5 prominent ridges on the back, 5-12 mm broad. Fls yellow, 6-petalled, with a *green band* on the outer side of each petal, starlike; in an umbel-like cluster with a pair of large lflike bracts at its base on an otherwise lfless stem; Mar-May. Open woods, scrub, grassland. T.

5 LEAST GAGEA *Gagea minima.* Low slender perennial. Lvs, 1 narrow basal, 1-2 broader opposite on stem below inflorescence. Fls yellow, with *pointed* petals, long-stalked, 1-7 in a cluster; Mar-May. Woods, grassy places, on lime. F, G, S, eastern. **5a Meadow Gagea** *G. pratensis* has a broader basal lf, always 2 broad opposite stem lvs, and 2-6 greener yellow fls with blunter petals. F, G.

6* WELSH STAR OF BETHLEHEM *Gagea bohemica. Very short* perennial. Lvs, 2 *thread-like, wavy* basal; 2 lanceolate alternate on stem. Fls *bright* yellow, usually solitary, sometimes 2-3; Jan-Mar. Rocks, dry stony places. B, F, G,

7 BELGIAN GAGEA *Gagea spathacea.* Short perennial with two very narrow lvs. Fls yellow-green, 2-5 on each spike, with hairless stalks and very broad bracts; Apr-May. Damp grassy places in woods. F, G. **7a** *G. arvensis* has downy stems and two more shaggily hairy bracts; Feb-Apr. Not on lime; also in cultivated places. F, G, S.

8* WILD TULIP *Tulipa sylvestris.* Slender short/medium hairless perennial. Lvs linear, grass-like, hooded at tip, with *no* prominent veins, grooves or ridges, all from roots. Fls yellow, tinged green or red outside, like a slender garden tulip, fragrant, solitary on lfless stalk; Apr-May. Dry grassy and bare places. (B), F, G.

On p. 269:

10* ROUND-HEADED LEEK *Allium sphaerocephalon* (see p. 269) Medium perennial. Lvs semi-cylindrical, grooved, hollow. Fls pink-purple, with *protruding stamens,* no bulbils and short papery bracts; in *globular heads;* June-Aug. Grassy places, rocks, dunes. B (rare), F, G.

3a

3

1

2

4

5

6

8

7

Lily Family (contd.)

LEEKS and GARLICS *Allium.* Hairless unbranched bulbous perennials, smelling *strongly* of garlic or onion. Fls usually bell-shaped, often mixed with bulbils, in a head or loose umbel, enclosed by one or more papery bracts.

1* RAMSONS *Allium ursinum.* Short/medium *carpeting* perennial; stem triangular. Lvs *broad* elliptical, all from roots, recalling Lily of the Valley (p. 264), but brighter green and garlic-smelling. Fls white, *starlike,* in an umbel; Apr-June. Woods, shady banks. T.

2* THREE-CORNERED LEEK *Allium triquetrum.* Short/medium perennial; stem 3-sided. Lvs linear, sharply keeled, all at base. Fls white, with a narrow *green line* on the back of each petal, no bulbils, in a drooping one-sided umbel-like head; Apr-June. Hedge-banks, stream-sides, shady places. (B), from Mediterranean. **2a* Few-flowered Leek** *A. paradoxum* is often colonial and has a single narrower lf and bulbils mixed with the smaller fls, which have unequal stalks; Apr-May. (B, F, G), from the Caucasus.

3* CROW GARLIC *Allium vineale.* Medium stiff perennial. Lvs semi-cylindrical, grooved, hollow. Fls greenish or pinkish, stamens *protruding,* bract shorter than fls; either long-stalked in a loose umbel, usually mixed with bulbils, or in a tight head of bulbils only; June-Aug. Grassy and cultivated places. T. **3a* Field Garlic** *A. oleraceum* □ has flatter lvs, looser umbels of fls with unequal stalks and stamens not protruding; form with bulbils only is rare. T.

4* CHIVES *Allium schoenoprasum.* Short/medium *tufted* perennial; a herb used for flavouring. Lvs cylindrical, hollow, greyish, *all from roots.* Fls purple-pink, short-stalked, in a head, stamens not protruding, papery bract short; June-Aug. Grassy places, rocks. T, rare in B.

5* KEELED GARLIC *Allium carinatum.* Medium perennial. Lvs linear, grooved, keeled beneath, up to middle of stem. Fls pink-purple, *blunt*-petalled, with *protruding* purple stamens and long bracts; long-stalked in a loose umbel, usually mixed with bulbils; July-Aug. Open woods, rocks, grassy places. T, but (B). **5a* Rosy Garlic** *A. roseum* has lvs only at base of stem, pinker fls, stamens not protruding, and shorter bracts; Apr-June. (B), from Mediterranean.

6* WELSH ONION *Allium fistulosum.* Medium/tall perennial. Lvs cylindrical, hollow. Fls yellowish-white, numerous, in umbel on stem swollen and hollow *above middle;* June-Sept. Escape from cultivation, occasionally naturalised. (T). **6a Onion** *A. cepa* is taller and stouter, with greenish-white fls on a stem swollen and hollow below middle.

7* SAND LEEK *Allium scorodoprasum.* Tall stiff perennial. Lvs linear, rough-edged. Fls red-purple, stamens *not* protruding, long-stalked in an umbel, mixed with *bulbils;* June-Aug. Sandy, grassy and cultivated places, hedge-banks. T. **7a** *A. rotundum* has lvs not finely toothed and a denser head of darker purple fls without bulbils. F, G.

8* WILD LEEK *Allium ampeloprasum.* Tall stout perennial, *to 2 m.* Lvs linear, greyish, keeled, finely toothed. Fls pale purple to whitish, with yellow anthers, slightly protruding stamens, and papery bract falling before flg; in large compact umbels *50-70 mm* across, sometimes with a few small bulbils; July-Aug. Hedge- and other banks, rocks, often near the sea. B (rare), F. **8a* Babington's Leek** *A. babingtonii* has loose umbels with fewer fls, numerous bulbils and sometimes also secondary umbels. B (rare). **8b* Cultivated Leek** *A. porrum* has broader lvs, white fls, reddish anthers and a persistent green bract. Escape from cultivation. (T).

9 GERMAN GARLIC *Allium senescens.* Short/medium perennial; stem angled. Lvs variable, linear, sometimes *twisted,* all from roots. Fls rose-pink, numerous in a dense umbel on a lfless stem; June-July. Dry, often grassy or sandy places. F, G, S, south-eastern. **9a** *A. angulosum* has keeled lvs. Marshy places. G. **9b** *A. suaveolens* and **9c** *A. strictum* have bright purple fls and round stems; 9b in damp meadows, 9c on rocks. G (rare). **9d** *A. victorialis* has broader lvs than 9c and yellowish-white fls. G, southern, rare.

10* ROUND-HEADED LEEK *Allium sphaerocephalon*: see p. 266.

Lily Family *(contd.)*

1* BLUEBELL *Endymion non-scriptus.* Short hairless carpeting perennial. Lvs linear, keeled, with a hooded tip, all from roots. Fls azure blue, occasionally white, elongated bell-shaped, fragrant, in a *long* one-sided spike, drooping at the tip, on a lfless stem; anthers creamy; Apr-June. Fr egg-shaped, seeds black. Woods, scrub, hedge-banks, also on sea cliffs and mountains. B, F, G.
1a* Spanish Bluebell *E. hispanicus* □ is stouter, with broader lvs, larger paler fls in an erect not one-sided spike, and blue anthers; rather later flg. (B, F), from S W Europe.

2* SPRING SQUILL *Scilla verna.* Low/short hairless perennial. Lvs narrow, grass-like, often curly, *all from roots,* appearing before fls. Fls sky-blue, rarely white. 6-petalled, starlike, with bluish *bracts,* in a cluster on a lfless stem; *Apr-June.* Grassy places, especially near the sea. B, F, S.

3 ALPINE SQUILL *Scilla bifolia.* Low/short hairless perennial. Lvs narrow lanceolate, usually two, less often 3-5, *all on stems.* Fls bright blue, rarely pink or white, 6-petalled, starlike, with *no* bracts, in a loose cluster; Mar-Aug. Woods, scrub, grassland, also on mountains. F, G.

4* AUTUMN SQUILL *Scilla autumnalis.* Low/short hairless perennial. Lvs narrow, grass-like, *all from roots,* appearing *after* fls. Fls purplish-blue, starlike, with *no* bracts, in a cluster on a lfless stem; *Aug-Oct.* Dry grassy places, often near the sea. B, F.

5* GRAPE HYACINTH *Muscari atlanticum.* Short hairless perennial. Lvs *1-3 mm* wide, *grooved* and semi-cylindrical, rather limp, all from roots. Fls dark blue, *egg-shaped,* with small whitish petal-lobes, in a tight elongated head on a lfless stem; Apr-May. Dry grassy and cultivated places. B (rare), F, G.
5a *M. neglectum* is larger overall with more fls. G, southern, rare.

6 SMALL GRAPE HYACINTH *Muscari botryoides.* Short hairless perennial. Lvs 3-7 mm wide, grass-like, broader *at tip,* all from roots. Fls pale blue-violet, *globular,* with small whitish petal-lobes, in a loose, more conical head than 5, on a lfless stem; Mar-May. Grassy places, woods. F, G, S. **6a* Tassel Hyacinth** *M. comosum* □ is larger and has broader lvs and larger spikes of longer stalked fls, with a "tassel" of purple sterile fls at the top. (B), F, G, southern.

7* MEADOW SAFFRON *Colchicum autumnale.* Short hairless perennial. Lvs *oblong-lanceolate,* bright green, in clumps, appearing in spring and dying *before* flg time. Fls pale rosy mauve, rarely white, *crocus-like,* on long weak white stalks (which are actually tubular and part of the fl), solitary; *Aug-Sept.* Fr egg-shaped, appearing in spring. Differs from Autumn Crocus (p. 276) both in its lvs and in its *orange* anthers and six stamens. Woods, damp meadows. T, southern.

Lily Family continued on p. 272.

Yam Family   Dioscoreaceae

8* BLACK BRYONY *Tamus communis.* Hairless perennial climber to 4 m, twining clockwise; no tendrils (unlike the completely unrelated White Bryony (p. 152)). Lvs *heart-shaped,* dark green, shiny, alternate. Fls yellow-green, tiny, *6-petalled,* in loose, sometimes branched spikes, stamens and styles on separate plants; May-Aug. Fr a red berry. Woods, scrub, hedges. B, F, G.

6a* Tassel Hyacinth, fruits × ½

7* Meadow Saffron, fruits × ½

271

Lily Family *(contd.)*

1* COMMON SOLOMON'S SEAL *Polygonatum multiflorum*. Medium patch-forming hairless perennial; stems rounded, arching. Lvs elliptical, alternate, along the stems. Fls greenish-white, bell-shaped, waisted, unscented, in hanging *clusters* of 1-3 at base of lvs; May-June. Fr a *blue-black* berry. Woods, scrub. T. **1a* Angular Solomon's Seal** *P. odoratum* □ has stems angled and fls fragrant, 1-2 together and not waisted; on lime. **1b*** *P.* × *hybridum,* the hybrid between 1 and 1a, is commonly grown in gardens and sometimes escapes. (T). **1c** *Streptopus amplexifolius* is like 1a but fls small, star-like, on long stalks. Mountain woods. G, southern, rare.

2* WHORLED SOLOMON'S SEAL *Polygonatum verticillatum*. Medium/tall hairless perennial. Lvs narrow lanceolate, in *whorls* of 4-5 up the stem. Fls white, tipped green, bell-shaped, unscented, 1-3 together at each whorl; June-July. Fr a *red* berry. Woods in hill and mountain districts. T, rare in B.

3* HERB PARIS *Paris quadrifolia*. Low/short hairless colonial perennial. Lvs pointed oval, unstalked, in a *whorl* of usually four (less often 3-8) together at top of otherwise lfless stem. Fls yellow-green, 4-6-petalled, starlike, with prominent stamens, solitary; May-June. Fr a black berry. Woods, on lime. T.

4* SPIKED STAR OF BETHLEHEM *Ornithogalum pyrenaicum*. Medium/tall hairless perennial; unbranched. Lvs linear, grass-like, greyish, all from roots, usually withered by flg time. Fls *greenish-white* or yellowish, 6-petalled, starlike, in a *stalked spike* on a lfless stem; May-July. Woods, hedge-banks, grassland. B, F, G.

5* DROOPING STAR OF BETHLEHEM *Ornithogalum nutans*. Short/medium hairless perennial. Lvs linear, grass-like, grooved and with a central *white stripe,* all from roots. Fls white, with a green stripe on the back of each petal, bell-shaped, nodding, in a one-sided spike on a lfless stem; Apr-May. Grassy and cultivated places. (T), southern, from S Europe.

6* COMMON STAR OF BETHLEHEM *Ornithogalum umbellatum*. Low/short hairless perennial. Lvs linear, grass-like, limp, grooved and with a central *white stripe,* all from roots. Fls white, with a green stripe on the back of each petal, 6-petalled, starlike, in an *umbel-like* cluster, on a lfless stem; May-June. Grassy and cultivated places. T, southern.

7* WILD ASPARAGUS *Asparagus officinalis*. Medium/tall perennial, hairless; well branched. Lvs (actually reduced stems) short, *needle-like,* in tufts. Fls yellowish or greenish-white, bell-shaped, 1-2 together at base of branches; stamens and styles on separate plants; June-Aug. Fr a red berry. Grassy and waste places; often by the sea, when may be prostrate, with fleshy greyish 'lvs' and male fls tinged red at base. Widely cultivated. T.

8* BUTCHER'S BROOM *Ruscus aculeatus*. Medium evergreen hairless bush. Lvs (actually flattened branches) oval, ending in a sharp *spine.* Fls whitish or greenish, tiny, 6-petalled, solitary or paired on upper surface of 'lvs; stamens and styles on separate plants; Jan-Apr. Fr a red berry. Woods, scrub, hedge-banks. B, F.

fruits of:

1a* Angular Solomon's Seal

2* Whorled Solomon's Seal

1* Common Solomon's Seal

273

Daffodil Family Amaryllidaceae.

Bulbous perennials, hairless. Lvs linear, all from roots. Fl buds enclosed in a sheath (spathe), which splits at flg; petals in two rings, each usually of three petals or petal-like sepals each. Fr a capsule.

1* WILD DAFFODIL *Narcissus pseudonarcissus.* Medium perennial. Lvs greyish, grooved. Fls yellow, the erect *trumpet-like* inner ring as long as but darker yellow than the spreading outer ring, nodding, solitary on lfless stem; Mar-Apr. Woods, grassland; numerous garden forms, including double ones, escape and appear on waysides and waste ground. T, southern. **1a* Tenby Daffodil** *N. obvallaris* has erecter uniformly deep yellow fls, the trumpet longer than the outer petals. B (Pembrokeshire), rare.

2* PRIMROSE PEERLESS *Narcissus × biflorus.* Probably originally a natural hybrid between 1 and 3. Fls *paired,* fragrant, the short trumpet deep yellow with a whitish crisped margin, the outer petals creamy or whitish; Apr-May. Widely naturalised in grassy places. (B, F).

3* POET'S NARCISSUS *Narcissus poeticus.* Medium perennial. Lvs greyish, grooved. Fls with a short yellow trumpet, the margin *red and crisped,* and white outer petals, fragrant, nodding, solitary; Apr-May. Grassy places. (B, F, G), from Mediterranean. **3a** *N. stellaris* has lvs less than 5 mm wide. G, southern, rare.

4* SPRING SNOWFLAKE *Leucojum vernum.* Short perennial. Lvs *bright* green. Fls white tipped green, bell-shaped, nodding, petals 20-25 mm, *solitary* or occasionally paired on lfless stems; anthers orange; Feb-Mar. Damp woods, copses, meadows. B (rare), F, G.

5* SUMMER SNOWFLAKE *Leucojum aestivum.* Short/medium *tufted* perennial. Lvs bright green. Fls like 4 but smaller (petals 15-20 mm) and 3-6 in a *cluster* on unequal stalks; Apr-May. Damp grassy places, by fresh water. B (rare), F, G.

6* SNOWDROP *Galanthus nivalis.* Low/short perennial. Lvs *greyish,* narrower than 4 and 5, grooved, keeled. Fls white, inner petals tipped green, outer row green on the back, bell-shaped, nodding, petals 20-25 mm long, *solitary* on lfless stems, anthers green; Jan-Mar. Damp woods, shady streamsides, meadows, and extensively naturalised. B, F, G.

Arum Family Araceae

7* LORDS AND LADIES *Arum maculatum.* Low/short hairless perennial. Lvs bluntly *arrow-shaped,* dark green, often *dark-spotted,* all from roots, appearing in January. Fls tiny, male above female, in dense whorls topped by the conspicuous purple finger-like spadix, fls and spadix both being enveloped by the even more conspicuous pale green hooded spathe, whose base conceals the fls; Apr-May. Fr a spike of bright orange berries. Woods, copses, shady banks. B, F, G. **7a* Large Cuckoo Pint** *A. italicum* ☐ has larger, more triangular lvs with creamy veins, appearing in autumn, and a yellow spadix. B, F. **7b* Bog Arum** *Calla palustris* ☐ is aquatic, with unspotted heart-shaped lvs, a shorter greenish spadix in a smaller white spathe, and red berries; June-Aug. Swamps, lake-sides. T, but (B).

N.B. **Adderstongue** *Ophioglossum vulgatum* ☐ and **Moonwort** *Botrychium lunaria* ☐ are two rather atypical ferns, whose spathe-like fronds respectively oval and pinnate, together with their spadix-like spike of spores, make them look not unlike diminutive arums.

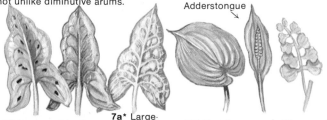

Adderstongue

7* Lords and Ladies   **7a*** Large Cuckoo Pint   **7b*** Bog Arum   Moonwort

Iris Family Iridaceae

IRISES *Iris*. Hairless perennials, often showing conspicuous rhizomes (underground stems) above ground. Lvs sword-shaped (except 5), flat, mostly from roots. Fls showy, with three spreading, sometimes bearded, outer petals (falls), three more or less erect and twisted inner ones (standards), all narrow at the base, and three large, almost petal-like stigmas with branched tips (crests). Fl buds enclosed in a sheath (spathe) which splits at flg time. Fr a 3-sided capsule.

N.B. Illustrations for this page are painted at half natural size.

1 SIBERIAN IRIS *Iris sibirica*. Slender tufted perennial. Lvs narrow, grass-like. Fls blue-violet, 60 mm, with an *all-brown* spathe. Grassy places. F, G, (S), southeastern.

2* PURPLE FLAG *Iris versicolor*. Medium/tall; little branched. Lvs broader than 3, slightly greyish and without raised midrib. Fls pale *pinkish*-purple, blade of falls as long as haft, crests nearly *white*, 1-3 together; June-July. Seeds dark brown. Wet places. (B, rare), from N America.

3* YELLOW IRIS *Iris pseudacorus*. Tall; stems branched. Lvs with a raised midrib. Fls *yellow*, 80-100 mm across, 1-3 together; June-Aug. Seeds brown. Marshes, by fresh water. T.

4* GARDEN IRIS *Iris germanica*. Stout medium/tall; branched. Lvs *greyish*. Fls fragrant, falls blue-violet with *yellow* beard, standards deep lilac, up to 100 mm across, 2-3 together; spathe green and brown; May-June. Grassy places, waysides. (T), from Mediterranean. **4a** *I. aphylla* is shorter, with lvs longer than stems. G, eastern. **4b** *I. sambucina* has lvs smelling strongly of Elder (p. 226) and fls darker. (G, southern).

5* BUTTERFLY IRIS *Iris spuria*. Medium; little branched. Lvs rather narrow, *unpleasant*-smelling when crushed. Fls pale blue-violet, falls with rounded blade half as long as haft and keel of haft *yellow*, crests violet, 40-50 mm across, unstalked, 1-3 together; May-June. Damp grassland and chalk downs. (B, rare), F, G.

6* STINKING IRIS *Iris foetidissima*. Medium, tufted. Lvs dark green, smelling *sickly sweet* when crushed. Fls grey-purple, occasionally yellowish, 80 mm across, 2-3 together; May-July. Seeds bright orange. Open woods, scrub, hedge-banks, sea cliffs and dunes. B, F.

7* GLADIOLUS *Gladiolus illyricus*. Medium/tall hairless perennial. Lvs sword-shaped, all from roots. Fls red-purple, 6-petalled, 3-8 in a spike on a lfless stem; June-July. Grassy and heathy places, scrub, marshes. B (rare), F. Several similar species occur as garden escapes. **7a** *G. palustris* has fls marked with white. G. **7b* Montbretia** *Crocosmia × crocosmiiflora* is shorter and patch-forming with a one-sided spike of smaller orange fls; July-Aug. A man-made hybrid widely established on cliffs and waste ground. (B, F).

8* SAND CROCUS *Romulea columnae*. Low hairless perennial, to 5 cm. Lvs threadlike, *curly*, all from roots. Fls pale purple, greenish outside, much smaller than 9 *(10 mm* across) and opening only in full sunshine; Apr. Sandy grassland, dunes. B (rare), F.

9* SPRING CROCUS *Crocus albiflorus*. Low hairless perennial. Lvs linear, grooved, keeled, with a *white midrib*, tufted. Fls like garden crocuses, purple, white or both, solitary on lfless stalk; styles orange; Mar-Apr. Grassy places. (B, F), G. **9a* Autumn Crocus** *C. nudiflorus* has lvs appearing in spring and dying before fls appear in Sept-Oct. Cf. Meadow Saffron (p. 270), which has 6 (not 3) stamens and much broader lvs. From S W Europe.

10* BLUE-EYED GRASS *Sisyrinchium bermudiana*. Low/short hairless perennial. Lvs linear, tufted, all from roots. Fls blue with a yellow centre, 6-petalled, *starlike*, 2-4 together on a *winged* lfless stem, opening only in sunshine; July. Grassy places. B, (F, G).

277

Orchid Family   Orchidaceae

pp. 278-90. Unbranched perennials, hairless or almost so, except *Epipactis* (p. 288). Lvs undivided, untoothed, often linear/lanceolate, keeled and rather fleshy. Fls of rather diverse shape, but two-lipped, the lower lip often developed to appear the predominant part of the fl; often spurred; in a stalked spike with a lflike bract at base of each fl. Fr egg-shaped to cylindrical.

1* LADY'S SLIPPER *Cypripedium calceolus.* Short/medium perennial. Lvs broad lanceolate, strongly ribbed, pale green. Fls maroon with a large hollow *yellow lip,* spotted red inside, solitary, the largest orchid fl of the region; May-June. Woods, scrub, in hill and mountain districts; mainly on lime. T, rare in B.

2* BEE ORCHID *Ophrys apifera.* Short/medium perennial. Lvs elliptical, pointed. Fls with sepals *pink,* petals *green* and lip red-brown, *rounded at tip,* patterned to appear like the rear of a small bumblebee apparently visiting the fl; June-July. Grassland, open scrub, dunes; only on lime. B, F, G. **2a* Wasp Orchid** *O. apifera* var. *trollii* □ has a point to the lip (in 2 this point is bent back under the lip) and also lacks the pale U-shaped mark usually found on the bee orchid's lip.

3* LATE SPIDER ORCHID *Ophrys fuciflora.* Shorter than 2 and differs in having both petals and sepals pink and a more elaborate pattern on the lip, supposedly resembling a spider, also a *heart-shaped* appendage at the tip; June-July. Grassy places; on lime. B (rare), F, G.

4* EARLY SPIDER ORCHID *Ophrys sphegodes.* Shorter than 2 and differs in having both sepals and petals *yellow-green,* and an *H- or X-shaped* mark on the lip, sometimes also likened to the Greek letter pi (π); Apr-June. Grassy places; on lime. B, F, G.

5* FLY ORCHID *Ophrys insectifera.* Slender short/medium perennial. Lvs narrow elliptical, shiny. Fls brown with green sepals, the narrow lip 3-lobed, the middle lobe forked, and with a *bluish patch* at its base; May-June. Woods, scrub, fens, grassy places; on lime. T.

6 BLACK VANILLA ORCHID *Nigritella nigra.* Low/short perennial. Lvs linear, pointed. Fls *blackish-purple,* rarely pink, spurred, *vanilla-scented,* in a tight head; June-Aug. Mountain meadows. F, G, S.

7 CALYPSO *Calypso bulbosa.* Low/short perennial. Lf elliptical, markedly veined, *solitary.* Fl purplish-pink with a large *hollow* yellow-blotched pale pink lip, solitary; May. Marshes, marshy woodland. S, a high northern plant.

5* Fly Orchid

2* Bee Orchid   **2a*** Wasp Orchid   **3*** Late Spider   **4*** Early Spider
                                                              Orchid   Orchid

279

Orchid Family (contd.)

1* EARLY PURPLE ORCHID *Orchis mascula.* Short/medium perennial. Lvs narrow oblong, pointed, usually *blotched* purplish-black. Fls purple, less often pink-purple or white, smelling of tom-cats; petals with one sepal forming a hood, the two other sepals *erect,* lip shallowly 3-lobed, spur long; Apr-June. Woods, scrub, grassland. T. **1a Pale-flowered Orchid** *O. pallens* has broader lvs and pale yellow fls. On lime. F, G, eastern.

2* GREEN-WINGED ORCHID *Orchis morio.* Short perennial. Lvs narrow oblong, pointed, unspotted. Fls purple, pink or white, fragrant, hood formed by all sepals (which are *green-veined)* and petals, lip shallowly 3-lobed, spur long; fewer and in more open spike than 1; May-June. Grassland, open scrub. T.

3* LADY ORCHID *Orchis purpurea.* Medium/tall perennial. Lvs broad oblong, shiny, unspotted, mostly in a basal rosette. Fls shaped like a *manikin,* all sepals and petals forming the hood or 'head', and the long lip being lobed narrowly to make the 'arms' and *broadly* to make the 'legs'; hood *dark* purple, and lip *pale* pink, fragrant, spur short; May-June. Woods, scrub; on lime. B(rare), F, G.

4* MILITARY ORCHID *Orchis militaris.* Short/medium perennial. Lvs broad oblong, unspotted. Fls manikin-shaped, like 3, but with a small central lobe between the 'legs', pinkish-grey with darker pink markings, especially inside the hood; spur short; May-June. Wood edges, scrub, grassland; on lime. B(rare), F, G.

5* BURNT ORCHID *Orchis ustulata.* Low/short perennial. Lvs oblong, pointed, unspotted. Fls like miniatures of 3, the hood *dark maroon* at first, becoming paler, the lip less markedly 4-lobed, white with pink dots, short-spurred, fragrant; May-June. Grassland; on lime. T, southern. **5a Toothed Orchid** *O. tridentata* ☐ has the 'legs' of the manikin poorly developed, the central lobe of the lip being itself shallowly 3-lobed, and having purplish-red spots. G.

6* MONKEY ORCHID *Orchis simia.* Short/medium perennial. Lvs broad oblong, shiny, unspotted. Fls like 4 but more delicate, and 'arms' and 'legs' of manikin *much thinner* with the central tooth accentuated to form the monkey's 'tail'; May-June. Grassland, open scrub; mainly on lime. B(rare), F, G.

7 BUG ORCHID *Orchis coriophora.* Short perennial. Lvs narrow lanceolate. Fls red-brown, all sepals (which are *green-veined)* and petals joined in a hood, the lip 3-lobed and dark wine-purple tipped red and green, spur short, *smelling of bed-bugs* (but a more southern darker red form with white bracts and a large central lobe to the lip is vanilla-scented); Apr-June. Damp grassland. F, G.

8 LOOSE-FLOWERED ORCHID *Orchis laxiflora.* Medium perennial. Lvs lanceolate, *unspotted.* Fls purple, in a rather loose spike, hood formed from petals and one sepal, the two other sepals *spreading,* lip often only *2-lobed,* spur short; bracts and stem often purplish; May-June. Damp meadows. F, G. **8a** *O. palustris* ☐ has paler fls with tip clearly 3-lobed. G, S.

5a Toothed Orchid   **8a** *Orchis palustris*

Orchid Family *(contd.)*

1* PYRAMIDAL ORCHID *Anacamptis pyramidalis.* Medium perennial. Lvs narrow lanceolate, unspotted. Fls *bright pink* or rose-purple, sepals spreading, petals hooded, lip 3-lobed, spur very *long,* slender and curved, often foxy-smelling, in a densely flattened *pyramidal* spike; June-Aug. Dry grassland, dunes; mainly on lime. T, southern.

MARSH AND SPOTTED ORCHIDS *Dactylorhiza.* Fls with sepals spreading outwards and forwards like a bird's wing, and a broad lip. Frequently hybridise with each other.

2* EARLY MARSH ORCHID *Dactylorhiza incarnata.* Short/medium perennial. Lvs oblong, keeled, *hooded at tip, yellowish-green,* unspotted. Fls very variable in colour, pink, purple, brick-red or yellow; lip with a small central tooth and looking very narrow because the sides soon fold *backwards,* spur straight; May-July. Damp grassland, marshes, fens, dune slacks. T. **2a* Flecked Marsh Orchid** *D. cruenta* is shorter and has stem, lvs and bracts streaked dark purple on both sides, and fls dark red-purple with untoothed lip. T, rare. **2b* Pugsley's Marsh Orchid** *D. traunsteineri* has narrower, usually spotted lvs and fewer dark rose-purple fls with a 3-lobed lip less markedly folded back but strongly marked with dark purple.

3* SOUTHERN MARSH ORCHID *Dactylorhiza praetermissa.* Short/medium perennial. Lvs dark green, not hooded at tip, unspotted. Fls rose-purple, lip broad, streaked and spotted darker, with a small central tooth, the sides *not* folding backwards; spur short, stout; bracts and upper stem often purplish; June-July. Damp grassland, marshes, fens, dune slacks. T, but rare in S. **3a* Northern Marsh Orchid** *D. purpurella* is shorter, with occasionally spotted lvs and wine-purple fls with lip either untoothed or shallowly 3-lobed, and a tapering spur. B, G, S.

4* BROAD-LEAVED MARSH ORCHID *Dactylorhiza majalis.* Medium/tall perennial. Lvs broad oblong, *bluish-green,* sometimes dark-spotted. Fls lilac-purple (bright purple or reddish in mountains), the shallowly 3-lobed lip with darker lines and spots, spur short; May-July. Damp meadows, marshes, on lime. T.

5* COMMON SPOTTED ORCHID *Dactylorhiza fuchsii.* Short/medium perennial. Lvs narrow lanceolate, keeled, usually *dark-spotted.* Fls pale pink, pale purple or white, dotted and lined with crimson or purple, lip markedly 3-lobed, spur long; June-Aug. Grassy places, open scrub; on lime. T. **5a* Heath Spotted Orchid** *D. maculata* □ has the lip with a wavy edge and a single small tooth. Heathy places on acid soils, including bogs.

6 ELDER-FLOWERED ORCHID *Dactylorhiza sambucina.* Short perennial. Lvs pale green, shiny, unspotted. Fls either *yellow* with a purple-spotted lip and pale *green* bracts or *red-purple* with a purple lip and *reddish* bracts, often mixed in same colony; lip very shallowly 3-lobed; not fragrant; Apr-June. Mountain grassland. F, G, S.

7* FRAGRANT ORCHID *Gymnadenia conopsea.* Medium perennial. Lvs narrow lanceolate, unspotted. Fls in an elongated spike, *pale* purplish-pink, *fragrant* (sometimes clove-scented), with spreading sepals, hooded petals, 3-lobed lip and *long* slender spur; June-July. Grassland, marshes, fens. T. **7a** *G. odoratissima* has narrower greyish lvs, pale pink or white vanilla-scented fls in a shorter tighter spike, and a shorter stouter spur. F, G, S.

Orchid Family *(contd.)*

1* MAN ORCHID *Aceras anthropophorum.* Short perennial. Lvs oblong lanceolate, keeled, shiny. Fls *greenish-yellow,* in long dense spikes, petals and sepals forming a hood and lip shaped as a manikin (cf. Lady Orchid, 280/3), often tinged red-brown, unspurred; May-June. Grassland, scrub; on lime. B, F, G.

2* LIZARD ORCHID *Himantoglossum hircinum.* Short/medium greyish perennial. Lvs oblong lanceolate, soon withering. Fls purplish grey-green, petals and sepals forming a hood, the *long straplike lip* 3-lobed, spur short; smelling strongly of billy goat; in a tangled, often elongated spike; June-July. Grassy banks, scrub, wood edges, dunes. B (rare), F, G.

3* DENSE-FLOWERED ORCHID *Neotinea intacta.* Short perennial. Lvs oblong elliptical, blunt, either unspotted or with *lines* of brownish spots. Fls either greenish-white or pink, sepals and petals hooded, lip shortly 3-lobed, spur short, vanilla-scented, in a short *tight* spike; Apr-June. May be more conspicuous in fr. Sandy grassland, dunes, limestone rocks. B (W Ireland, Isle of Man).

4 FALSE MUSK ORCHID *Chamorchis alpina.* Low perennial, hard to detect among other vegetation. Lvs linear, *grasslike.* Fls green tinged with violet or purple, sepals and petals hooded, lip yellowish, obscurely 3-lobed, unspurred, few in the spike; July-Aug. Damp grassland in *mountains*; on lime. S.

5* MUSK ORCHID *Herminium monorchis.* Low perennial. Lvs oval, yellow-green, 2-3 near *base* of stem. Fls greenish-yellow, petals and sepals spreading, lip 3-lobed, the middle lobe much the longest, spur very short or none; *honey-scented* (not musky); June-July. Grassland; on lime; in Britain prefers dry turf, on Continent damp places. T.

6* FROG ORCHID *Coeloglossum viride.* Low/short perennial. Lvs lanceolate, narrower up stem. Fls yellow-green tinged red-brown, sepals and petals hooded, lip strap-shaped, *forked* near tip with a *tooth* in the middle, spur very short; June-July. Grassland, often on mountains. T.

7* LESSER TWAYBLADE *Listera cordata.* Low slender perennial, sometimes only 5 cm. Lvs *heart-shaped,* shiny, *a single pair* at base of stem. Fls reddish-green, petals and sepals half-hooded, lip obscurely 3-lobed, the middle lobe conspicuously forked, unspurred; June-Aug. Moors, bogs, coniferous woods, often half-hidden among heather and moss or under bracken. T.

8* COMMON TWAYBLADE *Listera ovata.* Medium perennial. Lvs *broad oval,* not shiny, *a single pair* at base of stem. Fls yellow-green, petals and sepals half-hooded, lip forked, unspurred, in a long spike; May-July. Woods, scrub, grassland. T.

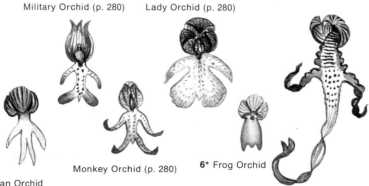

Military Orchid (p. 280) Lady Orchid (p. 280)

Monkey Orchid (p. 280) **6*** Frog Orchid

1* Man Orchid

2* Lizard Orchid

Orchid Family *(contd.)*

1* GREATER BUTTERFLY ORCHID *Platanthera chlorantha.* Medium perennial. Lvs broad elliptical, shiny, *a single pair* at base, much smaller and lanceolate up stem. Fls greenish-white, *vanilla-scented,* sepals and petals spreading, lip long, narrow and undivided, spur very long, in a rather loose spike; pollen masses diverging; June-July. Woods, open scrub, grassland. T. **1a* Lesser Butterfly Orchid** *P. bifolia* □ is shorter and smaller, with less broad basal lvs and pollen masses parallel. Also on moors and in marshes. **1b Northern Butterfly Orchid** *P. hyperborea* □ is shorter, with no basal lvs but 3-6 lanceolate lvs up stem; and small greenish or greenish-yellow, short-spurred fls. Damp peaty or grassy moors. Iceland only. **1c Arctic Butterfly Orchid** *P. obtusa* ssp. *oligantha* has only one basal lf and one very small stem lf, with few small short-spurred greenish-white fls. Arctic heaths on lime. S.

2* RED HELLEBORINE *Cephalanthera rubra.* Short/medium perennial, slightly tinged purple. Lvs narrow lanceolate, sometimes almost grasslike in deep shade. Fls bright *purple-pink,* not opening widely, with an unspurred whitish lip; June-July. Woods, especially of beech, and scrub; on lime. T, but rare in B.

3* WHITE HELLEBORINE *Cephalanthera damasonium.* Short/medium perennial. Lvs *broad lanceolate,* slightly bluish-green, narrower up stem. Fls creamy white, the yellow base of the unspurred lip usually hidden because fl scarcely opens, bracts conspicuous and *lflike;* May-July. Woods, especially of beech, shady banks; on lime. T.

4* NARROW-LEAVED HELLEBORINE *Cephalanthera longifolia.* Slenderer than 3. Lvs longer, *narrower,* parallel-sided and darker green. Fls pure white, opening more widely to reveal a much smaller orange spot at base of lip, with bracts too *small* to appear lflike; May-June. Woods. T.

5* WHITE FROG ORCHID *Pseudorchis albida.* Short perennial. Lvs narrow oblong, keeled, shiny, unspotted. Fls *creamy white, fragrant,* sepals and petals hooded, lip 3-lobed, short-spurred; May-June. *Upland grassland.* T.

6* AUTUMN LADY'S TRESSES *Spiranthes spiralis.* Low perennial. Lvs pointed oval, bluish-green, in a basal rosette which withers before flg, leaving only a few scale-like lvs up stem. Fls white, fragrant, the lip greenish, unspurred, with a frilled recurved margin, *spirally* in a *single* row up the spike; Aug-Sept. *Dry* grassland. B, F, G. **6a Summer Lady's Tresses** *S. aestivalis* □ has no basal rosette but linear-lanceolate lvs up the stem, fls all-white, slightly larger and less fragrant, and longer bracts; July-Aug. Marshy or boggy places, often by streams. Extinct in B; rare in F, G.

7* IRISH LADY'S TRESSES *Spiranthes romanzoffiana.* Low/short perennial. Lvs linear-lanceolate up the stem, no basal rosette. Fls white or creamy white, greenish at the base, fragrant, unspurred, in *three spiral* rows on a spike consequently much broader than 6; bracts conspicuous; Aug-Sept. *Wet* grassy places, peat marshes, bogs. B, western.

8* CREEPING LADY'S TRESSES *Goodyera repens.* Low/short creeping perennial, with *runners.* Lvs pointed oval conspicuously net-veined, in basal *rosettes,* smaller and scale-like up the stems. Fls white, fragrant (to some people unpleasantly so), in a *spiral* spike, the unspurred lip neither frilled nor recurved; July-Aug. *Woods,* especially coniferous, mainly in hills and mountains. T.

1

1a

1b

2

3

4

5

6a

6a

6

6a

7

8

287

Orchid Family *(contd.)*

HELLEBORINES *Epipactis.* Lvs usually broad elliptical, pointed. Fls in one-sided stalked spikes; lip unspurred, in two parts joined by a narrow waist, the base cupped, the tip more or less triangular.

1* MARSH HELLEBORINE *Epipactis palustris.* Medium perennial. Lvs narrow elliptical, keeled, folded. Fls with sepals purple or purple-brown, petals *crimson and white,* and lip white with cup crimson-streaked and tip with a yellow spot and a *notched frilly margin;* also sometimes fls yellowish-white with lip white; July-Aug. Fr downy. *Fens,* marshes, dune slacks. T.

2* DARK RED HELLEBORINE *Epipactis atrorubens.* Medium perennial, *tinged red-purple;* stem downy. Lvs in two rows up stem. *Fls dark wine-red,* fragrant, opening fully; June-July. Fr downy. Woods, grassy places, rocks, dunes; on lime. T.

3* BROAD-LEAVED HELLEBORINE *Epipactis helleborine.* Medium/tall perennial, often tinged purplish. Lvs sometimes short and almost rounded, alternating *spirally* up stem. Fls very, variable, greenish-yellow to purple-red, unscented, opening fully, lip appearing rounded because point of tip recurved *underneath;* July-Sept. Woods, shady banks, less often open hillsides, dune slacks. T. **3a** *E. microphylla* □ has much smaller lvs and fls greenish with purple margin, the lip showing a small point. Woods. F, G, S, rare. **3b** *E. muelleri* □ has pale green fls with the inside of the lip cup reddish. F, G.

4* NARROW-LIPPED HELLEBORINE *Epipactis leptochila.* Short/medium perennial, slightly downy, tinged *yellow.* Lvs in *two rows* up stem. Fls green, usually tinged white or yellow, *drooping,* sometimes not opening fully, the lip with cup mottled red and tip extended to a *point;* July-Aug. Woods, especially of beech, dunes; on lime. B, F, G.

5* VIOLET HELLEBORINE *Epipactis purpurata.* Medium perennial, purple-tinged; stems usually in a *clump.* Lvs narrow elliptical, spirally up stem. Fls pale greenish-white (purplish-green outside), lip with cup usually mottled violet and tip recurved; bracts often longer than fls; Aug-Sept. Woods, often of beech. B, F, G.

6* DUNE HELLEBORINE *Epipactis dunensis* agg. Medium perennial; stem downy above, purplish below. Lvs in *two rows* up stem. Fls pale yellow-green, *not fully opening,* lip with cup mottled red inside and tip recurved as in 3; lowest bracts *longer than fls;* June-July. Dune slacks, pine plantations. B (Lancashire, Anglesey).

7* GREEN-FLOWERED HELLEBORINE *Epipactis phyllanthes.* Medium perennial; stems hairless. Lvs variable, in *two rows* up stem. Fls green, often tinged yellow or purple, drooping, *scarcely opening,* lip cup usually whitish; lowest bracts longer than fls; July-Sept. Woods, dunes. B, F, G. **7a** *E. confusa* □ is slenderer and has fls opening fully with a pink lip. G, S.

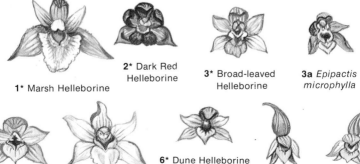

1* Marsh Helleborine

2* Dark Red Helleborine

3* Broad-leaved Helleborine

3a *Epipactis microphylla*

3b *E. muelleri*

5* Violet Helleborine

6* Dune Helleborine

7* Green-flowered Helleborine

7a *Epipactis confusa*

Orchid Family *(contd.)*

1* BIRDSNEST ORCHID *Neottia nidus-avis.* Short/medium perennial; stem *honey-coloured*. No lvs but scale-like bracts on stem. Fls honey-brown like stem, with a somewhat sickly fragrance, petals and sepals small, lip large, unspurred, *2-lobed* (Broomrapes (p. 222) have a 3-lobed lip); May-July. Woods, especially of beech, in deep shade. T.

2* CORALROOT ORCHID *Corallorhiza trifida.* Low/short perennial; stem yellow-green. No lvs but a few scale-like bracts on stem. Fls yellow-green or yellowish-white, sepals and petals spreading, lip obscurely 3-lobed, unspurred and marked with *red;* June-July. Woods, especially coniferous, in mountains, dune slacks. T.

3* GHOST ORCHID *Epipogium aphyllum.* Low perennial; stem *mauvish-yellow*. No lvs but a few tiny pale brown bracts. Fls pale mauve and pale yellow, petals and sepals spreading, lip recurved, undivided, spur whitish, solitary or 2-3 in a spike; June-Sept, flg most freely after a wet spring. Broad-leaved woods, especially of beech, in deep shade, often in damp places. T, but rare in B.

4* BOG ORCHID *Malaxis paludosa.* Low perennial, sometimes only 5 cm high. Lvs oval, *2-4* up the stems, usually with tiny bulbils on their margins. Fls yellow-green, sepals and petals spreading, the narrow lanceolate lip twisted round to appear at the top; July-Sept. Bogs and other wet *acid* places, almost always among *sphagnum* moss. T.

5 ONE-LEAVED BOG ORCHID *Malaxis monophyllos.* Low/short perennial. Lf *solitary*, more or less oval. Fls greenish-yellow, sepals and petals spreading, the narrowly triangular unspurred lip twisted round to appear at the top; July. Damp places, usually among *sphagnum* moss. G, S, rare.

6* FEN ORCHID *Liparis loeselii.* Low/short perennial. Lvs broad lanceolate, shiny, yellow-green, a *single pair* at base of stem. Fls yellow-green, petals and sepals spreading, the broad unspurred lip twisted round to appear at the top; June-July. *Fens,* marshes, dune slacks; *on lime.* T.

7 VIOLET BIRDSNEST ORCHID *Limodorum abortivum.* Medium perennial; stem violet. No lvs but scales on stem. Fls violet with a yellow tinge, *40 mm* across, petals and sepals spreading, lip undivided, spurred; May-July, flg most freely after a wet spring. Woods, especially coniferous, shady banks. F, G.

Ghost Orchid, Coralroot Orchid and the two Birdsnest Orchids, together with Yellow Birdsnest (p. 172) are all *saprophytes*, plants without chlorophyll (green colouring matter) which feed on rotting vegetation with the aid of a fungus partner. They are not to be confused with the other group of plants with no chlorophyll, the Broomrapes and Toothworts (p. 222), which are parasites on the roots of other plants.

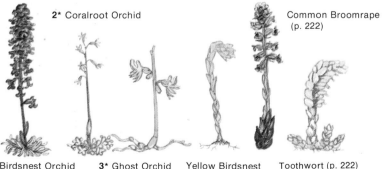

2* Coralroot Orchid

Common Broomrape (p. 222)

1* Birdsnest Orchid   **3*** Ghost Orchid   Yellow Birdsnest (p. 172)   Toothwort (p. 222)

Key to Waterweeds

(based on leaf-shape)

Includes all flowering plants in the book that grow actually in or under the water, but not waterside plants that just "get their feet wet". Those with conspicuous flowers are also included in the Main Key (p.13). The present key is based on leaf-shape, subdivided into floating or emergent and submerged leaves. Aquatic plants can be very confusing because their leaves may be of different shapes, depending on whether they are above the surface or submerged, and if submerged also on the speed of the water in which they grow.

Leaves oval or pointed oval

floating: Floating Water-plantain, Water Chestnut 262; Purslanes 294; Pondweeds 298

submerged: Purslanes 294; Pondweeds 298; Ivy-leaved Duckweed 300

Leaves pointed oblong

floating: Amphibious Bistort 40

submerged: Pondweeds 298

Leaves roundish

floating: Water-lilies 66; Fringed Water-lily 182; Frogbit 262; Duckweeds 300

submerged: Yellow Water-lily 66

Leaves short and narrow

emergent or floating: Blinks 42; Water Tillaea 102; Waterworts, Mudworts 294; Marestail, Water Starwort 296

submerged: Blinks 42; Water Tillaea 102; Waterworts, Mudworts 294; Marestail, Water Starwort, Canadian Waterweed 296; Naiads 300

Leaves quill-like, tufted (usually submerged)

Water Soldier 262; Pipewort 264; Awlwort, Shoreweed, Water Lobelia, Quillworts 294

Leaves long and grass like

floating: Bur-reeds 302

submerged: Ribbon-leaved Water-plantain 262; Tape-grass 296; Pondweeds 298; Tasselweeds, Eel-grasses 300; Bur-reeds 302

Compound leaves

emergent or floating: Water Crowfoots 70; Bogbean 72; Watercress 90; Fool's Watercress 166

submerged: Water Crowfoots 70; Watercress 90; Fool's Watercress, River Water Dropwort 166; Water Violet 180; Bladderworts 294; Hornworts, Water Milfoils 296

The waterweeds in the key opposite are those with inconspicuous flowers that are largely or completely submerged. Their text and illustrations are on pages 294-303. Water plants with conspicuous flowers will be found in the main text and in the main key on p.13. The diagram below shows the zonation of water plants from those growing on damp or marshy ground, through those that like to "keep their feet wet" to those that are completely submerged. The order in which plants are shown does not necessarily indicate the precise order in which they will occur at any given pond or lake, nor, of course, is the selection of plants in any way comprehensive.

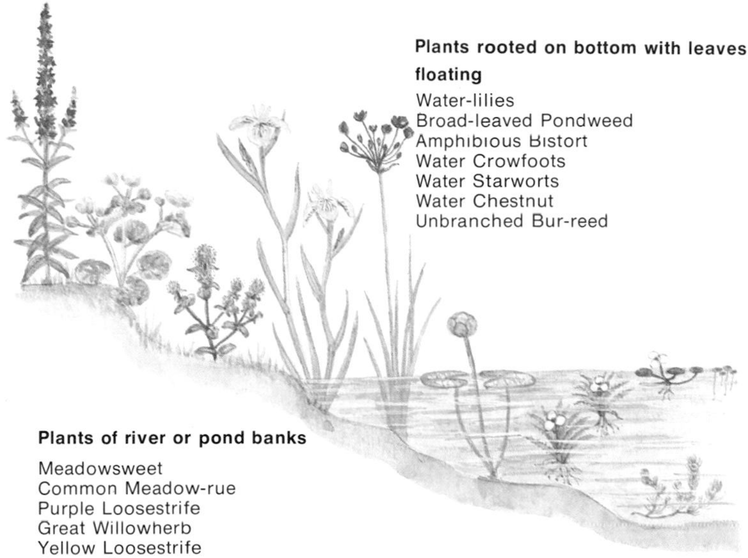

Plants rooted on bottom with leaves floating

Water-lilies
Broad-leaved Pondweed
Amphibious Bistort
Water Crowfoots
Water Starworts
Water Chestnut
Unbranched Bur-reed

Plants of river or pond banks

Meadowsweet
Common Meadow-rue
Purple Loosestrife
Great Willowherb
Yellow Loosestrife
Orange Balsam
Gipsywort
Water Mint
Marsh Woundwort
Hemp Agrimony

Plants of bare mud at water's edge

Celery-leaved Buttercup
Ivy-leaved Crowfoot
Brooklime, Shoreweed
Water Purslane
Common Water Starwort
Creeping and Marsh Yellowcress
Bur Marigolds, Blinks
Mudwort

Plants growing with their "feet wet"

Watercress, Monkey Flower
Fool's Watercress, Flowering Rush
Yellow Iris, Water Plantains
Water Dock, Arrowhead
Greater Spearwort, Great Yellowcress
Water Speedwell, Great Water Parsnip
Lesser Water Parsnip

Plants rooted on bottom with leaves submerged

Pondweeds
Water Starworts
Water Crowfoots (especially fast streams)
Water Milfoils
Hornwort
Awlwort
Water Lobelia
Pipewort

Plants not rooted on bottom with leaves floating

Frogbit
Duckweeds
Water Soldier (summer)
Bladderworts

Plants not rooted on bottom with leaves submerged

Ivy-leaved Duckweed
Water Soldier (winter)

Waterweeds

BLINKS *Montia fontana* (Portulacaceae): see p. 42.

1 AWLWORT *Subularia aquatica* (Cruciferae: see p. 84). Low annual, usually submerged. Lvs narrow, grass-like, pointed, all in a basal tuft. Fls small, white, with 4 narrow petals, in a *stalked spike*; June-Sept. Fr egg-shaped. Shallow pools in hills and mountains, not on lime. T, northern.

Crassula aquatica, C. vaillantii (Crassulaceae): see p. 102.

2* SIX-STAMENED WATERWORT *Elatine hexandra* (Elatineaceae). Low, creeping, often submerged annual. Lvs narrow oval, linear when submerged, *opposite*, reddening. Fls tiny, pinkish, *stalked, solitary* at base of lvs, with 3 *petals*, 3 sepals and 6 *stamens*. In or by shallow, often peaty fresh water, sometimes drying out in summer. T. **2a Three-stamened Waterwort** *E. triandra* has unstalked fls with only 3 stamens. F, G, S. **2b Eight-stamened Waterwort** *E. hydropiper* has un- or scarcely stalked fls with 4 petals, 4 sepals and 8 stamens. **2c Whorled Waterwort** *E. alsinastrum* may be perennial and is larger, with lvs whorled and fls like 2b. F, G.

3* WATER PURSLANE *Lythrum portula* (Lythraceae, see p. 154). More or less prostrate creeping hairless annual; stems often reddish. Lvs oval, tapering to stalk, opposite, sometimes reddish. Fls small, with 6 pinkish or no petals, *solitary*, unstalked at base of lvs; July-Aug. Fr roundish. In and by shallow water, often drying out in summer, and on damp ground in woods, not on lime. T. **3a** *L. borysthenicum* is usually hairy and erect, with lvs unstalked and usually alternate. F.

4* HAMPSHIRE PURSLANE *Ludwigia palustris* (Onagraceae, see pp. 152-54). Prostrate hairless perennial; stems reddish. Lvs pointed oval, glossy, red-veined, opposite. Fls petalless, green tinged red, unstalked *in pairs* at base of upper lvs; June-July. Fr surmounted by sepals. In and by fresh water. B (rare), F, G.

5* MUDWORT *Limosella aquatica* (Scrophulariaceae, see p. 212). Prostrate hairless annual with creeping *runners*. Lvs narrow, elliptical, long-stalked, all in a basal rosette. Fls small, pink or whitish, 4-5 petalled, stalked, *solitary*, at base of lvs; June-Sept. Fr oval, on down-turned stalks. Wet mud. T. **5a* Welsh Mudwort** *L. australis* has smaller, more linear lvs, larger fls, white with an orange tube, and rounder fr. Muddy estuaries. B (Wales), rare.

6* GREATER BLADDERWORT *Utricularia vulgaris* agg. (Lentibulariaceae, see p. 222). Floating *rootless* insectivorous perennial. Lvs *pinnately* divided into numerous thread-like green segments, with small *bladders* and tiny bristles. Fls *yellow*, *2-lipped*, with a blunt spur, on lfless stalks above the water; June-Aug. Fr roundish. Still fresh water, usually fairly deep. T. **6a* Intermediate Bladderwort** *U. intermedia* agg. is smaller, with toothed lflets and bladders not on the feathery green lvs but on colourless ones, often sunk in mud; spur of fl pointed. Prefers peaty water. **6b* Lesser Bladderwort** *U. minor* agg. is like 6a but has untoothed lflets, bladders on green lvs as well, and smaller pale yellow fls with a short blunt spur.

7* SHOREWEED *Littorella uniflora* (Plantaginaceae, see p. 224). Low hairless perennial, with rooting *runners*, often forming submerged swards. Lvs *semi-cylindrical*, flat on one side, all in a basal tuft. Fls whitish, 4-petalled, solitary, not usually appearing under water; male stalked and with *prominent stamens* separate from unstalked female; June-Aug. In and by shallow fresh water, not on lime. T.

8* WATER LOBELIA *Lobelia dortmanna* (Campanulaceae, see p. 230). Hairless perennial with submerged rosettes of untoothed linear lvs. Fls *pale lilac, 2-lipped*, in lfless spikes above surface; July-Aug. Lakes with acid water. T, western.

9* QUILLWORT *Isoetes lacustris* (Isoetaceae). Not a flowering plant, but its submerged tuft of cylindrical pointed dark green lvs could be confused with 1, 5a, 7 or 8 or Pipewort (p. 264). Spores in capsules at base of lvs. Still water in hill districts. T, northern. **9a* Spring Quillwort** *I. setacea* is smaller with paler green, more often spreading or recurved lvs. Also in lowlands.

PIPEWORT *Eriocaulon aquaticum*: see p. 264.

Waterweeds *(contd.)*

1* RIGID HORNWORT *Ceratophyllum demersum* (Ceratophyllaceae). Submerged rootless perennial. Lvs dark green, rather stiff, *forked*, with linear toothed segments, in whorls. Fls minute, *solitary*, unstalked, with 8 or more linear petals, whitish male and green female at base of separate lf-whorls; July-Sept. Fr a nut with 3 spines. Still or slow fresh water. T. **1a* Soft Hornwort** *C. submersum* has laxer, paler green lvs, and fr with 1 or no spines. Less frequent, often in brackish water. N.B. Some Stoneworts (Characeae) could be confused with Hornworts, but they are flowerless plants, Algae, recognisable by their grey, or translucent green colour and their foetid smell.

2* SPIKED WATER MILFOIL *Myriophyllum spicatum* (Haloragaceae). Submerged rooted perennial. Lvs *pinnate*, feathery, with linear segments, *in whorls* usually of 4. Fls small, pinkish or red, with 4 or no petals, male (anthers yellow) above female, in a *whorled spike* projecting above water; bracts small, undivided; June-Sept. Fr globular. Still or slow, fresh or brackish water. T. **2a* Whorled Water Milfoil** *M. verticillatum* is easily told by its conspicuous pinnate bracts; fls greenish-yellow. **2b* Alternate Water Milfoil** *M. alterniflorum* is slenderer and has a shorter spike of red-streaked yellow fls, the upper solitary or paired; not in limy water. 2a and 2b are only in fresh water. N.B. Some long trailing Water Crowfoots (p. 74) have rather similar feathery lvs.

3* MARESTAIL *Hippuris vulgaris* (Hippuridaceae). Unbranched perennial, erect but largely submerged in fast streams. Lvs *strap-like*, dark green, *in whorls*, longer and paler green when submerged. Fls petalless, pink, male and female separate, not or very shortly stalked at base of lvs; June-July. Fr a greenish nut. Fresh water. T. Not to be confused with Horsetails, which are not flowering plants.

4* COMMON WATER STARWORT *Callitriche stagnalis* agg. (Callitrichaceae). A difficult group, lf-shape varying with presence or depth of water and the ripe fr needed for sure identification often hard to find. Annuals/perennials, submerged or partly submerged in fresh water, or on wet mud. Lvs oval or elliptical, narrower and *opaque* when submerged, opposite, often forming a *floating rosette*. Fls tiny, green, petalless, male (anthers yellow) and female separate, solitary or paired at base of lvs; Apr-Sept. Fr 4-lobed. T. **4a* Intermediate Water Starwort** *C. hamulata* agg. has the submerged lvs linear with a deeply notched tip and the terrestrial lvs mostly elliptical. **4b* Autumnal Water Starwort** *C. hermaphroditica* agg. has all lvs submerged and transparent, linear but narrowing towards the notched tip. Fls May-Sept.

WATER-THYMES (Hydrocharitaceae, see p. 262). Submerged perennials. Fls 3-petalled, male and female on separate plants, floating on long thread-like stalks on the surface, but male fls (which often break and float away) are rare in Europe. Still and slow-moving fresh water.

5* CANADIAN WATERWEED or WATER-THYME *Elodea canadensis*. Lvs oblong, 10-15 mm long, *2 mm* wide, *blunt* at the tip, minutely toothed, dark green, in overlapping *whorls of 3*, or paired at base of stem, not collapsing when taken from water. Fls white or pale *purplish*; May-Sept. (T) from N America. **5a* Greater Water-thyme** *E. ernstiae* (*E. callitrichoides*) is larger, with lvs longer, to 25 mm, and tapering to a fine point, and larger white fls. (B,F) from S America. **5b* Nuttall's Water-thyme** *E. nuttallii* has paler green, longer and narrower lvs, the margins curving to a pointed tip, in whorls of 3 more widely spaced up the stem, and collapsing when taken from the water; fls smaller, white. (B,F,G) from N America, spreading fast and replacing 5 in some areas.

6* ESTHWAITE WATERWEED *Hydrilla verticillata*. Lvs in whorls of 3-8, *linear*, *distinctly toothed*. Fls transparent, red-streaked. Deep water. B (Ireland, rare; no longer at Esthwaite Water, Cumbria).

7* CURLY WATER-THYME *Lagarosiphon major*. Lvs narrow, densely crowded *spirally* up the stem, markedly *curved*, minutely toothed. Fls *pinkish*. (B,F) from S Africa.

8* LARGE-FLOWERED WATER-THYME *Egeria densa* see p. 302.
9* TAPE-GRASS *Vallisneria spiralis* see p. 302.

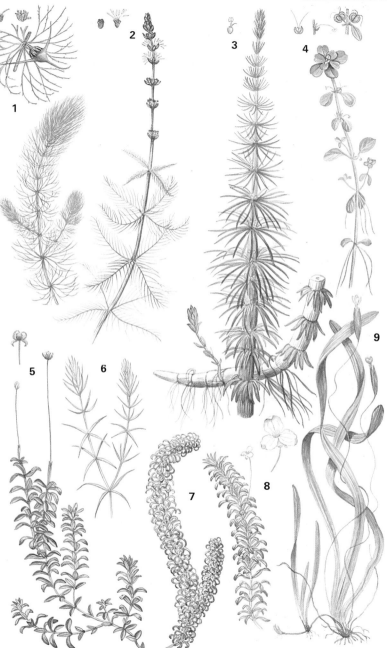

Pondweed Family   Potamogetonaceae

Submerged perennials with emergent fl-spikes, some with floating lvs. Often hard to identify because lf-shape varies with depth and speed of water; lvs alternate, with stipules or sheaths at their base. Fls small, green, petalless, 4-sepalled, in stalked spikes at base of lvs. Fr with 4 nutlets. Usually in still or slow fresh water. Hybrids are frequent, but only 2 × 3 has fr. A simplified treatment is given, based on lf and stipule characters.

1* BROAD-LEAVED PONDWEED *Potamogeton natans*. Floating lvs *broad oval*, to 120 mm long, with conspicuous stipules, *apparently jointed* at junction with long stalk (unlike 1a-1c); submerged lvs linear. Fl-stalks long, stout; May-Sept. T. **1a* Bog Pondweed** *P. polygonifolius* has floating lvs smaller and often reddish and submerged lvs lanceolate; often in bogs, flushes or very shallow water, not on lime. **1b* Fen Pondweed** *P. coloratus* ☐ has all lvs broad, translucent, net-veined and often reddish, the submerged ones longer and narrower; usually in limy water. **1c* Loddon Pondweed** *P. nodosus* ☐ has all lvs broad and opaque, the submerged ones longer, narrower and net-veined. B, F, G.

2* VARIOUS-LEAVED PONDWEED *Potamogeton gramineus*. May have oval *floating* lvs to 70 mm, with conspicuous stipules; submerged lvs narrow, *wavy-edged*, unstalked, to 80 mm long. Fl-stalks thickened at top; June-Aug. Not in limy water. T. **2a* Reddish Pondweed** *P. alpinus* ☐ is often tinged red and may have narrow net-veined floating lvs; submerged lvs longer. **2b* American Pondweed** *P. epihydrus* has flattened stems, narrower floating lvs and much longer linear submerged lvs. B, rare.

3* SHINING PONDWEED *Potamogeton lucens*. Lvs *all* submerged, *oblong lanceolate*, stalked, translucent, *wavy-edged*, shiny; stipules conspicuous. Fl-stalks thickened at top; June-Sept. Fairly deep water, especially on lime. **3a* Long-stalked Pondweed** *P. praelongus* ☐ has greener, less wavy, blunter-tipped unstalked lvs and longer fl-stalks not thickened at top. Cf. 2 and 2a when without floating lvs. Not especially on lime.

4* PERFOLIATE PONDWEED *Potamogeton perfoliatus*. Lvs *all* submerged, dark green, oval, *clasping* the stem; stipules small. Fls June-Sept. T.

5* CURLED PONDWEED *Potamogeton crispus*. Stems 4-angled. Lvs *all* submerged, narrow oblong, markedly *wavy*, translucent, finely toothed; stipules small. Fl spikes small; May-Sept. T.

6* SMALL PONDWEED *Potamogeton berchtoldii*. Stems slightly flattened. Lvs *all* submerged, narrow and *grass-like*, 3-veined, bluntly notched with a *tiny awn* and a *conspicuous gland* at the junctions; stipules small, *open*. Fl spike small; June-Sept. Not in limy water, unlike 6a-6d. T. **6a* Lesser Pondweed** *P. pusillus* ☐ is very similar, with a still tinier awn, no nodal glands, and stipules closed; prefers limy water, also in brackish water. **6b* Shetland Pondweed** *P. rutilus* ☐ has more finely pointed lvs, smaller nodal glands and closed stipules. Rare in B. **6c* Hair-like Pondweed** *P. trichoides* has almost hair-like lvs with a prominent midrib but no awn and smaller nodal glands. **6d* Flat-stalked Pondweed** *P. friesii* has stems more flattened, broader 5-veined lvs and stipules closed. 6b-6d in still water only.

7* BLUNT-LEAVED PONDWEED *Potamogeton obtusifolius*. Lvs grass-like, but twice as broad as 6, with *bluntly* pointed 3-veined lvs, *conspicuous* nodal glands and *open* stipules. Fl-stalks rather short; June-Sept. T. **7a* Sharp-leaved Pondweed** *P. acutifolius* ☐ has sharply pointed lvs with 3 main and many smaller veins and no nodal glands; still water only. **7b* Grasswrack Pondweed** *P. compressus* has blunter 5-veined lvs, no nodal glands and much longer fl-stalks.

8* FENNEL PONDWEED *Potamogeton pectinatus* see p. 302.

9* OPPOSITE-LEAVED PONDWEED *Groenlandia densa*. All lvs submerged, *opposite*, narrow oval, unstalked, sometimes very short. Fls in heads; May-Sept. Also in fast and lime-rich water. T.

299

1* HORNED PONDWEED *Zannichellia palustris* (Zannichelliaceae). Slender submerged perennial. Lvs *thread-like*, usually opposite. Fls tiny, petalless, green, male (1 stamen) and female separate, *solitary*, short- or unstalked at base of lvs; May-Aug. Fr beaked. Still or slow fresh or brackish water. T.

2* SLENDER NAIAD *Najas flexilis* (Najadaceae). Slender submerged perennial. Lvs narrow, grass-like, *in whorls* of 2-3, translucent. Fls small, green, 2-lipped, unstalked, *1-3 together* at base of lvs, male and female separate; Aug-Sept. Lakes. B, G, S.

3* HOLLY-LEAVED NAIAD *Najas marina*. Submerged perennial, with stems slightly and lvs strongly *toothed*. Fls like 2. Fresh or brackish water. B (rare), F, G. **3a** *N. minor* has lvs narrower and less sharply toothed and male and female lvs on same plant. F, G.

4* SPIRAL TASSELWEED *Ruppia cirrhosa* (Ruppiaceae). Slender submerged perennial. Lvs narrow, *thread-like*, dark green, alternate or opposite. Fls petalless, green, with 2 anthers, rising to surface on long stalks, *in pairs*, the individual stalks much shorter than the common one, which coils in fr; July-Sept. Shallow flowing salt or brackish water. T. **4a Beaked Tasselweed** *R. maritima* has paler green lvs, longer fl-stalks and the common stalk not coiling. Still water.

Eel-Grass Family   Zosteraceae

The only flowering plants that grow completely submerged by the sea.

5* EEL-GRASS *Zostera marina*. Perennial. Lvs more than 200 mm long, *grass-like*, 5-10 mm wide, blunt or pointed at tip, with 3 or more parallel veins. Fls minute, petalless, green, male and female separate, in *branched spikes*; June-Sept. Muddy, sandy and gravelly sea shores, near and below spring low water mark. T.

6* NARROW-LEAVED EEL-GRASS *Zostera angustifolia*. Like Eel-grass, but has lvs 150-300 mm long and 2-3 mm wide, 1-3-veined and notched at tip when mature. Muddy shores, mainly from half to low tide mark. B, G, S. **6b* Dwarf Eel-grass** *Z. noltii* has 1-veined lvs less than 200 mm long and 1 mm wide, and fl spikes unbranched.

Duckweed Family   Lemnaceae

Small floating annuals, with no stems or apparent lvs, but a frond or thallus, which may be a modification of one or both. Fls minute, petalless, rare in B; male and female separate but within same sheath. Still water, often making a complete green carpet. N.B. Water Fern *Azolla filiculoides* from tropical America (B, F, G), which appears like a feathery duckweed and turns red in autumn, is a fern, not a flowering plant.

7* ROOTLESS DUCKWEED *Wolffia arrhiza*. The smallest European flowering plant. with a rootless egg-shaped frond $\frac{1}{2}$-*1 mm* across. Fl not found in N W Europe. B, F, G.

8* IVY-LEAVED DUCKWEED *Lemna trisulca*. Fronds floating just *below surface*, lanceolate, translucent, 7-12 mm, *at right angles* to each other, each with one root; fls May-July. T.

9* COMMON DUCKWEED *Lemna minor*. Fronds *rounded*, 1$\frac{1}{2}$-4 mm, nearly flat on both sides, 1-5-veined, floating *on surface*; one root. Fls May-July. T. **9a* Greater Duckweed** *Spirodela polyrhiza* □ has fronds 5-8 mm, 5-11-veined, often purplish beneath, with a tuft of roots. **9b* Fat Duckweed** *L. gibba* □ has fronds 3-5 mm and markedly swollen. Also in brackish and polluted water. **9c Lesser Duckweed** *L. miniscula* (*L. valdiviana*) has smaller, often oblong or curved 1-veined fronds. (B, F), from Americas.

Bur-reed Family   Sparganiaceae

Aquatic perennials. Fls greenish, in rounded heads, male and female separate on same plant.

1* BRANCHED BUR-REED *Sparganium erectum. Tall,* to 150 cm or more. Lvs stout, iris-like, keeled, 3-sided near the base. Inflorescence *branched,* the heads *unstalked;* July-Aug. Freshwater *margins.*

2* UNBRANCHED BUR-REED *Sparganium emersum.* Shorter than 1, usually growing *in the water,* with both emergent and *floating* lvs, the floating ones long and *ribbon-like,* not keeled. Inflorescence *unbranched,* the lower heads *stalked;* lower bracts erect, longer than inflorescence; July-Aug. **2a** *S. gramineum* may be branched at the base of the inflorescence; lower bracts floating. S.

3* LEAST BUR-REED *Sparganium minimum.* Lvs floating, *flat, translucent.* Inflorescence short, *unbranched,* with *unstalked* flhds (only 1 male) and the lowest bract *scarcely exceeding* the inflorescence; June-July. Still water with no lime. T. **3a* Floating Bur-reed** *S. angustifolium* is intermediate between 2 and 3, with usually 2 male flhds and the lowest bract twice as long as the inflorescence; Aug-Sept. Peaty pools. **3b** *S. glomeratum* has lvs keeled, sometimes 2 male flhds and the lowest bract 3 times the length of the inflorescence. S. **3c** *S. hyperboreum* has lower female flhds stalked and the lowest bract longer than the inflorescence. 3a-3c have opaque lvs.

Bulrush Family   Typhaceae

4* BULRUSH or CATTAIL *Typha latifolia.* Tall (to 200 cm), stout, aggressively creeping perennial, forming *extensive patches.* Lvs long, flat, greyish, 10-20 mm wide. Fls in two tight contiguous spikes, 10-15 cm long, the straw-coloured male *immediately above* the much stouter chocolate-brown, *sausage-shaped* female; July-Aug. Swamps, freshwater margins, often invading open water and eliminating ponds. T. **4a** *T. shuttleworthii* has mature female spikes silvery grey. F, G.

5* LESSER BULRUSH *Typha angustifolia.* Slenderer than Bulrush and differing in its narrower lvs, 5 mm wide, and the male and female spikes being *well apart.* Similar places to 4. T. **5a Least Bulrush** *T. minima* ☐ is shorter, with still narrower lvs, 1-2 mm wide, and shorter inflorescence with male and female spikes sometimes contiguous. F, G.

6* SWEET FLAG *Acorus calamus* (Araceae, p. 274). Tall, stout, creeping perennial, to 1 m. Lvs iris-like with one or both edges *crinkled, sweet-smelling* when crushed. The tiny green fls are packed into a markedly phallic spike up the stem; June-July. By fresh water. (T), from Asia.

On p. 297.

8* LARGE-FLOWERED WATER-THYME *Egeria densa.* (See p. 297). Lvs *long,* to 15-40 mm, and up to *5 mm* wide, densely packed in whorls of 3-5, minutely toothed. Fls white, 14–20 mm across. (B, F, G) from S America.

9* TAPE-GRASS *Vallisneria spiralis.* (See p. 297). Submerged perennial. Lvs all from roots, long and *ribbon-like,* with 3-5 veins, slightly toothed at tip. Fls small, pinkish-white, male (several in a head) separate from female (solitary, with stalks twisted after flg); June-Oct. Fresh water, usually *artificially heated* canals. (B, F), increasing, from tropics.

On p. 299.

8* FENNEL PONDWEED *Potamogeton pectinatus.* (See p. 299). Lvs *all* submerged, dark green, linear, often *rounded* like a bootlace, *pointed,* with no nodal glands and *open, white-edged* stipules. Fl spikes short, interrupted in fr; June-Sept. Also in fast fresh, brackish and slightly polluted water; likes lime. **8a* Slender-leaved Pondweed** *P. filiformis* is smaller, with flattened stems, still narrower lvs with blunt tips, closed not white-edged stipules, and long-stalked fl spikes, the individual whorls still more widely spaced. Usually near the sea, often in brackish water. **8b** *P. vaginatus* has blunt-tipped lvs and stipules not white-edged. S. Usually in brackish water.

Appendix 1 Introduced Shrubs

So many ornamental or cultivated shrubs have been introduced to Europe during the past three or four centuries and have been self-sown, bird-sown or otherwise naturalised that it is not possible to deal with them in as great detail as the native trees or shrubs. Many will anyway be familiar from gardens. The following list gives brief details of the most important.

As elsewhere, an asterisk indicates a plant to be found wild in Britain.

*Fig *Ficus carica* (Moraceae). Lvs palmately lobed. Fls small, green. Fr pear-shaped.

*Shrubby Orache *Atriplex halimus* (Chenopodiaceae). Lvs broad, silvery. Fls small, green.

*Barberries *Berberis* (Berberidaceae). Fls yellow: *B. darwinii*, lvs evergreen, holly-like, fr dark purple; *B. glaucocarpa*, lvs evergreen, leathery, fr violet-blue; *B. x stenophylla*, lvs evergreen, narrow, spine-tipped, whitish beneath, hybrid so no fr; *B. wilsonae*, lvs deciduous, glaucous beneath, fr red.

*Mock orange or **Syringa** *Philadelphus coronarius* (Hydrangeaceae). Lvs lanceolate, fls large, white, fragrant; fr dry.

*Flowering currant *Ribes sanguineum* (Grossulariaceae). Lvs palmately lobed. Fls reddish-pink. Fr blue-black.

*Escallonia *Escallonia macrantha* (Escalloniaceae). Lvs oval, Fls pink.

*Pittosporum *Pittosporum crassifolium* (Pittosporaceae). Lvs oval, evergreen. Fls purple-black.

Sorbaria *Sorbaria sorbifolia* (Rosaceae). Lvs pinnate. Fls white.

*Physocarpus *Physocarpus opulifolius* (Rosaceae). Lvs roundish. Fls white.

*Spiraea *Spiraea salicifolia* agg. (Rosaceae). Lvs elliptical. Fls pink.

*Purple-flowering Raspberry *Rubus odoratus* (Rosaceae). Lvs palmately lobed. Fls pink.

*Wineberry *R. phoenicolasius*. Lvs trefoil. Fls pink.

*Rubus *R. spectabilis*. Lvs trefoil. Fls bright pink.

Roses **Rosa sempervirens, R. foetida, *R. virginiana, *R. rugosa* (Rosaceae). Lvs pinnate. Fls pink.

Quince *Cydonia oblonga* (Rosaceae). Lvs pointed oval. Fls pink.

*Cotoneasters *Cotoneaster* (Rosaceae). Large shrubs, fls white, fr red: *C. frigida*, lvs lanceolate, matt; *C. salicifolia*, lvs narrower, glossier.

*Firethorn *Pyracantha coccinea* (*Rosaceae*). Spiny. Lvs elliptical. Fls white.

*Medlar *Mespilus germanica* (Rosaceae). Sometimes a small tree and spiny. Lvs lanceolate. Fls white. Fr brown.

Aronia *Aronia melanocarpa* (Rosaceae). Lvs pointed oval, dark glossy green. Fls white. Fr black.

*Staghorn Sumach *Rhus typhina* (Anacardiaceae). Lvs pinnate. Fls greenish. Fr red.

Box Elder *Acer negundo* (Aceraceae). Lvs trefoil or pinnate. Fls green, in hanging clusters, before lvs.

*Evergreen Spindle *Euonymus japonicus* (Celastraceae). Lvs evergreen, glossy, leathery. Fls greenish. Fr pink.

*Vine *Vitis vinifera* (Vitaceae). Woody climber. Lvs palmately lobed. Fls green.

Virginia Creepers **Parthenocissus quinquefolia, P. inserta* (Vitaceae). Woody climbers. Lvs with 5 elliptical Lflets. Fls green.

*Fuchsia *Fuchsia magellanica* (Onagraceae). Lvs oval, toothed. Fls bell-shaped, purple and red.

*Rhododendron *Rhododendron ponticum* (Ericaceae). Lvs lanceolate, dark green. Fls large, purple.

*Yellow Rhododendron *R. luteum*. Similar but fls yellow.

*Shallon *Gaultheria shallon* (Ericaceae). Carpeting evergreen undershrub. Lvs broad, leathery. Fls pink, bell-shaped. Fr black.

*Prickly Heath *Pernettya mucronata* (Ericaceae). Suckers. Lvs toothed, sharp-tipped. Fls pink, bell-shaped. Fr pink.

*Lilac *Syringa vulgaris* (Oleaceae). Lvs pointed oval. Fls in spikes, lilac or white.

*Butterfly Bush *Buddleja davidii* (Buddlejaceae). Lvs lanceolate. Fls mauve, in long dense spikes.

*Duke of Argyll's Tea-tree *Lycium barbarum* (Solanaceae). Sparsely spiny. Lvs lanceolate. Fls pale purple.

*Snowberry *Symphoricarpos albus* (Caprifoliaceae). Suckering. Lvs rounded or lobed. Fls small, pink. Fr large, white.

*Flowering Nutmeg *Leycesteria formosa* (Caprifoliaceae). Lvs ace-of-spades. Bracts lf-like. Fls purplish.

Appendix 2 **Aggregates**

Where the abbreviation "agg." follows a scientific name, several closely similar species or microspecies are included under the one heading, as follows:

Salix caprea (p. 26) inc. *S. coaetanea.*
S. cinerea inc. *S. atrocinerea.*
S. nigricans inc. *S. borealis.*
S. phylicifolia inc. *S. bicolor, S. hibernica.*
S. starkeana inc. *S. xerophila.*
S. glauca inc. *S. stipulifera.*
S. lanata inc. *S. glandulifera.*
S. repens inc. *S. arenaria, S. rosmarinifolia.*
Polygonum minus (p. 40) inc. *P. foliosum.*
Atriplex hastata (p. 46) inc. *A. glabriuscula, A. longipes*
Salicornia europaea (p. 48): too numerous to list
Sagina intermedia (p. 56) inc. *S. caespitosa.*
Ranunculus montanus (p. 68) inc. *R. oreophilus.*
Fumaria officinalis (p. 78) *F. rostellata.*
F. vaillantii inc. *F. schrammii.*
Papaver radicatum (p. 80) inc. *P. dahlianum, P. laestadianum, P. lapponicum.*
Draba alpina (p. 84) inc. *D. glacialis.*
Biscutella laevigata inc. *B. apricorum, B. divoniensis, B. guilloni, B. neustriaca,* all very local in F.
Cardamine pratensis (p. 90) inc. *C. nymanii, C. palustris, C. matthiolii.*
Cakile maritima (p. 92) inc. *C. edentula.*
Cochlearia officinalis (p. 92) inc. *C. pyrenaica, C. alpina, C. micacea, C. scotica.*
Arabis hirsuta (p. 94) inc. *A. brownii.*
Draba norvegica (p. 96) inc. *D. cacuminum.*
D. daurica inc. *D. cinerea.*
Saxifraga rosacea (p. 106) inc. *S. hartii.*
Alchemilla vulgaris (p. 108); too numerous to list.
Rosa canina (p. 110): too numerous to list.
R. tomentosa inc. *R. mollis, R. scabriuscula, R. sherardii, R. villosa.*
R. rubiginosa inc. *R. agrestis, R. elliptica, R. micrantha.*
Rubus fruticosus (p. 110): too numerous to list.
Potentilla argentea (p. 114) inc. *P. neglecta.*
P. nivea inc. *P. chamissonis.*
Sorbus aria (p. 116): too numerous to list.
S. latifolia: too numerous to list.
Vicia sativa (p. 126) inc. *V. angustifolia, V. segetalis.*
Linum perenne (p. 136) inc. *L. austriacum, L. leonii.*
Geranium bohemicum (p. 138) inc. *G. lanuginosum.*
Ribes rubrum (p. 144) inc. *R. spicatum.*
Hypericum perforatum (p. 148) inc. *H. elegans.*
Epilobium montanum (p. 154) inc. *E. duriaei.*
Anthriscus sylvestris (p. 160) inc. *A. nitida.*
Monotropa hypopitys (p. 172) inc. *M. hypophegea.*
Limonium binervosum (p. 182) inc. *L. paradoxum, L. recurvum, L. transwallianum.*
Galium mollugo (p. 190) inc. *G. album.*
G. palustre inc. *G. elongatum.*
G. pumilum inc. *G. oelandicum, G. suecicum, G. valdepilis.*
G. sterneri inc. *G. normanii.*
G. fleurotii inc. *G. timeroyi.*
Myosotis laxa (p. 192) inc. *M. baltica.*
Lamium amplexicaule (p. 204) inc. *L. moluccellifolium.*

Galeopsis tetrahit (p. 204) inc. *G. bifida.*
G. angustifolia inc. *G. ladanum.*
Thymus serpyllum (p. 208): too numerous to list.
Solanum sarrachoides (p. 210): inc. *S. nitidibaccum.*
Euphrasia officinalis (p. 220): too numerous to list.
Rhinanthus minor (p. 220): too numerous to list.
Valeriana officinalis (p. 226) inc. *V. sambucifolia.*
Valerianella locusta (c. 226) inc. *V. carinata, V. dentata, V. eriocarpa, V. rimosa.*
Erigeron alpinus (p. 236) inc. *E. borealis*
Artemisia campestris (p. 242) inc. *A. borealis.*
Senecio squalidus (p. 246) inv. *S. cambrensis.*
Arctium minus (p. 248) inc. *A. nemorosum, A. pubens.*
Centaurea rhenana (p. 252) inc. *C. maculosa.*
Taraxacum spp. (p. 258): too numerous to list.
Hieracium spp. (p. 260): too numerous to list.
Epipactis dunensis (p. 288) inc. *E. youngiana.*
Utricularia vulgaris (p. 294) inc. *U. australis.*
U. intermedia inc. *U. ochroleuca.*
U. minor inc. *U. bremii.*
Callitriche stagnalis (p. 296) inc. *C. cophocarpa, C. obtusangula, C. palustris,*
    C. platycarpa.
C. hamulata inc. *C. brutia.*
C. hermaphroditica inc. *C. truncata.*
Spiraea salicifolia (p. 304) inc. *S. alba, S. douglasii, S. tomentosa.*

Plant Ecology

Habitat is often an important aid to identifying an unknown plant or in distinguishing between two similar ones, particularly in those groups such as the Umbellifers which many people find 'difficult'. It is useful, therefore, to gain an understanding of the habitat preferences of plants or, to use a now familiar word, their ecology. You will, anyway, want to know more about plant ecology out of simple curiosity as your knowledge of plant identification increases: it is fascinating to ask oneself why there should be so many kinds of, often very similar, plants and why, for example, Heath Bedstraw (*Galium saxatile*) is found on heaths and moors and acid grasslands, while its close relative Limestone Bedstraw *G. sterneri* is confined to soils formed on chalk and limestone. That, incidentally, is an excellent example of a case where professional botanists do their identification on habitat, and not by looking to see which way the prickles on the leaf margin face.

We cannot here give a proper introduction to the complex and fast-developing scientific discipline of plant ecology, but only indicate briefly some of the factors that control plant distribution, and give some idea of the sorts of plants to be found in different habitats.

First, we have to consider the range of any plant species: some are exceedingly widespread, found in a few cases on all continents, though usually thanks to the intervention of human transport. Others are very restricted, either within a large range (Snowdon Lily *Lloydia serotina*, for example, occurs in North Wales, the Alps, the Himalayas, Arctic Russia and Alaska) or literally to a single area. Really localised species, such as Lundy Cabbage *Rhynchosinapis wrightii*, which only occurs on Lundy Island in the Bristol Channel, are usually either recently evolved species (effectively isolated populations of more widespread ones), or more ancient species approaching extinction. One has to be careful in assuming that the physical factors of the environment are the only limits to the range of a species. Human activity may be equally important, and where ranges are spreading or contracting they may be hard to define.

Nevertheless, all plants are constrained by their physiological make-up to survive within certain physical limits, and prime amongst these is **temperature**. Within the area covered by this book temperature produces a north–south gradient, and the differences in the floras of Edinburgh and Paris can largely be ascribed to this. Of course altitude has almost as strong an effect as latitude, so that a plant such as Mountain Avens *Dryas octopetala* is a mountain plant in Central Europe, an upland plant in northern England (albeit confined to north-facing cliffs) but descends to sea-level in northern Scotland and western Ireland. Gardeners will know that this same plant survives happily in 'captivity' at sea-level almost anywhere in Europe, which only goes to show that in the wild many other factors must be operating.

One of these is **rainfall**, or rather the excess of rainfall over evaporation. Whereas temperature provides north–south gradient, rainfall is high in the west and low in the east, thanks to the Atlantic Ocean. The two together therefore neatly quarter our area, giving a cool moist north-west, a cool dry north-east, a warm moist south-west and warm dry south-east This is, of course, an over-simplification (most statements in ecology are) and hides the fact that the 'continental' climate of the south-eastern quadrant may have the warmest summers but also suffers very cold winters, being far from the ameliorating influence of the westerly winds. The most continental climate in England is found in the low uplands of East Anglia: a whole group of plants characteristic of Central Europe is found there and only there in Britain, Spring Speedwell *Veronica verna*, Spanish Catchfly *Silene otites* and Field Wormwood *Artemisia campestris* to name three. A much larger group of plants in our flora is known as the Atlantic element, and these are restricted (probably by sensitivity to frost) to the Atlantic seaboard of Europe. Most heathers *Erica* spp. fall into this category, and so does Gorse *Ulex europaeus*. On exposed western coasts wind can be an important influence on the growth of trees and shrubs, Blackthorn *Prunus spinosa*, for instance, being reduced from its normal 3–4 m to less than 1 m.

Even in an area so well defined that no large climatic differences occur, however, there will be obvious differences in the vegetation from place to place, and these can often be traced to differences in **soil type**. The contrast between the lush green meadows of the Yorkshire Dales and the darker moors of the central Pennines; between the Dark Peak and the White Peak in Derbyshire; or between the flower-studded grassy slopes of the North Downs and the gorse and heather-covered heaths a few miles away around Bagshot in Surrey are all the result of the presence or absence of a single mineral: limestone, which includes chalk. Soils rich in calcium carbonate – chalks and limestones – are said to be base-rich (calcium, like magnesium, is in chemical terms a base, the opposite of an acid), while soils lacking it or some other base are said to be base-poor or acid. These base-rich soils lack the toxins that characterise acid soils, and so a much richer flora flourishes on them. Many orchids are wholly confined to calcareous or other base-rich soils, including the striking Man Orchid *Aceras anthropophorum*, and many pairs of related plant species occur one on acid and the other on base-rich soils – for example, Heath and Common Spotted Orchids *Dactylorhiza maculata* (acid) and *D. fuchsii* (calcareous). This distinction is so important to the field botanist that on the map on p. 6 we have shown the major areas of limestone soils in north-western Europe.

The other major habitat distinction based on soils is that between wet and dry soils. Dry soils pose obvious problems of desiccation and the plants that grow on them have many adaptive features to minimise water loss. Where there is too much water, however, and the soil becomes waterlogged, quite different problems occur, for here it is lack of air and the presence of noxious gases that affect plants: as a result plants of wet soils are again a very distinctive group. In really waterlogged soils the dead vegetation of each year does not rot but accumulates as peat, again forming distinctive habitats with a characteristic flora.

Peatlands (or mires) may be either acid or, if the surrounding rocks are limestones, base-rich. Base-rich mires are called fens and have a marvellously rich flora. If the surrounding rocks are acid or if bog-mosses *Sphagnum* have taken hold, the peat will be acid and have a quite different and much poorer flora. Such mires are termed bogs, with sundews *Drosera* and other characteristic plants of their own.

These two soil factors – acidity and wetness – between them account for much of the habitat variation to be found in our flora, though of course in special situations, such as the coast, where salty soils occur, with their highly specialised flora, including Sea Aster *Aster tripolium*, other factors will be important. In the absence of human activity such a simple soil classification might be sufficient, but in that case much of the land would be wooded, instead of the small proportion that is today. Man is the main creator of open habitats and so introduces a new dimension – that from open to wooded conditions. The lists on the following pages are constructed with these factors in mind, but first let us quickly survey some of the more important habitats.

Woodland formed of deciduous trees has a flora of plants that come into leaf and flower early and catch the light before the leaves on the trees cut it out. Plants such as Early Purple Orchid *Orchis mascula* and Bluebell *Endymion non-scriptus* are characteristic. Most of the low-lying part of the area covered in this book would naturally be covered by oakwood. Beech trees often intercept so much light that in some beechwoods no plants can survive except those such as the Birdsnest Orchid *Neottia nidus-avis,* living on dead leaves. On the Continent beechwood is found mainly in the hills and has a richer flora, but in Britain it is on the edge of its range and is found mainly in the south.

Pinewoods have leaves on the trees all year, but natural pinewoods have well-scattered trees and ample light, and seem very different from the sombre plantations. The flora normally comprises dwarf shrubs, mainly in the heather family (Ericaceae), of which the commonest is heather *Calluna vulgaris.* When the trees are cleared, the same areas form the vast tracts of **heather moorland,** so characteristic of the British hills. These moors have very few plants in their flora.

Grasslands have very different flowers depending on whether they are on acid, neutral, or lime-rich soils, and whether they are cut for hay or grazed. Chalk and limestone grassland have especially rich floras, with many rare or decreasing plants. In grazed pastures the plants are often low-growing, with their buds low down and protected. Hay-meadows, usually neutral grassland, contain taller plants, flowering in spring and early summer, but they too have their buds low down.

Salt-marshes formed in estuaries are often grass-covered, but they have a very distinctive flora, including many members of the Goosefoot family. Many of the plants, such as the Glassworts *Salicornia,* are succulent and flower late in the year. Succulence enables them to counteract the salt in the soil.

Similarly the plants of other coastal habitats are often succulent. The floras of **shingle, sandy, and rocky shores** are all quite different, and *Glaucium flavum,* the Yellow Horned Poppy, is a typical plant of shingle banks. Behind the shore, dunes of lime-rich shell sand are often found and these have a flora somewhat resembling that of chalk grassland. Sea-shore plants are often covered with a thick greyish layer of protective wax, which prevents them drying out in their very exposed habitat. For the same reason they are also often cushion-shaped, and these characteristics are shared by many plants of **mountain tops,** another exposed habitat. Indeed several plants, such as Thrift *Armeria maritima* are found both on sea-cliffs and on mountain tops.

In addition there are many other habitats, some of which are widespread, such as those of **freshwater** or those, such as **arable fields,** that are directly the result of human activity. These habitats have a flora of annual plants, but many of the typical cornfield weeds such as Cornflower *Centaurea cyanus* are now scarce owing to the widespread use of modern herbicides.

Finally one can mention the more specialised habitats which have a particular fascination for botanists as they are the home of many rare and beautiful plants. These are the **limestone rocks** and inland cliffs, especially in the mountains; these habitats are protected from grazing, and often from collectors too. In addition to the mountain plants such as Alpine Gentian *Gentiana nivalis,* many lowland woodland plants also occur here.

The following lists are intended to give an impression of the sorts of plants to be found in certain typical habitats. They are based on a grid system so that reading across the page increasingly wet soils are found, while down the page the lists proceed from the pioneer species of open ground to the plants associated with the climax woodland community. They are divided firstly by the different soils – chalk and limestone, neutral soils (clays, for example), and acid sandy and peaty soils. In addition a list for salty soils by the sea, divided on the distance from the sea and whether the ground is muddy, sandy or rocky is included.

By identifying first the soil type and then finding the correct position on the grid in terms of wetness and the stage of development of the community (open, scrub or wooded) the typical species of that habitat can be located. Few localities will contain all of those mentioned, nor is the list exhaustive, but they may be regarded as indicator species, suggesting the others that are likely to be found as well.

Reading down the columns, as the habitats become more wooded, groups of species are separated by gaps. Species from one stage may often survive through to later stages, but the later plants are unlikely to appear earlier. In addition species in one column are likely to appear in the immediately adjacent column.

Calcareous Soils

	Dry *Rocks*	**Moister** *Grassland*	**Wet** *Fen*
Open	Sticky Catchfly Whitlow grasses Rock stonecrop Buckler mustard Wallflower Dark red helleborine Pellitory of the wall Spring sandwort Alpine penny-cress Spiked speedwell Hairy rock-cress Stone bramble	Wild candytuft Autumn gentian Rockroses Clustered bellflower Common spotted orchid Pyramidal orchid Fragrant orchid Bee orchid Salad burnet Horseshoe vetch Fairy flax Hairy violet Greater hawkbit Dwarf thistle	Marsh pea Marsh valerian Marsh helleborine Early marsh orchid Milk parsley Common comfrey Marsh cinquefoil Greater spearwort Fen orchid Marsh bedstraw Yellow loosestrife Cabbage thistle Water chickweed
Scrub	Traveller's joy Hazel Blackthorn	Hawthorn Dogwood Wayfaring tree Wild privet Spindle tree Sweet briar	Alder buckthorn Buckthorn Grey willow Boat willow Guelder rose Blackcurrant
Wooded	Whitebeams Yew Wych elm	Beech Ash Lime Crab apple Wild cherry Field maple	Alder Birches Hawthorn

Understory	Herb robert
	Mountain
	  St. John's wort
	Wall lettuce
	Jacob's ladder
	Enchanter's
	  nightshade
	Herb bennet
	Wild strawberry
	Broad-leaved
	  willowherb
		Spurge laurel
		Dog's mercury
		Sanicle
		Woodruff
		Wood dog violet
		Fly orchid
		Birdsnest orchid
		White helleborine
		Hellebores
		Alpine ragwort
		Wood sorrel
			Hedge bindweed
			Tufted vetch
			Yellow iris
			Dewberry
			Comfrey
			Hop
			Hemp agrimony

Neutral Soils

	Dry *Waste Ground*	Moister *Grassland*	Wet *Marsh*

Open	Red dead-nettle
	Broad-leaved dock
	Common field
	  speedwell
	Groundsel
	Shepherd's purse
	Dandelions
	Common stitchwort
	Poppies
	Fumitories
	Charlock
	Hoary cinquefoil
		Birdsfoot trefoil
		Great burnet
		Pepper saxifrage
		Dyer's greenweed
		Saw-wort
		Greater butterfly
		  orchid
		Common mouse-ear
		Knapweeds
		Meadow buttercup
		Ribwort plantain
		Yarrow
			Broad-leaved
			  pondweed
			Water-lilies
			Amphibious bistort
			
			Marsh marigold
			Branched bur-reed
			Yellow iris
			Purple loosestrife
			Meadowsweet
			Water forgetmenots
			Lesser spearwort
			Great willow-herb

Scrub	Goat willow Elder	Hazel Midland hawthorn Holly Blackthorn	Goat willow Grey willow
Wood	Common oak	Durmast oak Wych elm Wild service tree Field maple	Black poplar Crack willow White willow Alder

Understory	Nettle (and see
	  next column)
	Barren strawberry
	Bramble
	Rosebay willowherb
		Bugle, Wood anemone
		Hepatica, Ramsons
		Primrose, Common
		  dog violet
		Yellow pimpernel
		Red campion
		Early purple orchid
		Wolfsbane
		Purple lettuce
			Cuckoo flower
			Ragged robin
			Water avens
			Creeping buttercup
			Bugle
			Hedge woundwort
			Creeping jenny
			Marsh thistle

Acid soils

	Dry *Heath*	Moister *Moor*	Wet *Bog*

Open	Spring beauty
	Sand spurrey
	Changing
	  forgetmenot
	Annual knawel
	Cudweeds
	
	Thyme
	Harebell
	Petty whin
	Maiden pink
	Tormentil
	Heath milkwort
	Heath dog violet
		Shepherd's cress
		Sheep's sorrel
		
		Lousewort
		Heath spotted orchid
		Lesser butterfly
		  orchid
		Coralroot orchid
		Foxgloves
		Heath bedstraw
			Bog pondweed
			Bogbean
			
			Marsh asphodel
			Sundews
			Grass of Parnassus
			Butterworts
			Bog orchids
			Marsh lousewort
			Marsh pennywort
			Marsh St John's wort
			Marsh violet

Dwarf Shrub	Bell heather Gorse Broom Bilberry	Heather Bearberry Crowberry Dwarf cornel	Cross-leaved heath Bog rosemary Cranberry Bog myrtle Creeping willow

	Dry *Heath*	Moister *Moor*	Wet *Bog*
Scrub	Rowan Downy rose	Juniper Aspen	Gorse
Wood	Birches Durmast oak	Scots pine	Norway spruce Birch

Understory	Common cow-wheat
	Lesser stitchwort
	Trailing St John's
	  wort
	Lesser periwinkle
	Climbing corydalis
		Wintergreens
		Twinflower
		Bitter vetch
		Cowberry
		Creeping lady's
		  tresses
			Labrador tea
			Chickweed
			  wintergreen

Marine

	Muddy *Dry saltmarsh*	Sandy *Fixed dunes*	Rocky *Cliffs*
Upper Shore	Grass-leaved orache Sea plantain Sea aster Sea arrow-grass Sea wormwood Scurvy-grasses Marsh mallow Wild celery Sea clover Dittander	Sea buckthorn Burnet rose Bloody cranesbill Purple milk-vetch Yellow-wort Wild pansy Centauries Crow garlic Common rest-harrow	Thrift Rock samphire Squills Goldilocks aster Rock spurrey English stonecrop Rock sea-lavender Wild cabbage Stocks
	Lower saltmarsh Sea milkwort Common sea lavender Sea purslane Common seablite Sea spurreys	*Slack* Creeping willow Marsh helleborine Yellow iris Bog pimpernel Marsh pennywort Dewberry	*Fixed shingle* Tree mallow Nottingham catchfly Sea campion Curled dock Goosegrass Wall-pepper Lady's fingers Yellow vetch Bittersweet Herb robert
		Mobile dune Houndstongue Sea bindweed Viper's bugloss Ragwort Sea spurge Common whitlow- grass Common storksbill Rue leaved saxifrage	
Lower Shore	*Mud* Glassworts Eelgrasses	*Foreshore* Frosted orache Sea rocket Saltwort Sea beet Sea sandwort	*Shingle banks* Sea kale Yellow horned poppy Sea holly Sea mayweed Sea pea Shrubby seablite Oyster plant

Glossary

Some Basic Terms to Learn

The following botanical terms are used very frequently in the text, since it is not possible to "translate" them. If they are unfamiliar to you, it will simplify your use of this book to memorise them:

Anther, Bract, Petal, Pinnate, Sepal, Stamen, Stigma, Stipule, Style.

Alternate

Anther

Appressed

Acid soils have very few basic minerals and are formed on rocks such as sandstone. Peaty soils are usually acid since plant humus is often so. See also Ecology (p. 306)

Aggregate: group of closely related species, often microspecies (q.v.): see Appendix 2 (p. 305).

Alternate: neither opposite nor whorled.

Anther: the tip of the stamen producing the male pollen.

Annual plants live for a year or less. They are usually shallow rooted and never woody.

Appressed: flattened against the stem.

Bracts

Bulbil

Berry: fleshy fruit.

Biennials: live for two years. Usually the first year's growth produces a leaf-rosette, the second the flowers.

Bog: a habitat on wet, acid peat.

Bracts are small, usually leaf-like organs just below the flowers, and sometimes, as in Daisies, numerous and overlapping.

Bulbs are underground storage organs, composed of fleshy leaves.

Bulbils are small bulb-like organs at the base of the leaves or in place of the flowers, breaking off to form new plants.

Calyx

Catkins

Calyx refers to the sepals as a whole, usually used when they are joined.

Casual: plant appearing irregularly, without fixed localities.

Catkins are hanging flower-spikes, the individual flowers of which are usually rather inconspicuous.

Cluster: loose group of flowers.

Composite: member of the Daisy Family, Compositae.

Corms are bulb-like underground storage organs, comprising a swollen stem.

Coppice: trees or shrubs cut to the ground and growing from the old stools.

Corolla refers to the petals as a whole, usually when they are joined.

Crests: see p. 276.

Crucifer: member of Cabbage Family, Cruciferae.

Cluster

Composite

Corolla

Deciduous: with leaves falling in autumn.

Deflexed: bent downwards.

Disc Florets: see p. 234.

312

Dunes are areas of wind-blown, usually lime-rich shell sand near the sea, with intervening damp hollows, termed *slacks*. See also Ecology (p. 306).

Epicalyx is a ring of sepal-like organs just below the true sepals (calyx). Common in Rose Family.

Deflexed

Falls: see p. 276
Female flowers contain styles only, no stamens.
Fen: a habitat on wet, lime-rich peat, not acid as in a bog (q.v.).
Florets are small flowers, part of a compound head.
Flower: the reproductive structure of a plant.
Flower parts comprise petals, sepals, stamens, styles, and sometimes other organs.
Fruits are composed of the seeds and structures surrounding them.

Disc florets

Head: used when flowers are grouped in more or less compact terminal groups.
Heath: area, often dominated by heathers or related shrubs, on acid soils.
Hips are usually brightly coloured false fruits, characteristic of roses.
Hoary: greyish with short hairs.
Honey-leaf; see Nectary.

Epicalyx

Introduced plants are not native to the area, but brought in by man.

Female flower

Keel: see p. 122.

Labiate: belonging to the Labiate Family, Labiatae.
Lanceolate: spear-shaped, narrowly oval and pointed.
Lax: of a flower-head with the flowers well spaced; not dense.
Lime: strictly the product (Calcium oxide) of the burning of limestone rock (Calcium carbonate), but here used loosely to include limestone and chalk, and also soils formed on them; the opposite of acid. Lime-rich soils are those formed on limestone.
Linear: almost parallel-sided.
Lobed: of lvs deeply toothed, but not formed of separate leaflets. Cf *pinnatifid*.

Lobed

Male flowers contain stamens only, no styles.
Marsh: a community on wet ground, but not on peat.
Microspecies: species produced by complex reproductive processes which result in a large number of biologically distinct units, only distinguished with difficulty, and often on microscopic characters. They are here grouped into Aggregates (q.v.).
Midrib: central vein of a leaf, usually thick and raised.
Moor: usually heather covered, upland areas.
Morphology: the study of the shape, form or appearance of plants.

Male flower

Midrib

Nectar: sugary substance attractive to insects and secreted by many flowers.

Nectary

Nectary: organ in the flower producing nectar.
Net-veined: of a leaf with the veins not all parallel.
Node: point of origin of leaves on the stem.

Net-veined

Ochrea: see p.40.
Opposite: of leaves arising opposite each other on the stem.

Nodes

Palmate: with finger-like lobes or leaflets.
Parasites are plants, usually without green colouring, that obtain nutriment from other plants.
Peat is a soil composed of undecayed plant matter, often acid. See also Ecology (p. 306).
Perennial: plant surviving more than 2 years; often stouter than annuals and sometimes woody.
Petals: usually conspicuous organs above the sepals, and surrounding the reproductive parts of the flower. See Figure.
Pinnate, Pinnatifid: see Figure.
Pod: fruit, usually long and cylindrical and never fleshy; as in peas.

Pinnate

Rhizome

Ray: see p. 158.
Ray-floret: see p. 234.
Rhizome: horizontal underground stem, and therefore bearing leaf-scars.
Rosette: flattened, rose-like group of leaves at the base of the stem.
Runners are horizontal above-ground stems, often rooting at the nodes.

Rosette

Runners

Samara: a winged key-shaped fruit.
Scale: small appendage not resembling a leaf. Normally small, brown, and papery.
Sepals form a ring immediately below the petals and are usually green or brown and less conspicuous.
Shrub: much-branched woody plant, shorter than a tree.
Shy flowerer: plant often passing whole years without flowering and for which other characters are therefore important for identification.
Silicula and Siliqua: see p.84.
Spadix: see p. 274.
Spathe: see p. 274.
Species: the basic unit of classification; see Introduction, p.10.
Spike: flower-head with flowers arranged up a central axis in a cylinder. Stalked spike has individual fls stalked.
Spine: straight, sharp-tipped appendage; cf. thorn.
Spreading: standing out horizontally or at a wide angle from the stem.
Stamens: the male organs in a flower, comprising a filament and a pollen sac, the anther (see Figure). Distinguished from styles (q.v.) by lying outside the centre of the flower usually in a ring, and by the usually coloured anthers.
Standard: see pp. 122 and 276.
Stigma: the surface receptive to pollen at the tip of the style.
Stipules are scale- or leaf-like organs at the base of the leaf-stalk.

Samara

Scales

Sepals

Spike

Styles: the columns of filaments leading from the female organs to the stigma (q.v.); see Figure. Distinguished from stamens (q.v.) by lying in the centre of the flower, within the ring of stamens.

Subspecies: the division in the classification of organisms immediately below the species. Subspecies are morphologically distinct from each other, but interbreed freely and are therefore included in the same species.

Stamens

Stigma

Tendrils are twisted filaments forming part of a leaf or stem and used for climbing.

Thallus: plant body not differentiated into stems and lvs.

Thorn: sharp-tipped, woody appendage, straight or curved.

Tree: tall, woody plant, with a single woody stem at the base.

Trefoil, Trifoliate: with three leaflets.

Stipules

Umbel: a flowerhead with the flowers in a spike but the lower branches longer so that all are level. **Umbellifers** are members of the Umbelliferae (pp. 158-70), characterised by having flowers in umbels.

Undershrub: low, often creeping woody perennial, often quite unlike taller shrubs.

Style

Valve: see Docks, p. 40.

Variety: a distinct form of a plant, of even lower rank than a subspecies.

Tendrils

Waste places: areas much disturbed by man, but not cultivated.

Whorl: where several organs arise at the same point on a stem. Cf opposite.

Winged: with a flange running down the stem or stalk.

Wings: see p. 122.

Thorn

Winged fruit

Winged stem

Winged flower

Whorl

315

Plant Photography

It was once the custom for botanists to pick and press plants for their own herbaria, and this practice caused certain rare plants to become extinct or nearly so. Nowadays the botanist can keep a permanent photographic record which is more responsible, more informative, and more attractive. Here are a few hints.

Ideally use a single-lens reflex camera, enabling you to see your proposed picture through the lens. This is especially important for close-up work, for which you can obtain extension tubes or use a macro lens.

Use slow speeds to give narrow apertures (high f values) and good depth of focus. A fast film helps here and a tripod may be useful.

Modern cameras have through-the-lens metering, but fully automatic may yield poor results where the flower is much brighter than its background; if necessary move up half-a-stop or a stop on the f-scale to avoid over-exposure.

If you need to move surrounding vegetation to take your picture, be sure to replace it afterwards to protect the plant.

And look where you tread!

Societies to Join

Botanical Society of the British Isles, c/o Natural History Museum, Cromwell Road, London S.W.7. The leading national society for serious students of the British flora, both amateur and professional.

Wild Flower Society, Rams Hill House, Horsmonden, Tonbridge, Kent, is for amateur lovers of wild flowers, who enjoy keeping annual diaries, and enter for various flower-hunting and recording competitions.

Your local library will supply addresses of local natural history societies, many of which have botanical sections, and almost all of which run field meetings where beginners are helped to learn the local wild flowers.

The Royal Society for Nature Conservation, The Green, Nettleham, Lincoln, will supply the address of your local County Naturalists' Trust, which all lovers of British wild flowers who are seriously interested in their conservation ought to join.

Further Reading

Britain

Flora of the British Isles, by A. R. Clapham, T. G. Tutin, and E. F. Warburg (Cambridge) is the standard British flora for serious workers.

The Pocket Guide to Wild Flowers, by David McClintock and R. S. R. Fitter (Collins) is a complete field guide, fully illustrated.

Finding Wild Flowers, by Richard Fitter (Collins). A comprehensive key to the British flora, with details of good sites for flower hunting.

The Concise British Flora in Colour, by W. Keble Martin (Ebury Press/ Michael Joseph). A tour-de-force, all British wild flowers painted by one man. Text rather brief. A modern equivalent is *The Wild Flowers of the British Isles* by I. Garrard and D. T. Streeter (Macmillan).

The New Naturalist Series (Collins) includes *Wild Flowers,* by John Gilmour and Max Walters; *Wild Orchids of Britain,* by V. S. Summerhayes; *Wild Flowers of Chalk and Limestone,* by J. E. Lousley; *Flowers of the Coast,* by Ian Hepburn; *Mountain Flowers,* by John Raven and Max Walters; and *British Plant Life,* by W. B. Turrill.

Europe

Flora Europaea, edited by T. G. Tutin and others (Cambridge) is the standard European flora, in five volumes.

Flowers of Europe: a field guide, by Oleg Polunin (Oxford). Covers a wider area less thoroughly than the present volume. Illustrated by colour photographs.

Index of English Names

Abraham, Isaac and Jacob, 196
Acacia, False, 124
Aconite, Winter, 66
Adderstongue, 274
Agrimony, 108
    Bastard, 108
    Fragrant, 108
    Hemp, 234
Alder, 30
    Green, 30
    Grey, 30
Alexanders, 170
    Biennial, 170
Alison, Hoary, 96
    Mountain, 84
    Small, 84
    Sweet, 96
Alkanet, 196
    Green, 196
    Large Blue, 196
    Yellow, 192
Allseed, 136
    Four-leaved, 58
Alyssum, Golden, 84
Amaranth, Common, 48
    Green, 48
    White, 48
Amelanchier, 116
Amsinckia, 192
Androsace, Northern, 180
Anemone, Blue, 76
    Wood, 76
    Yellow, 76
Angelica, 164
    Garden, 164
Apple, Crab, 116
    Cultivated, 116
Apple of Peru, 210
Archangel, Yellow, 204
Arnica, 240
Aronia, 304
Arrow-grass, Marsh, 224
    Sea, 224
Arrowhead, 262
    Canadian, 262
Artichoke, Jerusalem, 240
Arum, Bog, 274
Asarabacca, 38
Ash, 36
    Flowering 36
    Mountain, 116
Asparagus, Wild, 272
Aspen, 30
Asphodel, Bog, 266
    German, 264
    Scottish, 264
Aster, Sea, 236
    Stink, 234
Astrantia, 158
Auricula, 178
Avens, Alpine, 112
    Mountain, 112
    Water, 112
Awlwort, 294
Azalea, Wild, 176

Balm, 206
    Bastard, 202
Balsam, Himalayan, 142

Orange, 142
    Small, 142
    Touch-me-not, 142
Baneberry, 72
Barberry, 78, 304
Bartsia, Alpine, 216
    Red, 216
    Yellow, 216
Basil, Cow, 62
    Wild, 208
Bayberry, 28
Bearberry, Alpine, 176
    Black, 176
Bedstraw, Cheddar, 190
    Fen, 190
    Heath, 190
    Hedge, 190
    Lady's, 190
    Limestone, 190
    Marsh, 190
    Northern, 190
    Slender, 190
    Slender Marsh, 190
    Wall, 190
Beech, 32
Beet, Sea, 46
Beggar Ticks, 242
Bellflower, Bearded, 230
    Clustered, 230
    Creeping, 230
    Giant, 230
    Ivy-leaved, 230
    Nettle-leaved, 230
    Peach-leaved, 230
    Rampion, 230
    Spreading, 230
Bermuda Buttercup, 136
Betony, 206
Bilberry, 176
    Northern, 176
Bindweed, Black, 42
    Copse, 42
    Field, 188
    Great, 188
    Hairy, 188
    Hedge, 188
    Sea, 188
Birch, Downy, 30
    Dwarf, 30
    Silver, 30
Birdsfoot, 132
    Orange, 132
Birdsfoot Trefoil, 132
    Greater, 132
    Hairy, 132
    Narrow-leaved, 132
    Slender, 132
Birdsnest, Yellow, 172
Birthwort, 38
Bistort, 40
    Alpine, 40
    Amphibious, 40
Bittercress, Coralroot, 90
    Hairy, 94
    Large, 90
    Narrow-leaved, 90
    Wavy, 90
Bittersweet, 210
Blackberry, 110
Black-eyed Susan, 240

Blackthorn, 118
Bladder-nut, 144
Bladderseed, 162
Bladderwort, Greater, 294
    Intermediate, 294
    Lesser, 294
Blinks, 42
Blood-drop Emlets, 216
Bluebell, 270
    Spanish, 270
Blue-eyed Grass, 276
Blue-eyed Mary, 194
Bogbean, 182
Bog Myrtle, 28
Bog Rosemary, 174
Borage, 196
Box, 144
Bramble, 110
    Arctic, 110
    Stone, 110
Briar, Sweet, 110
Brooklime, 220
Brookweed, 180
Broom, 124
    Black, 124
    Butcher's, 272
    Clustered, 124
    Winged, 124
Broomrape, Branched, 222
    Carrot, 224
    Clove-scented, 222
    Common, 224
    Greater, 222
    Ivy, 224
    Knapweed, 222
    Ox-tongue, 224
    Purple, 222
    Thistle, 224
    Thyme, 224
Bryony, Black, 270
    White, 152
Buckthorn, 144
    Alder, 144
    Sea, 144
Buckwheat, 40
Bugle, 200
    Blue, 200
    Pyramidal, 200
Bugloss, 196
    Purple Viper's, 196
    Viper's, 196
Bullace, 118
Bulrush, 302
    Lesser, 302
Burdock, Greater, 248
    Least, 302
    Lesser, 248
Bur Marigold, Nodding, 242
    Trifid, 242
Burnet, Great, 108
    Salad, 108
Burnet Saxifrage, 162
    Greater, 162
Burning Bush, 142
Bur Parsley, Greater, 160
    Knotted, 160
    Small, 160
    Spreading, 160
Bur-reed, Branched, 302
    Floating, 302

Least, 302
Unbranched, 302
Butcher's Broom, 272
Butterbur, 242
White, 242
Buttercup, Bermuda, 136
Bulbous, 68
Celery-leaved, 70
Corn, 70
Creeping, 68
Glacier, 70
Goldilocks, 70
Hairy, 68
Jersey, 68
Large White, 70
Meadow, 68
Small-flowered, 70
Woolly, 68
Butterfly Bush, 304
Butterwort, Alpine, 222
Common, 222
Large-flowered, 222
Pale, 222
Buttonweed, 242

Cabbage, Bastard, 88
Hare's-ear, 86
Isle of Man, 88
Lundy, 88, 306
St. Patrick's, 104
Wallflower, 88
Warty, 86
Wild, 88
Calamint, Common, 208
Lesser, 208
Caltrop, 136
Calypso, 278
Campion, Bladder, 58
Red, 60
Sea, 58
White, 60
Candytuft, Wild, 96
Canterbury Bell, 230
Caraway, 160
Whorled, 160
Carnation, 64
Carrot, Moon, 162
Wild, 160
Cassiope, 176
Catchfly, Alpine, 60
Berry, 62
Flaxfield, 58
Forked, 58
Italian, 58
Moss, 58
Night-flowering, 60
Northern, 58
Nottingham, 58
Rock, 58
Sand, 62
Small-flowered, 62
Spanish, 58
Sticky, 60
Sweet William, 58
White Sticky, 58
Catmint, 202
Hairless, 202
Catsear, Common, 258
Smooth, 258
Spotted, 258
Cattail, 302
Cedar, Western Red, 24
Celandine, Greater, 80
Lesser, 68
Celery, Wild, 166
Centaury, Common, 184
Guernsey, 184
Lesser, 184
Perennial, 184

Seaside, 184
Slender, 184
Yellow, 184
Chaffweed, 180
Chamomile, Corn, 236
Lawn, 236
Stinking, 236
Yellow, 240
Charlock, 88
Cherry, Bird, 118
Choke, 118
Cornelian, 158
Dwarf, 118
Rum, 118
St. Lucie's, 118
Wild, 118
Cherry, Laurel, 118
Cherry-plum, 118
Chervil, Bur, 160
Garden, 160
Golden, 160
Rough, 160
Chestnut, Horse, 36
Sweet, 32
Water, 262
Chickweed, Common, 54
Greater, 54
Lesser, 54
Umbellate, 54
Upright, 52
Water, 54
Wintergreen, 180
Chicory, 254
Chives, 268
Chondrilla, 256
Cinquefoil, Alpine, 114
Creeping, 114
Grey, 114
Hoary, 114
Marsh, 112
Norwegian, 114
Rock, 112
Shrubby, 114
Spring, 114
Sulphur, 114
White, 112
Clary, Meadow, 206
Whorled, 206
Wild, 206
Cleavers, Common, 190
Corn, 190
False, 190
Clematis, Alpine, 72
Cloudberry, 110
Clover, Alsike, 134
Burrowing, 134
Clustered, 134
Crimson, 134
Haresfoot, 134
Knotted, 134
Long-headed, 134
Mountain, 134
Mountain Zigzag, 134
Red, 134
Reversed, 134
Rough, 134
Sea, 134
Strawberry, 134
Suffocated, 134
Sulphur, 134
Twin-flowered, 134
Upright, 134
Western, 134
White, 134
Zigzag, 134
Club-rush, Common, 300
Cnidium, 168
Cocklebur, Rough, 234
Spiny, 234

Stinking, 234
Coltsfoot, 244
Purple, 244
Columbine, 74
Comfrey, Common, 192
Dwarf, 192
Rough, 192
Russian, 192
Soft, 192
Tuberous, 192
Compass Plant, 256
Cone Flower, 240
Coral Necklace, 58
Coriander, 160
Corn Cockle, 60
Cornel, Dwarf, 158
Cornelian Cherry, 158
Cornflower, 252
Perennial, 252
Cornsalad, 226
Corydalis, Bulbous, 78
Climbing, 78
Yellow, 78
Cotoneaster, 304
Himalayan, 118
Small-leaved, 118
Wall, 118
Wild, 118
Cottonweed, 242
Cowbane, 164
Cow Basil, 62
Cowberry, 176
Cow Parsley, 160
Cowslip, 178
Cow-wheat, Common, 220
Crested, 220
Field, 220
Small, 220
Crab Apple, 116
Cranberry, 176
Small, 176
Cranesbill, Bloody, 138
Cut-leaved, 138
Dovesfoot, 138
Dusky, 138
French, 138
Hedgerow, 138
Knotted, 138
Long-stalked, 138
Marsh, 138
Meadow, 138
Pencilled, 138
Round-leaved, 138
Shining, 138
Small-flowered, 138
Wood, 138
Creeping Jenny, 178
Cress, Garden, 98
Hoary, 98
Mitre, 84
Shepherd's, 96
Thale, 94
Crocus, Autumn, 276
Sand, 276
Spring, 276
Crosswort, 190
Crowberry, 176
Crowfoot, Common Water, 70
Ivy-leaved, 70
Round-leaved, 70
Three-lobed, 70
Cuckoo Flower, 90
Cuckoo Pint, Large, 274
Cucumber, Prickly, 152
Cudweed, Broad-leaved, 238
Cape, 238
Common, 238
Dwarf, 238
Heath, 238

Highland, 238
Jersey, 238
Marsh, 238
Narrow-leaved, 238
Red-tipped, 238
Small, 238
Upright, 238
Currant, Black, 144
Flowering, 304
Mountain, 144
Red, 144
Cyclamen, 180
Cyphel, 52
Cypress, Lawson's, 24

Daffodil, Tenby, 274
Wild, 274
Daisy, 236
Michaelmas, 236
Ox-eye, 244
Shasta, 244
Dame's Violet, 90
Dandelion, 258
Lesser, 258
Marsh, 258
Red-veined, 258
Dead-nettle, Cut-leaved, 204
Henbit, 204
Red, 204
Spotted, 204
White, 204
Dewberry, 110
Diapensia, 172
Dill, 170
Dittander, 92
Dock, Argentine, 44
Broad-leaved, 44
Clustered, 44
Curled, 44
Fiddle, 44
Golden, 42
Marsh, 42
Northern, 44
Patience, 44
Scottish, 44
Shore, 44
Water, 44
Willow-leaved, 44
Wood, 44
Dodder, Common, 188
Flax, 188
Greater, 188
Dogwood, 158
Dragon's Teeth, 132
Dropwort, 108
Water, 166
Duckweed, Common, 300
Fat, 300
Greater, 300
Ivy-leaved, 300
Lesser, 300
Rootless, 300
Dusty Miller, 54

Eel-grass, 300
Dwarf, 300
Narrow-leaved, 300
Elder, 226
Box, 304
Dwarf, 226
Ground, 162
Red-berried, 226
Elecampane, 240
Elm, English, 34
Fluttering, 34
Small-leaved, 34
Wych, 34
Enchanter's Nightshade, 152
Alpine, 152

Upland, 152
Eryngo, Field, 158
Escallonia, 304
Evening Primrose,
Common, 152
Fragrant, 152
Large-flowered, 152
Small-flowered, 152
Everlasting, Mountain, 238
Pearly, 238
Eyebright, 220
Irish, 220

Fat Hen, 46
Felwort, Marsh, 186
Fennel, 170
Hog's, 168
Fenugreek, 134
Classical, 130
Feverfew, 244
Fig, 304
Hottentot, 42
Figwort, Balm-leaved, 212
Common, 212
French, 212
Green, 212
Water, 212
Yellow, 212
Fir, Douglas, 24
Silver, 24
Firethorn, 304
Flag, Purple, 276
Sweet, 302
Flax, Common, 136
Pale, 136
Perennial, 136
Purging, 136
White, 136
Yellow, 136
Fleabane, Alpine, 236
Blue, 236
Canadian, 234
Common, 240
Irish, 240
Mexican, 236
Small, 238
Fleawort, Field, 246
Marsh, 246
Flixweed, 86
Fluellen, Round-leaved, 214
Sharp-leaved, 214
Fool's Parsley, 162
Forget-me-not, Alpine, 194
Bur, 194
Changing, 194
Creeping, 194
Early, 194
Field, 194
Jersey, 194
Pale, 194
Tufted, 194
Water, 194
Wood, 194
Foxglove, 216
Fairy, 216
Large Yellow, 216
Small Yellow, 216
Fringe Cups, 104
Fritillary, 266
Frogbit, 262
Fuchsia, 304
Fumana, Common, 152
Fumitory, Common, 78
Ramping, 78
Small, 78
Wall, 78

Gagea, Belgian, 266
Least, 266

Meadow, 266
Gallant Soldier, 236
Garlic, Crow, 268
Field, 268
German, 268
Keeled, 268
Rosy, 268
Gean, 118
Gentian, Alpine, 186
Autumn, 186
Bladder, 186
Chiltern, 186
Cross, 186
Dune, 186
Early, 186
Field, 186
Fringed, 186
Great Yellow, 184
Marsh, 186
Northern, 184
Purple, 186
Slender, 186
Spotted, 186
Spring, 186
Willow, 186
Germander, Cut-leaved,
200
Mountain, 200
Wall, 200
Water, 200
Gipsywort, 208
Gladiolus, 276
Glasswort, 48
Perennial, 48
Globeflower, 68
Globularia, 228
Goatsbeard, 254
Goatsbeard Spiraea, 108
Goat's Rue, 122
Golden-rod, 234
Canadian, 234
Gold of Pleasure, 84
Goldilocks, 234
Good King Henry, 46
Gooseberry, 144
Goosefoot, Fig-leaved, 46
Green, 46
Grey, 46
Many-seeded, 46
Maple-leaved, 46
Nettle-leaved, 46
Oak-leaved, 46
Red, 46
Small Red, 46
Stinking, 46
Upright, 46
Gorse, 124
Dwarf, 124
Western, 124
Grape-hyacinth, 270
Small, 270
Grass of Parnassus, 100
Grass Poly, 154
Gratiola, 216
Greenweed, Dyer's, 124
German, 124
Hairy, 124
Gromwell, Common, 192
Corn, 192
Purple, 192
Ground Elder, 162
Ground Ivy, 202
Ground-pine, 200
Groundsel, 246
Heath, 246
Sticky, 246
Guelder Rose, 226
Gypsophila, Annual, 62
Fastigiate, 62

Harebell, 230
Hare's-ear, Sickle, 170
  Slender, 168
  Small, 170
Hartwort, 168
Hawkbit, Autumn, 258
  Lesser, 258
  Rough, 258
Hawksbeard, Beaked, 260
  Bristly, 260
  French, 260
  Marsh, 260
  Northern, 260
  Rough, 260
  Smooth, 258
  Stinking, 260
Hawkweed, Alpine, 260
  Few-leaved, 260
  Leafy, 260
  Mouse-ear, 258
  Orange, 260
  Spotted, 260
Hawthorn, 116
  Midland, 116
Hazel, 32
Heath, Cornish, 174
  Cross-leaved, 174
  Dorset, 174
  Irish, 174
  Mackay's, 174
  Mountain, 174
  Prickly, 304
  St. Dabeoc's, 174
Heather, 174
  Bell, 174
Hedge Parsley, Upright, 160
Helichrysum, 238
Heliotrope, 192
  Winter, 242
Hellebore, Green, 66
  Stinking, 66
Helleborine, Broad-leaved,
  288
  Dark Red, 288
  Dune, 288
  False, 264
  Green-flowered, 288
  Marsh, 288
  Narrow-leaved, 286
  Narrow-lipped, 288
  Red, 286
  Violet, 288
  White, 288
Hemlock, 164
  Western, 24
Hemp, 38
Hemp-nettle, Common, 204
  Downy, 204
  Large-flowered, 204
  Red, 204
Hen and Chickens Houseleek,
  102
Henbane, 210
Hepatica, 76
Herb Bennet, 112
Herb Paris, 272
Herb Robert, 138
Hog's Fennel, 168
Hogweed, 164
  Giant, 164
Holly, 36
  Sea, 158
Hollyhock, 146
Honesty, 90
  Perennial, 90
Honewort, 162
Honeysuckle, 226
  Black-berried, 226
  Blue, 226

Fly, 226
  Perfoliate, 226
Honeywort, Lesser, 192
Hop, 38
Horehound, Black, 202
  White, 202
Hornbeam, 32
Horned-poppy, Red, 80
  Yellow, 80
Hornwort, Rigid, 296
  Soft, 296
Horse-radish, 92
Hottentot Fig, 42
Houndstongue, 192
  Green, 192
Houseleek, Hen and
  Chickens, 102
Hutchinsia, 96
Hyacinth, Tassel, 270
Hyssop, 202

Iris, Butterfly, 276
  Garden, 276
  Siberian, 276
  Stinking, 276
  Yellow, 276
Ivy, 158
  Ground, 202

Jack-go-to-bed-at-noon, 254
Jacob's Ladder, 188
Joint Pine, 24
Juneberry, 116
Juniper, 24
Jupiter's Distaff, 206

Kale, Sea, 92
Knapweed, Black, 252
  Brown, 252
  Greater, 252
  Jersey, 252
  Slender, 252
Knawel, Annual, 56
  Perennial, 56
Knotgrass, 40
  Ray's, 40
  Sea, 40
Knotweed, Giant, 42
  Himalayan, 42
  Japanese, 42

Labrador Tea, 176
Laburnum, 124
Lady's Mantle, 108
  Alpine, 108
Lady's Slipper, 278
Lady's Tresses, Autumn, 286
  Creeping, 286
  Irish, 286
  Summer, 286
Larch, European, 24
  Japanese, 24
Larkspur, 74
  Eastern, 74
  Forking, 74
Laurel, Cherry, 118
  Portugal, 118
  Spurge, 142
Leatherleaf, 174
Leek, Babington's, 268
  Cultivated, 268
  Few-flowered, 268
  Round-headed, 266
  Sand, 268
  Three-cornered, 268
  Wild, 268
Lentil, Wild, 122
Leopardsbane, 240
  Green, 240

Lettuce, Blue, 254
  Great, 256
  Least, 256
  Pliant, 256
  Prickly, 256
  Purple, 254
  Wall, 256
Lilac, 304
Lily, Kerry, 264
  Martagon, 266
  May, 264
  Pyrenean, 266
  St. Bernard's, 264
  Snowdon, 264, 306
Lily of the Valley, 264
Lime, Common, 36
  Large-leaved, 36
  Small-leaved, 36
Liquorice, Wild, 122
Little Robin, 138
Lobelia, Heath, 230
  Water, 294
London Pride, 104
Longleaf, 164
Loosestrife, Dotted, 178
  Purple, 154
  Tufted, 178
  Yellow, 178
Lords and Ladies, 274
Lousewort, 220
  Leafy, 220
  Marsh, 220
Lovage, 170
  Scots, 168
Love in a Mist, 66
Love Lies Bleeding, 48
Lucerne, 130
Lungwort, 194
  Narrow-leaved, 194
Lupin, Annual, 124
  Bitter Blue, 124
  Garden, 124
  Sweet, 124
  Tree, 124
  White, 124
  Wild, 124

Madder, Field, 188
  Wild, 190
Madwort, 194
Mallow, Chinese, 146
  Common, 146
  Dwarf, 146
  Least, 146
  Marsh, 146
  Musk, 146
  Rough, 146
  Small, 146
  Smaller Tree, 146
  Tree, 146
Maple, Field, 34
  Italian, 34
  Montpelier, 34
  Norway, 34
Marestail, 296
Marigold, 244
  Corn, 244
  Garden, 244
  Marsh, 68
  Nodding Bur, 242
  Trifid Bur, 242
Marjoram, 208
Marshwort, Creeping, 166
  Lesser, 166
Masterwort, 168
Mayweed, Pineapple, 236
  Scented, 236
  Scentless, 236
  Sea, 236

Meadow-rue, Alpine, 72
  Common, 72
  Greater, 72
  Lesser, 72
  Small, 72
Meadow Saffron, 270
Meadowsweet, 108
Medick, Black, 132
  Bur, 132
  Spotted, 132
  Toothed, 132
Medlar, 304
Melilot, Ribbed, 130
  Small, 130
  Tall, 130
  White, 130
Mercury, Annual, 140
  Dog's, 140
Mezereon, 142
Michaelmas Daisy, 230
Mignonette, Corn, 100
  White, 100
  Wild, 100
Milfoil, Alternate Water, 296
  Spiked Water, 296
  Whorled Water, 296
Milk-parsley, Cambridge, 164
Milk-vetch, Alpine, 122
  Hairy, 122
  Mountain, 122
  Northern, 122
  Purple, 122
  Yellow, 122
  Yellow Alpine, 122
Milkwort, Chalk, 142
  Common, 142
  Dwarf, 142
  Heath, 142
  Sea, 180
  Shrubby, 142
  Tufted, 142
Mind Your Own Business, 38
Mint, Corn, 208
  Round-leaved, 208
  Spear, 208
  Water, 208
Mistletoe, 38
Mock Orange, 304
Moneywort, Cornish, 220
Monkey Flower, 216
Monkshood, 74
Monk's Rhubarb, 44
Montbretia, 276
Moon Carrot, 162
Moonwort, 274
Moor-king, 220
Moschatel, 222
Motherwort, 204
  False, 204
Mouse-ear, Alpine, 54
  Arctic, 54
  Common, 54
  Dwarf, 54
  Field, 54
  Grey, 54
  Little, 54
  Sea, 54
  Starwort, 54
  Sticky, 52
Mousetail, 74
Mudwort, 294
  Welsh, 294
Mugwort, 244
  Chinese, 244
Mullein, Dark, 212
  Great, 212
  Hoary, 212
  Moth, 212
  Orange, 212

  Purple, 212
  Twiggy, 212
  White, 212
Musk, 216
Mustard, Ball, 84
  Black, 88
  Buckler, 84
  Garlic, 90
  Hedge, 86
  Hoary, 88
  Tower, 94
  Treacle, 86
  White, 88
  White Ball, 92
Myrobalan, 118
Myrtle, Bog, 28

Naiad, Holly-leaved, 300
  Slender, 300
Narcissus, Poet's, 274
Navelwort, 100
Nettle, Stinging, 38
  Annual, 38
Nightshade, Black, 210
  Deadly, 210
  Enchanter's, 152
  Green, 210
  Hairy, 210
Nipplewort, 256
Nonea, 192
Nutmeg, Flowering, 305

Oak, Durmast, 32
  Evergreen, 32
  Holm, 32
  Pedunculate, 32
  Red, 32
  Sessile, 32
  Turkey, 32
  White, 32
Odontites, Yellow, 216
Old Man's Beard, 72
Onion, 268
  Welsh, 268
Orache, Common, 46
  Frosted, 46
  Grass-leaved, 46
  Shrubby, 304
  Spear-leaved, 46
  Stalked, 48
Orchid, Bee, 278
  Arctic Butterfly, 286
  Birdsnest, 290
  Black Vanilla, 278
  Bog, 290
  Broad-leaved Marsh, 282
  Bug, 280
  Burnt, 280
  Common Spotted, 282
  Coral-root, 290
  Dense-flowered, 284
  Early Marsh, 282
  Early Purple, 280
  Early Spider, 278
  Elder-flowered, 282
  False Musk, 284
  Fen, 290
  Flecked Marsh, 282
  Fly, 278
  Fragrant, 282
  Frog, 284
  Ghost, 290
  Greater Butterfly, 286
  Green-winged, 280
  Heath Spotted, 282
  Lady, 280
  Late Spider, 278
  Lesser Butterfly, 286
  Loose-flowered, 280

  Lizard, 284
  Man, 284
  Marsh, 282
  Military, 280
  Monkey, 280
  Musk, 284
  Northern Butterfly, 286
  Northern Marsh, 282
  One-leaved Bog, 290
  Pale-flowered, 280
  Pugsley's Marsh, 282
  Pyramidal, 282
  Southern Marsh, 282
  Spotted, 282
  Toothed, 280
  Violet Birdsnest, 290
  Wasp, 278
  White Frog, 286
Oregon Grape, 78
Orpine, 102
Osier, 26
  Golden, 26
  Red, 158
Oxalis, Pink, 136
  Upright Yellow, 136
  Yellow, 136
Ox-eye, Large Yellow, 240
  Yellow, 240
Ox-eye Daisy, 244
Oxlip, 178
  False, 178
Ox-tongue, Bristly, 260
  Hawkweed, 260
Oyster Plant, 196
  Spanish, 252

Pansy, Dwarf, 146
  Field, 146
  Mountain, 146
  Wild, 146
Parsley, Bur, 160
  Cambridge Milk, 164
  Corn, 166
  Cow, 160
  Fool's, 162
  Greater Bur, 160
  Knotted Bur, 160
  Milk, 168
  Small Bur, 160
  Spreading Bur, 160
  Stone, 166
  Upright Hedge, 160
  Wild, 166
Parsley Piert, 108
Parsnip, Wild, 170
Pasque Flower, 76
  Eastern, 76
  Pale, 76
  Small, 76
Pea, Black, 128
  Broad-leaved Everlasting, 128
  Marsh, 128
  Narrow-leaved Everlasting, 128
  Sea, 128
  Spring, 128
  Tuberous, 128
Pear, Cultivated, 116
  Plymouth, 116
  Wild, 116
Pearlwort, Alpine, 56
  Annual, 56
  Heath, 56
  Knotted, 56
  Procumbent, 56
  Sea, 56
  Snow, 56
Pellitory of the Wall, 38

321

Pennycress, Alpine, 98
  Field, 98
  Garlic, 98
  Mountain, 98
  Perfoliate, 98
Pennyroyal, 208
Pennywort, Marsh, 158
Peppermint, 208
Pepperwort, Field, 98
  Least, 98
  Narrow-leaved, 98
  Poor Man's, 98
  Smith's, 98
  Tall, 98
Periwinkle, Greater, 182
  Lesser, 182
Persicaria, Pale, 40
Petty Whin, 124
Pheasant's Eye, 74
  Large, 74
  Summer, 74
Physocarpus, 304
Pick-a-back Plant, 104
Pignut, 162
  Great, 162
Pimpernel, Blue, 180
  Bog, 180
  Scarlet, 180
  Yellow, 178
Pine, Austrian, 24
  Joint, 24
  Maritime, 24
  Scots, 24
Pink, Carthusian, 64
  Cheddar, 64
  Childing, 62
  Clove, 64
  Deptford, 64
  Jersey, 64
  Large, 64
  Maiden, 64
  Proliferous, 62
  Wild, 64
Pipewort, 264
Pirri-pirri Bur, 108
Pittosporum, 304
Plane, London, 34
Plantain, Branched, 224
  Buckshorn, 224
  Greater, 224
  Hoary, 224
  Ribwort, 224
  Sea, 224
Pleurospermum, 164
Ploughman's Spikenard, 234
Plum, Wild, 118
Poet's Narcissus, 274
Polycnemum, 48
Pondweed, American, 298
  Blunt-leaved, 298
  Bog, 298
  Broad-leaved, 298
  Curled, 298
  Fen, 298
  Fennel, 298
  Flat-stalked, 298
  Grasswrack, 298
  Hair-like, 298
  Horned, 300
  Lesser, 298
  Loddon, 298
  Long-stalked, 298
  Opposite-leaved, 298
  Perfoliate, 298
  Reddish, 298
  Sharp-leaved, 298
  Shetland, 298
  Shining, 298
  Small, 298

Various-leaved, 298
Poplar, Balsam, 30
  Black, 30
  Grey, 30
  Italian, 30
  White, 30
Poppy, Arctic, 80
  Babington's, 80
  Common, 80
  Long-headed, 80
  Opium, 80
  Prickly, 80
  Red Horned, 80
  Rough, 80
  Welsh, 80
  Yellow Horned, 80
Primrose, 178
  Birdseye, 178
  Evening, 152
  Scottish, 178
Primrose Peerless, 274
Privet, Garden, 182
  Wild, 182
Purslane, Hampshire, 294
  Iceland, 42
  Pink, 42
  Sea, 48
  Water, 294

Quillwort, 294
  Spring, 294
Quince, 304

Radish, Garden, 88
  Sea, 88
  Wild, 88
Ragged Robin, 60
Ragweed, 234
Ragwort, 246
  Alpine, 246
  Broad-leaved, 246
  Fen, 246
  Hoary, 246
  Marsh, 246
  Oxford, 246
  Silver, 246
Rampion, Round-headed, 228
  Spiked, 228
Ramsons, 268
Rannoch Rush, 224
Rape, 88
Raspberry, 110
  Purple-flowering, 304
Rattle, Yellow, 220
Redshank, 40
Reed, 300
Reed Sweetgrass, 300
Reed-mace see Bulrush
Rest-harrow, 130
  Large Yellow, 130
  Small, 130
  Spiny, 130
Rhododendron, 304
  Arctic, 174
  Yellow, 304
Rhubarb, Monk's, 44
River Beauty, 154
Rockcress, Alpine, 94
  Annual, 94
  Bristol, 94
  Garden, 94
  Hairy, 94
  Northern, 94
  Tall, 94
Rocket, Eastern, 86
  False London, 86
  Hairy, 88
  London, 86
  Sea, 92

Tall, 86
Rock-rose, Common, 152
  Hoary, 152
  Spotted, 152
  White, 152
Rose, 304
  Burnet, 110
  Dog, 110
  Downy, 110
  Field, 110
  Guelder, 226
  Provence, 110
Rosemary, Bog, 174
Rose of Sharon, 148
Roseroot, 100
Rowan, 116
Rubus, 304
Rue, Goat's, 122
Rupturewort, Fringed, 56
  Hairy, 56
  Smooth, 56
Rush, Flowering, 262
  Rannoch, 224

Safflower, 252
Sage, Wild, 206
  Wood, 200
Sainfoin, 130
St. John's Wort, Flax-leaved,
  148
  Hairy, 148
  Imperforate, 148
  Irish, 148
  Marsh, 148
  Pale, 148
  Perforate, 148
  Slender, 148
  Square-stalked, 148
  Trailing, 148
  Wavy, 148
St. Patrick's Cabbage, 104
Sallow, Eared, 26
  Great, 26
  Grey, 26
Salsify, 254
Saltwort, Prickly, 48
Samphire, Golden, 234
  Rock, 170
Sandwort, Arctic, 52
  Balearic, 52
  Curved, 52
  Fine-leaved, 52
  Fringed, 52
  Mossy, 52
  Mountain, 52
  Sea, 52
  Spring, 52
  Teesdale, 52
  Three-veined, 52
  Thyme-leaved, 52
Sanicle, 158
Savory, Winter, 202
Sawwort, 252
  Alpine, 248
Saxifrage, Alternate
  leaved Golden, 100
  Arctic, 106
  Burnet, 162
  Drooping, 106
  Greater Burnet, 162
  Hawkweed, 106
  Highland, 106
  Irish, 106
  Kidney, 104
  Livelong, 104
  Marsh, 106
  Meadow, 104
  Mossy, 106
  Opposite-leaved Golden, 100

Pepper, 170
Purple, 106
Pyrenean, 104
Round-leaved, 104
Rue-leaved, 106
Starry, 104
Tufted, 106
Yellow, 106
Scabious, Devilsbit, 228
Field, 228
Sheepsbit, 228
Small, 228
Wood, 228
Yellow, 228
Scorpion Senna, 124
Scorpion Vetch, 124
Small, 124
Scurvy-grass, Common, 92
Early, 92
English, 92
Greenland, 92
Sea-beet, 46
Seablite, Annual, 48
Hairy, 48
Shrubby, 48
Sea-heath, 152
Sea Holly, 158
Sea Kale, 92
Sea-lavender, Common, 182
Lax-flowered, 182
Matted, 182
Rock, 182
Sea Spurrey, Greater, 56
Greek, 56
Lesser, 56
Rock, 56
Self-heal, 202
Cut-leaved, 202
Large, 202
Senna, Bladder, 124
Scorpion, 124
Sermountain, 168
Service Tree, Wild, 116
Shaggy Soldier, 236
Shallon, 304
Shepherd's Cress, 96
Shepherd's Needle, 162
Shepherd's Purse, 96
Shoreweed, 294
Sibbaldia, 114
Silverweed, 114
Skullcap, 200
Lesser, 200
Spear-leaved, 200
Sloe, 118
Snapdragon, 214
Lesser, 214
Sneezewort, 242
Snowbell, Alpine, 178
Snowberry, 305
Snowdrop, 274
Snowdrop Windflower, 76
Snowflake, Spring, 274
Summer, 274
Soapwort, 60
Soldier, Gallant, 236
Shaggy, 236
Water, 262
Solomon's Seal, Angular, 272
Common, 272
Whorled, 272
Sorbaria, 304
Sorrel, Common, 44
French, 44
Mountain, 42
Sheep's, 44
Wood, 136

Sowbread, 180
Sow-thistle, Alpine, 254
Blue, 254
Marsh, 256
Perennial, 256
Prickly, 256
Smooth, 256
Spearwort, Adderstongue, 68
Creeping, 68
Greater, 68
Lesser, 68
Speedwell, Alpine, 218
American, 216
Blue Ivy-leaved, 218
Breckland, 216
Common Field, 218
Fingered, 216
French, 216
Germander, 218
Green Field, 218
Grey Field, 218
Heath, 218
Large, 218
Lilac Ivy-leaved, 218
Marsh, 218
Pink Water, 218
Rock, 218
Slender, 218
Spiked, 218
Spring, 216
Thyme-leaved, 218
Wall, 216
Water, 218
Wood, 218
Spikenard, Ploughman's, 234
Spignel, 162
Spindle, Evergreen, 304
Spindle-tree, 144
Spiraea, 304
Goatsbeard, 108
Spring Beauty, 42
Spruce, Norway, 24
Sitka, 24
Spurge, Broad-leaved, 140
Caper, 140
Cypress, 140
Dwarf, 140
Hairy, 140
Irish, 140
Leafy, 140
Marsh, 140
Petty, 140
Portland, 140
Purple, 140
Sea, 140
Sun, 140
Sweet, 140
Upright, 140
Wood, 140
Spurge Laurel, 142
Spurrey, Corn, 56
Greater Sea, 56
Greek Sea, 56
Lesser Sea, 56
Pearlwort, 56
Rock Sea, 56
Sand, 56
Squill, Alpine, 270
Autumn, 270
Spring, 270
Squinancywort, 188
Star-fruit, 262
Star of Bethlehem, Common, 272
Drooping, 272
Spiked, 272
Welsh, 266
Yellow, 266
Star-thistle, Cockspur, 252

Red, 252
Rough, 252
Yellow, 252
Starwort, 296
Autumnal Water, 296
Common Water, 296
Intermediate Water, 296
Stink Aster, 234
Stitchwort, Bog, 54
Greater, 54
Lesser, 54
Marsh, 54
Wood, 54
Stock, Hoary, 92
Sea, 92
Stonecrop, Alpine, 102
Annual, 102
Biting, 102
Caucasian, 102
English, 102
Hairy, 102
Mossy, 102
Reflexed, 102
Rock, 102
Tasteless, 102
Thick-leaved, 102
White, 102
Storksbill, Common, 136
Musk, 136
Sea, 136
Strapwort, 56
Strawberry, Barren, 112
Garden, 112
Hautbois, 112
Pink Barren, 112
Wild, 112
Strawberry Tree, 176
Succory, Lamb's, 256
Sumach, Staghorn, 304
Sundew, Common, 100
Great, 100
Oblong-leaved, 100
Sunflower, 240
Perennial, 240
Swede, 88
Sweet Briar, 110
Sweet Cicely, 160
Sweet Gale, 28
Sweet William, 64
Swinecress, 98
Lesser, 98
Sycamore, 34
Syringa, 304

Tamarisk, 144
Tansy, 244
Tape-grass, 296
Tare, Hairy, 126
Slender, 126
Smooth, 126
Tarragon, 244
Tasselweed, Beaked, 300
Spiral, 300
Teasel, 228
Small, 228
Tea-tree, Duke of Argyll's, 305
Thistle, Cabbage, 248
Carline, 248
Cotton, 248
Creeping, 250
Dwarf, 250
Globe, 248
Great Marsh, 250
Marsh, 250
Meadow, 250
Melancholy, 248
Milk, 250
Musk, 250
Slender, 250

Spear, 250
Star, 252
Stemless Carline, 248
Tuberous, 250
Welted, 250
Woolly, 250
Yellow Melancholy, 248
Thorn-apple, 210
Thorow-wax, 170
Narrow, 170
Thrift, 182
Jersey, 182
Thyme, Basil, 208
Large, 208
Wild, 208
Thymelaea, Annual, 142
Tillaea, Water, 102
Toadflax, Alpine, 214
Bastard, 38
Common, 214
Daisy-leaved, 214
Ivy-leaved, 214
Jersey, 214
Pale, 214
Prostrate, 214
Purple, 214
Sand, 214
Small, 214
Tobacco Plant, 210
Small, 210
Tomato, 210
Toothwort, 222
Purple, 222
Tormentil, 114
Trailing, 114
Towercress, 94
Traveller's Joy, 72
Tree of Heaven, 36
Trefoil, Birdsfoot, 132
Hop, 132
Large Hop, 132
Lesser, 132
Slender, 132
Tulip, Wild, 266
Tunic Flower, 62
Turnip, Wild, 88
Tutsan, 148
Stinking, 148
Tall, 148
Twayblade, Common, 284
Lesser, 284
Twinflower, 226

Valerian, Common, 226
Marsh, 226
Red, 226
Venus's Looking Glass, 230
Large, 230
Vervain, 196
Vetch, Bithynian, 126
Bush, 126
Common, 126
Crown, 122
Danzig, 126
Fine-leaved, 126
Fodder, 126
Horseshoe, 132
Kidney, 130
Scorpion, 124
Small Scorpion, 124
Spring, 126
Tufted, 126
Upright, 126
Wood, 126
Yellow, 126
Vetchling, Bitter, 128
Grass, 128
Hairy, 128
Meadow, 128

Yellow, 128
Vincetoxicum, 182
Vine, 304
Violet, Common Dog, 150
Dame's, 90
Early Dog, 150
Fen, 150
Hairy, 150
Heath Dog, 150
Marsh, 150
Meadow, 150
Northern, 150
Pale Dog, 150
Sweet, 150
Teesdale, 150
Water, 180
White, 150
Yellow Wood, 150
Viper's Grass, 254
Purple, 254
Virginia Creeper, 304

Wallflower, 86
Wall-rocket, Annual, 84
Perennial, 84
Walnut, 30
Water Chestnut, 262
Watercress, 90
Fool's, 166
One-rowed, 90
Water Crowfoot, Brackish, 70
Chalk-stream, 70
Common, 70
Fan-leaved, 70
Pond, 70
River, 70
Thread-leaved, 70
Water Dropwort,
Corky-fruited, 166
Fine-leaved, 166
Hemlock, 164
Narrow-leaved, 166
Parsley, 166
River, 166
Tubular, 166
Water-lily, Fringed, 182
Least, 66
White, 66
Yellow, 66
Water Milfoil, Alternate, 296
Spiked, 296
Whorled, 296
Water Parsnip, Greater, 164
Lesser, 166
Water-pepper, 40
Small, 40
Tasteless, 40
Water-plantain, Common, 262
Floating, 262
Lesser, 262
Narrow-leaved, 262
Parnassus-leaved, 262
Ribbon-leaved, 262
Water Soldier, 262
Water-thyme, Canadian, 296
Curly, 296
Greater, 296
Large-flowered, 296
Nuttall's, 296
Water Tillaea, 102
Water Violet, 180
Waterweed, Canadian, 296
Esthwaite, 296
Waterwort, Eight-stamened,
294
Six-stamened, 294
Three-stamened, 294
Whorled, 294
Wayfaring Tree, 226

Weld, 100
Whitebeam, 116
Broad-leaved, 116
Swedish, 116
Whitlow-grass, Common, 96
Rock, 96
Twisted, 96
Wall, 96
Yellow, 84
Willow, Almond, 26
Bay, 26
Bluish, 28
Crack, 26
Creeping, 28
Cricket-bat, 26
Dark-leaved, 28
Downy, 28
Dwarf, 28
Eared, 26
Goat, 26
Grey, 26
Mountain, 28
Net-leaved, 28
Purple, 26
Tea-leaved, 28
Violet, 26
Weeping, 26
White, 26
Whortle-leaved, 28
Woolly, 28
Willowherb, Alpine, 154
American, 154
Broad-leaved, 154
Chickweed, 154
Great, 154
Hoary, 154
Marsh, 154
New Zealand, 154
Pale, 154
Rosebay, 154
Short-fruited, 154
Spear-leaved, 154
Square-stemmed, 154
Windflower, Snowdrop, 76
Wineberry, 304
Wintercress, American, 86
Common, 86
Medium-flowered, 86
Small-flowered, 86
Wintergreen, Chickweed,
180
Common, 172
Intermediate, 172
One-flowered, 172
Round-leaved, 172
Toothed, 172
Umbellate, 172
Yellow, 172
Woad, 86
Wolfsbane, 74
Northern 74
Woodruff, 188
Blue, 188
Dyer's, 188
Wood-sorrel, 136
Wormwood, 244
Field, 244
Scottish, 242
Sea, 242
Woundwort, Annual, 206
Downy, 206
Field, 206
Hedge, 206
Limestone, 206
Marsh, 206
Yellow, 206

Yarrow, 242
Yellowcress, Austrian, 86

Creeping, 84
Great, 86
Iceland, 84
Marsh, 84
Yellow-wort, 184
Yew, 24

Index of Scientific Names

Abies alba, 24
  grandis, 24
  procera, 24
Acaena anserinifolia, 108
Acer campestre, 34
  monspessulanus, 34
  negundo, 304
  opalus, 34
  platanoides, 34
  pseudoplatanus, 34
Aceras anthropophorum, 284
Achillea collina, 242
  millefolium, 242
  nobilis, 242
  pannonica, 242
  ptarmica, 242
  setacea, 242
Acinos arvensis, 208
Aconitum firmum, 74
  napellus, 74
  septentrionale, 74
  variegatum, 74
  vulparia, 74
Acorus calamus, 302
Actaea erythrocarpa, 72
  spicata, 72
Adonis aestivalis, 74
  annua, 74
  flammea, 74
  vernalis, 74
Adoxa moschatellina, 222
Aegopodium podagraria, 162
Aesculus hippocastanum, 36
Aethusa cynapium, 162
  ssp. *cynapioides*, 162
Agrimonia eupatoria, 108
  procera, 108
Agrostemma githago, 60
Ailanthus altissima, 36
Ajuga chamaepitys, 200
  genevensis, 200
  pyramidalis, 200
  reptans, 200
Alcea rosea, 146
Alchemilla alpina, 108
  conjuncta, 108
  vulgaris, 108, 305
Alisma gramineum, 262
  lanceolatum, 262
  plantago-aquatica, 262
Alliaria petiolata, 90
Allium ampeloprasum, 268
  angulosum, 268
  babingtonii, 268
  carinatum, 268
  cepa, 268
  fistulosum, 268
  oleraceum, 268
  paradoxum, 268
  porrum, 268
  roseum, 268
  rotundum, 268
  schoenoprasum, 268
  scorodoprasum, 268
  senescens, 268
  sphaerocephalon, 266
  strictum, 268
  suaveolens, 268
  triquetrum, 268
  ursinum, 268

  victorialis, 268
  vineale, 268
Alnus glutinosa, 30
  incana, 30
  viridis, 30
Althaea hirsuta, 146
  officinalis, 146
Alyssum alyssoides, 84
  montanum, 84
  saxatile, 84
Amaranthus albus, 48
  blitoides, 48
  bouchonii, 48
  caudatus, 48
  cruentus, 48
  graecizans, 48
  hybridus, 48
  lividus, 48
  retroflexus, 48
Ambrosia artemisiifolia, 234
Amelanchier grandiflora, 116
  ovalis, 116
  spicata, 116
Ammi majus, 160
  visnaga, 160
Amsinckia intermedia, 192
Anacamptis pyramidalis, 282
Anagallis arvensis, 180
  foemina, 180
  minima, 180
  tenella, 180
Anaphalis margaritacea, 238
Anarrhinum bellidifolium, 214
Anchusa arvensis, 196
  azurea, 196
  ochroleuca, 192
  officinalis, 196
Andromeda polifolia, 174
Androsace elongata, 180
  lactea, 180
  maxima, 180
  septentrionalis, 180
Anemone apennina, 76
  narcissiflora, 76
  nemorosa, 76
  ranunculoides, 76
  sylvestris, 76
Anethum graveolens, 170
Angelica archangelica, 164
  palustris, 164
  sylvestris, 164
Antennaria alpina, 238
  carpatica, 238
  dioica, 238
  porsildii, 238
Anthemis arvensis, 236
  austriaca, 240
  cotula, 236
  tinctoria, 240
Anthericum liliago, 264
  ramosum, 264
Anthriscus caucalis, 160
  cerefolium, 160
  nitida, 305
  sylvestris, 160
Anthyllis vulneraria, 130
Antirrhinum majus, 214
Aphanes arvensis, 110
Apium graveolens, 166
  inundatum, 166

  nodiflorum, 166
  repens, 166
Aposeris foetida, 258
Aquilegia vulgaris, 74
Arabidopsis thaliana, 94
  suecica, 94
Arabis alpina, 94
  brownii, 305
  caucasica, 94
  glabra, 94
  hirsuta, 94
  pauciflora, 94
  recta, 94
  stricta, 94
  turrita, 94
Arbutus unedo, 176
Arctium lappa, 248
  minus, 248
  nemorosum, 306
  pubens, 306
  tomentosum, 248
Arctostaphylos alpina, 176
  uva-ursi, 176
Aremonia agrimonioides, 108
Arenaria balearica, 52
  ciliata, 52
  gothica, 52
  humifusa, 52
  norvegica, 52
  serpyllifolia, 52
Aristolochia clematitis, 38
Armeria alliacea, 182
  maritima, 182
  maritima ssp. *elongata*, 182
Armoracia rusticana, 92
Arnica angustifolia, 240
  montana, 240
Arnoseris minima, 256
Aronia melanocarpa, 304
Artemisia absinthium, 244
  austriaca, 244
  borealis, 306
  campestris, 244
  dracunculus, 244
  laciniata, 244
  maritima, 242
  norvegica, 242
  pontica, 244
  rupestris, 244
  verlotiorum, 244
  vulgaris, 244
Arthrocnemum perenne, 48
Arum italicum, 274
  maculatum, 274
Aruncus dioicus, 108
Asarum europaeum, 38
Asparagus officinalis, 272
Asperugo procumbens, 194
Asperula arvensis, 188
  cynanchica, 188
  occidentalis, 188
  tinctoria, 188
Aster alpinus, 236
  amellus, 236
  linosyris, 234
  michelii, 236
  novae-angliae, 236
  novi-belgii, 236
  × *salignus*, 236
  tripolium, 236

Astragalus alpinus, 122
  arenarius, 122
  cicer, 122
  danicus, 122
  frigidus, 122
  glycyphyllos, 122
  norvegicus, 122
  penduliflorus, 122
Astrantia major, 158
Athamanta cretensis, 162
Atriplex calotheca, 46
  glabriuscula, 305
  halimus, 304
  hastata, 46
  laciniata, 46
  littoralis, 46
  longipes, 305
  oblongifolia, 46
  patula, 46
  rosea, 46
Atropa bella-donna, 210
Azolla filiculoides, 300

Baldellia ranunculoides, 262
  repens, 262
Ballota nigra, 202
Barbarea intermedia, 86
  stricta, 86
  verna, 86
  vulgaris, 86
Bartsia alpina, 216
Bassia hirsuta, 48
Bellis perennis, 236
Berberis darwinii, 304
  glaucocarpa, 304
  × stenophylla, 304
  vulgaris, 78
  wilsonae, 304
Berteroa incana, 96
Berula erecta, 166
Beta vulgaris maritima, 46
Betula humilis, 30
  nana, 30
  pendula, 30
  pubescens, 30
Bidens cernua, 242
  connata, 242
  frondosa, 242
  radiata, 242
  tripartita, 242
Bifora radians, 160
Bilderdykia convolvulus, 42
  dumetorum, 42
Biscutella apricorum, 305
  divoniensis, 305
  guilloni, 305
  laevigata, 84
  neustriaca, 305
Blackstonia perfoliata, 184
  ssp. serotina, 184
Bombycilaena erecta, 238
Borago officinalis, 196
Botrychium lunaria, 274
Brassica juncea, 88
  napus var. napobrassica,
  88
  napus, 88
  nigra, 88
  oleracea, 88
  rapa, 88
Braya linearis, 94
Bryonia alba, 152
  cretica, 152
Buddleja davidii, 304
Buglossoides arvensis, 192
  purpuro-caerulea, 192
Bunias erucago, 86
  orientalis, 86
Bunium bulbocastanum, 162

Bupleurum baldense, 170
  falcatum, 170
  gerardi, 168
  lancifolium, 170
  longifolium, 170
  rotundifolium, 170
  tenuissimum, 168
Buphthalmum salicifolium,
  240
Butomus umbellatus, 262
Buxus sempervirens, 144

Cakile edentula, 305
  maritima, 92
Calamintha nepeta, 208
  sylvatica, 208
Caldesia parnassifolia, 262
Calendula arvensis, 244
  officinalis, 244
Calepina irregularis, 92
Calla palustris, 274
Callitriche brutia, 306
  cophocarpa, 306
  hamulata, 296
  hermaphroditica, 296
  obtusangula, 306
  palustris, 306
  platycarpa, 306
  stagnalis, 296
  truncata, 306
Calluna vulgaris, 174
Caltha palustris, 68
Calycocorsus stipitatus, 256
Calypso bulbosa, 278
Calystegia pulchra, 188
  sepium, 188
  silvatica, 188
  soldanella, 188
Camelina alyssum, 84
  macrocarpa, 84
  microcarpa, 84
  sativa, 84
Campanula barbata, 230
  baumgartenii, 230
  bononiensis, 230
  cervicaria, 230
  cochleariifolia, 230
  glomerata, 230
  latifolia, 230
  medium, 230
  patula, 230
  persicifolia, 230
  rapunculoides, 230
  rapunculus, 230
  rhomboidalis, 230
  rotundifolia, 230
  scheuchzeri, 230
  trachelium, 230
  uniflora, 230
Cannabis sativa, 38
Capsella bursa-pastoris, 96
  rubella, 96
Cardamine amara, 90
  bellidifolia, 90
  bulbifera, 90
  enneaphyllos, 90
  flexuosa, 94
  heptaphylla, 90
  hirsuta, 94
  impatiens, 90
  matthiolii, 305
  nymanii, 305
  palustris, 305
  parviflora, 90
  pentaphyllos, 90
  pratensis, 90
  raphanifolia, 90
Cardaminopsis arenosa, 94
  halleri, 94

  petraea, 94
Cardaria draba, 98
Carduus acanthoides, 250
  crispus, 250
  defloratus, 250
  nutans, 250
  palustre, 250
  personata, 250
  tenuiflorus, 250
Carlina acaulis, 248
  vulgaris, 248
Carpinus betulus, 32
Carpobrotus edulis, 42
Carthamus lanatus, 252
  tinctorius, 252
Carum carvi, 160
  verticillatum, 160
Cassiope hypnoides, 176
  tetragona, 176
Castanea sativa, 32
Caucalis platycarpos, 160
Centaurea aspera, 252
  calcitrapa, 252
  cyanus, 252
  debeauxii ssp. nemoralis,
  252
  jacea, 252
  maculosa, 306
  melitensis, 252
  montana, 252
  nigra, 252
  paniculata, 252
  rhenana, 252
  scabiosa, 252
  solstitialis, 252
Centaurium erythraea, 184
  littorale, 184
  pulchellum, 184
  scilloides, 184
  tenuiflorum, 184
Centranthus ruber, 226
Cephalanthera damasonium,
  286
  longifolia, 286
  rubra, 286
Cerastium alpinum, 54
  arcticum, 54
  arvense, 54
  brachypetalum, 54
  cerastoides, 54
  diffusum, 54
  dubium, 54
  fontanum, 54
  glomeratum, 52
  pumilum, 54
  semidecandrum, 54
  tomentosum, 54
Ceratophyllum demersum,
  296
  submersum, 296
Cerinthe minor, 192
Chaenorhinum minus, 214
Chaerophyllum aromaticum,
  160
  aureum, 160
  bulbosum, 160
  hirsutum, 160
  temulentum, 160
Chamaecyparis lawsoniana, 24
Chamaecytisus
  ratisbonensis, 124
  supinus, 124
Chamaedaphne calyculata,
  176
Chamaemelum nobile, 236
Chamaespartium
  sagittale, 124
Chamomilla suaveolens, 236
  recutita, 236

Chamorchis alpina, 284
Cheiranthus cheiri, 86
Chelidonium majus, 80
Chenopodium album, 46
  berlandieri, 46
  bonus-henricus, 46
  botryodes, 46
  ficifolium, 46
  glaucum, 46
  hybridum, 46
  murale, 46
  opulifolium, 46
  polyspermum, 46
  rubrum, 46
  suecicum, 46
  urbicum, 46
  × variabile, 46
  vulvaria, 46
Chimaphila umbellata, 172
Chondrilla juncea, 256
Chrysanthemum segetum, 244
Chrysosplenium
  alternifolium, 100
  oppositifolium, 100
  tetrandrum, 100
Cicendia filiformis, 184
Cicerbita alpina, 254
  macrophylla, 254
  plumieri, 254
  sibirica, 254
  tatarica, 254
Cichorium intybus, 254
Cicuta virosa, 164
Circaea alpina, 152
  × intermedia, 152
  lutetiana, 152
Cirsium acaule, 250
  arvense, 250
  dissectum, 250
  eriophorum, 250
  eristhales, 248
  helenioides, 250
  oleraceum, 248
  rivulare, 250
  spinosissimum, 248
  tuberosum, 250
  vulgare, 250
Clematis alpina, 72
  recta, 72
  vitalba, 72
Clinopodium vulgare, 208
Cnidium dubium, 168
Cochlearia aestuaria, 92
  alpina, 305
  anglica, 92
  danica, 92
  fenestrata, 92
  groenlandica, 92
  micacea, 305
  officinalis, 92
  pyrenaica, 305
  scotica, 305
Coeloglossum viride, 284
Colchicum autumnale, 270
Colutea arborescens, 124
Conioselinum tataricum, 164
Conium maculatum, 164
Conopodium majus, 162
Conringia orientalis, 86
Consolida ambigua, 74
  orientalis, 74
  regalis, 74
Convallaria majalis, 264
Convolvulus arvensis, 188
Conyza canadensis, 234
Corallorhiza trifida, 290
Coriandrum sativum, 160
Corispermum leptopterum, 48
  marschalii, 48

Cornus alba, 158
  mas, 158
  sanguinea, 158
  sericea, 158
  suecica, 158
Coronilla coronata, 124
  emerus, 124
  minima, 124
  vaginalis, 124
  varia, 122
Coronopus didymus, 98
  squamatus, 98
Corrigiola litoralis, 56
Corydalis bulbosa, 78
  claviculata, 78
  intermedia, 78
  lutea, 78
  ochroleuca, 78
  pumila, 78
  solida, 78
Corylus avellana, 32
Cotoneaster frigida, 304
  horizontalis, 118
  integerrimus, 118
  microphyllus, 118
  nebrodensis, 118
  niger, 118
  salicifolia, 304
  simonsii, 118
Cotula coronopifolia, 242
Crambe maritima, 92
Crassula aquatica, 102
  tillaea, 102
  vaillantii, 102
Crataegus calycina, 116
  laevigata, 116
  monogyna, 116
Crepis biennis, 260
  capillaris, 258
  foetida, 260
  mollis, 260
  nicaeensis, 260
  paludosa, 260
  praemorsa, 260
  pulchra, 260
  sancta, 260
  setosa, 260
  tectorum, 260
  vesicaria, 260
Crithmum maritimum, 170
Crocosmia × crocosmiiflora,
  276
Crocus albiflorus, 276
  nudiflorus, 276
Cruciata glabra, 190
  laevipes, 190
Cucubalus baccifer, 62
Cuscuta epilinum, 188
  epithymum, 188
  europaea, 188
Cyclamen hederifolium, 180
  purpurascens, 180
Cydonia oblonga, 304
Cymbalaria muralis, 214
Cynoglossum germanicum,
  192
  officinale, 192
Cypripedium calceolus, 278
Cytisus scoparius, 124

Daboecia cantabrica, 174
Dactylorhiza cruenta, 282
  fuchsii, 282
  incarnata, 282
  maculata, 282
  majalis, 282
  praetermissa, 282
  purpurella, 282
  sambucina, 282

  traunsteineri, 282
Daphne laureola, 142
  mezereum, 142
Damasonium alisma, 262
Datura stramonium, 210
Daucus carota, 160
Descurainia sophia, 86
Dianthus arenarius, 64
  armeria, 64
  barbatus, 64
  carthusianorum, 64
  caryophyllus, 64
  deltoides, 64
  gallicus, 64
  gratianopolitanus, 64
  plumarius, 64
  seguieri, 64
  superbus, 64
Diapensia lapponica, 172
Dictamnus albus, 142
Digitalis grandiflora, 216
  lutea, 216
  purpurea, 216
Diplotaxis muralis, 84
  tenuifolia, 84
  viminea, 84
Dipsacus fullonum, 228
  laciniatus, 228
  pilosus, 228
Dittrichia graveolens, 234
Doronicum pardalianches,
  240
  plantagineum, 240
Dorycnium pentaphyllum,
  134
Draba aizoides, 84
  alpina, 84
  cacuminum, 305
  cinerea, 305
  crassifolia, 84
  daurica, 96
  fladnizensis, 96
  glacialis, 305
  incana, 96
  muralis, 96
  nemorosa, 96
  nivalis, 96
  norvegica, 96
Dracocephalum ruyschiana,
  202
  thymiflorum, 202
Drosera anglica, 100
  intermedia, 100
  rotundifolia, 100
Dryas octopetala, 112

Echinocystis lobata, 152
Echinops exaltatus, 248
  sphaerocephalus, 248
Echium plantagineum, 196
  vulgare, 196
Egeria densa, 302
Elatine alsinastrum, 294
  hexandra, 294
  hydropiper, 294
  triandra, 294
Elodea callitrichoides, 296
  canadensis, 296
  ernstiae, 296
  nuttallii, 296
Empetrum nigrum, 176
Endymion hispanicus, 270
  nonscriptus, 270
Ephedra distachya, 24
Epilobium adenocaulon, 154
  alpestre, 154
  alsinifolium, 154
  anagallidifolium, 154
  angustifolium, 154

brunnescens, 154
collinum, 154
davuricum, 154
dodonaei, 154
duriaei, 305
glandulosum, 154
hirsutum, 154
hornemanii, 154
komarovianum, 154
lactiflorum, 154
lanceolatum, 154
latifolium, 154
montanum, 154
nutans, 154
obscurum, 154
palustre, 154
parviflorum, 154
pedunculare, 154
roseum, 154
tetragonum, 154
Epipactis atrorubens, 288
confusa, 288
dunensis, 288
helleborine, 288
leptochila, 288
microphylla, 288
muelleri, 288
palustris, 288
phyllanthes, 288
purpurata, 288
youngiana, 306
Epipogium aphyllum, 290
Eranthis hyemalis, 66
Erica carnea, 174
ciliaris, 174
cinerea, 174
erigena, 174
mackaiana, 174
tetralix, 174
vagans, 174
Erigeron acer, 236
alpinus, 236
annuus, 236
borealis, 306
humilis, 236
karvinskianus, 236
uniflorus, 236
Erinus alpinus, 216
Eriocaulon aquaticum, 264
Erodium cicutarium, 136
maritimum, 136
moschatum, 136
Erophila verna, 96
Erucastrum gallicum, 88
nasturtiifolium, 88
Eryngium campestre, 158
maritimum, 158
planum, 158
Erysimum cheiranthoides, 86
crepidifolium, 86
hieracifolium, 86
odoratum, 86
repandum, 86
Escallonia macrantha, 304
Euonymus europaeus, 144
japonicus, 304
latifolius, 144
Eupatorium cannabinum, 234
Euphorbia amygdaloides, 140
brittingeri, 140
cyparissias, 140
dulcis, 140
esula, 140
exigua, 140
helioscopia, 140
hyberna, 140
lathyris, 140
lucida, 140
palustris, 140

paralias, 140
peplis, 140
peplus, 140
platyphyllos, 140
portlandica, 140
seguieriana, 140
serrulata, 140
villosa, 140
Euphrasia officinalis, 220,
306
salisburgensis, 220
Exaculum pusillum, 184

Fagopyrum esculentum, 40
tataricum, 40
Fagus sylvatica, 32
Falcaria vulgaris, 164
Ficus carica, 304
Filaginella uliginosa, 238
Filago lutescens, 230
pyramidata, 238
vulgaris, 238
Filipendula ulmaria, 108
vulgaris, 108
Foeniculum vulgare, 170
Fragaria × ananassa, 112
moschata, 112
vesca, 112
viridis, 112
Frangula alnus, 144
Frankenia laevis, 152
Fraxinus excelsior, 36
ornus, 36
Fritillaria meleagris, 266
Fuchsia magellanica, 304
Fumana procumbens, 152
Fumaria bastardii, 78
capreolata, 78
densiflora, 78
martinii, 78
muralis, 78
occidentalis, 78
officinalis, 78
parviflora, 78
purpurea, 78
rostellata, 305
schleicheri, 78
schrammii, 305
vaillantii, 78

Gagea arvensis, 266
bohemica, 266
lutea, 266
minima, 266
pratensis, 266
spathacea, 266
Galanthus nivalis, 274
Galega officinalis, 122
Galeopsis angustifolia, 204
bifida, 306
ladanum, 306
pubescens, 204
segetum, 204
speciosa, 204
tetrahit, 204
Galinsoga quadriradiata, 236
parviflora, 236
Galium album, 305
aparine, 190
boreale, 190
debile, 190
elongatum, 305
fleurotii, 190
glaucum, 188, 190
mollugo, 190
normanii, 305
odoratum, 188
oelandicum, 305
palustre, 190

parisiense, 190
pumilum, 190
rotundifolium, 190
saxatile, 190
spurium, 190
sterneri, 190
suecicum, 305
sylvaticum, 190
timeroyi, 305
tricornutum, 190
trifidum, 190
triflorum, 190
uliginosum, 190
valdepilis, 305
verum, 190
Gaultheria shallon, 304
Genista anglica, 124
germanica, 124
pilosa, 124
tinctoria, 124
Gentiana asclepiadea, 186
cruciata, 186
lutea, 184
nivalis, 186
pneumonanthe, 186
punctata, 186
purpurea, 186
utriculosa, 186
verna, 186
Gentianella amarella, 186
anglica, 186
aspera, 186
aurea, 186
campestris, 186
ciliata, 186
detonsa, 186
germanica, 186
tenella, 186
uliginosa, 186
Geranium bohemicum, 138
columbinum, 138
dissectum, 138
divaricatum, 138
endressii, 138
lanuginosum, 305
lucidum, 138
molle, 138
nodosum, 138
palustre, 138
phaeum, 138
pratense, 138
purpureum, 138
pusillum, 138
pyrenaicum, 138
robertianum, 138
rotundifolium, 138
sanguineum, 138
sylvaticum, 138
versicolor, 138
Geum hispidum, 112
× intermedium, 112
montanum, 112
rivale, 112
urbanum, 112
Gladiolus illyricus, 276
palustris, 276
Glaucium corniculatum, 80
flavum, 80
Glaux maritima, 180
Glechoma hederacea, 202
Globularia vulgaris, 228
Gnaphalium luteoalbum, 238
undulatum, 238
Goodyera repens, 286
Gratiola officinalis, 216
Groenlandia densa, 298
Guizotia abyssinica, 240
Gymnadenia conopsea, 282
odoratissima, 282

Gypsophila fastigiata, 62
  muralis, 62
  repens, 62

Halimione pedunculata, 48
  portulacoides, 48
Hedera helix, 158
Hedysarum hedysaroides,
  130
Helianthemum apenninum,
  152
  canum, 152
  nummularium, 152
  oelandicum, 152
Helianthus annuus, 240
  rigidus, 240
  tuberosus, 240
Helichrysum arenarium, 238
  stoechas, 238
Heliotropium europaeum, 192
Helleborus foetidus, 66
  viridis, 66
Hepatica nobilis, 76
Heracleum mantegazzianum,
  164
  sphondylium, 164
Herminium monorchis, 284
Herniaria ciliolata, 56
  glabra, 56
  hirsuta, 56
Hesperis matronalis, 90
Hieracium, 260, 306
  alpinum, 260
  aurantiacum, 260
  maculatum, 260
  murorum, 260
  lactucella, 258
  pilosella, 258
  umbellatum, 260
Himantoglossum hirçinum,
  284
Hippophae rhamnoides, 144
Hippocrepis comosa, 132
Hippuris vulgaris, 296
Hirschfeldia incana, 88
Holosteum umbellatum, 54
Homogyne alpina, 244
Honkenya peploides, 52
Hornungia petraea, 96
Hottonia palustris, 180
Humulus lupulus, 38
Hydrilla verticillata, 296
Hydrocharis morsus-ranae,
  262
Hydrocotyle vulgaris, 158
Hymenolobus procumbens,
  96
Hyoscyamus niger, 210
Hypericum androsaemum,
  148
  calycinum, 148
  canadense, 148
  elegans, 305
  elodes, 148
  hircinum, 148
  hirsutum, 148
  humifusum, 148
  inodorum, 148
  linarifolium, 148
  maculatum, 148
  majus, 148
  montanum, 148
  mutilum, 148
  perforatum, 148
  pulchrum, 148
  tetrapterum, 148
  undulatum, 148
Hypochaeris glabra, 258
  maculata, 258

  radicata, 258
Hyssopus officinalis, 202

Iberis amara, 96
  intermedia, 96
  umbellata, 96
Ilex aquifolium, 36
Illecebrum verticillatum, 58
Impatiens capensis, 142
  glandulifera, 142
  noli-tangere, 142
  parviflora, 142
Inula britannica, 240
  conyza, 234
  crithmoides, 234
  ensifolia, 240
  germanica, 240
  helenium, 240
  helvetica, 240
  hirta, 240
  salicina, 240
Iris aphylla, 276
  foetidissima, 276
  germanica, 276
  pseudacorus, 276
  sambucina, 276
  sibirica, 276
  spuria, 276
  versicolor, 276
Isatis tinctoria, 86
Isoetes lacustris, 294
  setacea, 294

Jasione laevis, 228
  montana, 228
Jovibarba sobolifera, 102
Juglans regia, 30
Juniperus communis, 24
  ssp. nana, 24
Jurinea cyanoides, 252

Kernera saxatilis, 92
Kickxia elatine, 214
  spuria, 214
Knautia arvensis, 228
  dipsacifolia, 228
Kochia laniflora, 48
  scoparia, 48
Koenigia islandica, 42

Laburnum anagyroides, 124
Lactuca perennis, 254
  quercina, 256
  saligna, 256
  serriola, 256
  viminea, 256
  virosa, 256
Lagarosiphon major, 296
Lamiastrum galeobdolon, 204
Lamium album, 204
  amplexicaule, 204
  hybridum, 204
  maculatum, 204
  moluccellifolium, 305
  purpureum, 204
Lappula deflexa, 194
  squarrosa, 194
Lapsana communis, 256
Larix decidua, 24
  kaempferi, 24
Laser trilobum, 168
Laserpitium latifolium, 168
  prutenicum, 168
  siler, 168
Lathraea clandestina, 222
  squamaria, 222, 290
Lathyrus aphaca, 128
  heterophyllus, 128
  hirsutus, 128

  japonicus, 128
  latifolius, 128
  montanus, 128
  montanus var. tenuifolius,
    128
  niger, 128
  nissolia, 128
  palustris, 128
  pannonicus, 128
  pratensis, 128
  sphaericus, 128
  sylvestris, 128
  tuberosus, 128
  vernus, 128
Lavatera arborea, 146
  cretica, 146
  thuringiaca, 146
Ledum palustre, 176
Legousia hybrida, 230
  speculum-veneris, 230
Lembotropis nigricans, 124
Lemna gibba, 300
  minor, 300
  minuscula, 300
  trisulca, 300
  valdiviana, 300
Leontodon autumnalis, 258
  hispidus, 258
  hispidus ssp. hyoseroides,
    258
  incanus, 258
  pyrenaicus, 258
  taraxacoides, 258
Leonurus cardiaca, 204
  marrubiastrum, 204
Lepidium bonariense, 98
  campestre, 98
  densiflorum, 98
  divaricatum, 98
  graminifolium, 98
  heterophyllum, 98
  latifolium, 92
  neglectum, 98
  perfoliatum, 98
  ruderale, 98
  sativum, 98
  virginicum, 98
Leucanthemum maximum, 244
  vulgare, 244
Leucojum aestivum, 274
  vernum, 274
Levisticum officinale, 170
Leycesteria formosa, 305
Ligusticum mutellinum, 168
  scoticum, 168
Ligustrum ovalifolium, 182
  vulgare, 182
Lilium bulbiferum, 266
  martagon, 266
  pyrenaicum, 266
  pyrenaicum ssp. bulbiferum,
    266
  pyrenaicum ssp. croceum,
    266
Limodorum abortivum, 290
Limonium auriculae-
  ursifolium, 182
  bellidifolium, 182
  binervosum, 182
  humile, 182
  paradoxum, 305
  recurvum, 305
  transwallianum, 305
  vulgare, 182
Limosella aquatica, 294
  australis, 294
Linaria alpina, 214
  arenaria, 214
  arvensis, 214

genistifolia, 214
incarnata, 214
pelisseriana, 214
purpurea, 214
repens, 214
supina, 214
vulgaris, 214
Lindernia procumbens, 214
Linnaea borealis, 226
Linum austriacum, 305
bienne, 136
catharticum, 136
flavum, 136
le nii, 305
perenne, 136
suffruticosum, 136
tenuifolium, 136
usitatissimum, 136
viscosum, 136
Liparis loeselii, 290
Listera cordata, 284
ovata, 284
Lithospermum officinale, 192
Littorella uniflora, 294
Lloydia serotina, 264, 306
Lobelia dortmanna, 294
urens, 230
Lobularia maritima, 96
Logfia arvensis, 238
gallica, 238
minima. 238
neglecta, 238
Loiseleuria procumbens, 176
Lomatogonium rotatum, 186
Lonicera alpigena, 226
caerulea, 226
caprifolium, 226
nigra, 226
periclymenum, 226
xylosteum, 226
Lotus angustissimus. 132
corniculatus, 132
subbiflorus, 132
tenuis, 132
uliginosus, 132
Ludwigia palustris, 294
Lunaria annua, 90
rediviva, 90
Lupinus albus, 124
angustifolius, 124
arboreus, 124
luteus, 124
micranthus, 124
nootkatensis, 124
polyphyllus, 124
Luronium natans, 262
Lychnis alpina, 60
flos-cuculi, 60
viscaria, 60
Lycium barbarum, 305
Lycopersicon esculentum, 210
Lycopus europaeus, 208
Lysimachia nemorum, 178
nummularia, 178
punctata, 178
thyrsiflora, 178
vulgaris, 178
Lythrum borysthenicum, 294
hyssopifolia, 154
portula, 294
salicaria, 154

Maianthemum bifolium, 264
Mahonia aquifolium, 78
Malaxis monophyllos, 290
paludosa, 290
Malus domestica, 116
sylvestris, 116

Malva alcea, 146
moschata, 146
neglecta, 146
parviflora, 146
pusilla, 146
sylvestris, 146
verticillata, 146
Marrubium vulgare, 202
Matricaria maritima, 236
perforata, 236
Matthiola incana, 92
sinuata, 92
Meconopsis cambrica, 80
Medicago arabica, 132
lupulina, 132
minima, 132
polymorpha, 132
sativa, 130
sativa ssp. falcata, 130
sativa ssp. sativa, 130
Melampyrum arvense, 220
cristatum, 220
nemorosum, 220
pratense, 220
sylvaticum, 220
Melilotus alba, 130
altissima, 130
dentata, 130
indica, 130
officinalis, 130
Melissa officinalis, 204
Melittis melissophyllum, 202
Mentha aquatica, 208
arvensis, 208
× piperita, 208
pulegium, 208
spicata, 208
suaveolens, 208
× verticillata, 208
× villosa var. alopecuroides, 208
Menyanthes trifoliata, 182
Mercurialis annua, 140
ovata. 140
perennis, 140
Mertensia maritima, 196
Mespilus germanica, 304
Meum athamanticum, 162
Mimulus guttatus, 216
luteus, 216
moschatus, 216
Minuartia biflora, 52
hybrida, 52
mediterranea, 52
recurva, 52
rubella, 52
rubra, 52
sedoides, 52
setacea, 52
stricta, 52
verna, 52
viscosa, 52
Misopates orontium, 214
Moehringia muscosa, 52
trinervia, 52
Moenchia erecta, 52
Moneses uniflora, 172
Monotropa hypophegea, 305
hypopitys, 172, 290
Montia fontana, 42
perfoliata, 42
sibirica, 42
Muscari atlanticum, 270
botryoides, 270
comosum, 270
neglectum, 270
Myagrum perfoliatum, 84
Mycelis muralis, 256
Myosotis alpestris, 194

arvensis, 194
baltica, 305
discolor, 194
laxa, 194
ramosissima, 194
secunda, 194
scorpioides. 194
sicula, 194
sparsiflora, 194
stolonifera, 194
stricta, 194
sylvatica, 194
Myosoton aquaticum, 54
Myosurus minimus, 74
Myrica carolinensis, 28
gale, 28
Myricaria germanica, 144
Myriophyllum alterniflorum, 296
spicatum, 296
verticillatum, 296
Myrrhis odorata, 160

Najas flexilis, 300
marina, 300
minor, 300
Narcissus × biflorus, 274
obvallaris, 274
poeticus, 274
pseudonarcissus, 274
stellaris, 274
Narthecium ossifragum, 266
Nasturtium microphyllum, 90
officinale, 90
Neotinea intacta, 284
Neottia nidus-avis, 290
Nepeta cataria, 202
nuda, 202
Neslia paniculata, 84
Nicandra physalodes, 210
Nicotiana rustica, 210
tabacum, 210
Nigella arvensis, 66
damascena, 66
Nigritella nigra, 278
Nonea pulla, 192
rosea, 192
Nuphar lutea, 66
pumila, 66
Nymphaea alba, 66
candida, 66
Nymphoides peltata, 182

Odontites jaubertiana, 216
lutea, 216
verna, 216
Oenanthe aquatica, 166
crocata, 164
fistulosa, 166
fluviatilis, 166
lachenalii, 166
peucedanifolia, 166
pimpinelloides, 166
silaifolia, 166
Oenothera biennis, 152
cambrica, 152
erythrosepala, 152
fallax, 152
rubricaulis, 152
salicifolia, 152
stricta, 152
Omalotheca norvegica, 238
supina, 238
sylvatica, 238
Omphalodes scorpioides, 194
verna, 194
Onobrychis arenaria, 130
viciifolia, 130
Ononis arvensis, 130

natrix, 130
pusilla, 130
reclinata, 130
repens, 130
spinosa, 130
Onopordum acanthium, 248
Onosma arenaria, 196
Ophioglossum vulgatum, 274
Ophrys apifera, 278
apifera var. trollii, 278
fuciflora, 278
insectifera, 278
sphegodes, 278
Orchis coriophora, 280
laxiflora, 280
mascula, 280
militaris, 280, 284
morio, 280
pallens, 280
palustris, 280
purpurea, 280, 284
simia, 280, 284
tridentata, 280
ustulata, 280
Origanum vulgare, 208
Orlaya grandiflora, 160
Ornithogalum nutans, 272
pyrenaicum, 272
umbellatum, 272
Ornithopus perpusillus, 132
pinnatus, 132
Orobanche alba, 224
alsatica, 222
amethystea, 222
arenaria, 222
caryophyllacea, 222
elatior, 222
flava, 222
gracilis, 224
hederae, 224
loricata, 224
maritima, 224
minor, 224, 290
purpurea, 222
ramosa, 222
rapum-genistae, 222
reticulata, 224
teucrii, 224
Orthilia secunda, 172
Otanthus maritimus, 242
Oxalis acetosella, 136
articulata, 136
corniculata, 136
corymbosa, 136
europaea, 136
latifolia, 136
pes-caprae, 136
stricta, 136
Oxyria digyna, 42
Oxytropis campestris, 122
deflexa, 122
halleri, 122
lapponica, 122
pilosa, 122

Papaver argemone, 80
dahlianum, 305
dubium, 80
hybridum, 80
laestadianum, 305
lapponicum, 305
lecoqii, 80
radicatum, 80
rhoeas, 80
somniferum, 80
Parentucellia viscosa, 216
Parietaria judaica, 38
officinalis, 38
Paris quadrifolia, 272

Parnassia palustris, 100
Parthenocissus inserta, 304
quinquefolia, 304
Pastinaca sativa, 170
Pedicularis flammea, 220
foliosa, 220
hirsuta, 220
lapponica, 220
oederi, 220
palustris, 220
sceptrum-carolinum, 220
sylvatica, 220
Pentaglottis sempervirens, 196
Pernettya mucronata, 304
Petasites albus, 242
fragrans, 242
frigidus, 242
hybridus, 242
japonicus, 242
spurius, 242
Petrorhagia nanteulii, 62
prolifera, 62
saxifraga, 62
Petroselinum crispum, 166
segetum, 166
Peucedanum alsaticum, 168
carvifolia, 168
cervaria, 168
gallicum, 168
lancifolium, 168
officinale, 168
oreoselinum, 168
ostruthium, 168
palustre, 168
Philadelphus coronarius, 304
Phyllodoce caerulea, 174
Physocarpus opulifolius, 304
Physospermum cornubiense,
162
Phyteuma nigrum, 228
orbiculare, 228
spicatum, 228
Picea abies, 24
sitchensis, 24
Picris echioides, 260
hieracioides, 260
Pimpinella major, 162
saxifraga, 162
Pinguicula alpina, 222
grandiflora, 222
lusitanica, 222
villosa, 222
vulgaris, 222
Pinus contorta, 24
mugo, 24
nigra, 24
pinaster, 24
radiata, 24
strobus, 24
sylvestris, 24
Pittosporum crassifolium,
304
Plantago arenaria, 224
coronopus, 224
lanceolata, 224
major, 224
maritima, 224
media, 224
Platanthera bifolia, 286
chlorantha, 286
hyperborea, 286
obtusa ssp. oligantha, 286
Platanus hybrida, 34
Pleurospermum austriacum,
164
Polemonium acutiflorum, 188
caeruleum, 188
Polycarpon diphyllum, 58
tetraphyllum, 58

Polycnemum arvense, 48
majus, 48
Polygala amarella, 142
calcarea, 142
chamaebuxus, 142
comosa, 142
serpyllifolia, 142
vulgaris, 142
Polygonatum × hybridum, 272
multiflorum, 272
odoratum, 272
verticillatum, 272
Polygonom amphibium, 40
arenastrum, 40
aviculare, 40
bistorta, 40
boreale, 40
foliosum, 305
hydropiper, 40
lapathifolium, 40
maritimum, 40
minus, 40
mite, 40
oxyspermum, 40
patulum, 40
persicaria, 40
polystachyum, 42
rurivagum, 40
viviparum, 40
Populus alba, 30
× canadensis, 30
canescens, 30
gileadensis, 30
nigra, 30
nigra var. italica, 30
tremula, 30
Portulaca oleracea, 42
Potamogeton acutifolius, 298
alpinus, 298
berchtoldii, 298
coloratus, 298
compressus, 298
crispus, 298
epihydrus, 298
filiformis, 302
friesii, 298
gramineus, 298
lucens, 298
natans, 298
nodosus, 298
obtusifolius, 298
pectinatus, 302
perfoliatus, 298
polygonifolius, 298
praelongus, 298
pusillus, 298
rutilus, 298
trichoides, 298
vaginatus, 302
Potentilla alba, 112
anglica, 114
anserina, 114
argentea, 114
aurea, 114
chamissonis, 305
cinerea, 114
collina, 114
crantzii, 114
erecta, 114
fruticosa, 114
heptaphylla, 114
hyparctica, 114
inclinata, 114
intermedia, 114
micrantha, 112
montana, 112
multifida, 114
neglecta, 305
nivea, 114

norvegica, 114
palustris, 112
pusilla, 114
recta, 114
reptans, 114
rupestris, 112
sterilis, 112
supina, 114
tabernaemontani, 114
thuringiaca, 114
Prenanthes purpurea, 254
Primula auricula, 178
elatior, 178
farinosa, 178
nutans, 178
scandinavica, 178
scotica, 178
stricta, 178
veris, 178
veris × vulgaris, 178
vulgaris, 178
Prunella grandiflora, 202
laciniata, 202
vulgaris, 202
Prunus avium, 118
cerasifera, 118
cerasus, 118
domestica, 118
Prunus domestica ssp.
insititia, 118
fruticosa, 118
laurocerasus, 118
lusitanica, 118
mahaleb, 118
padus, 118
serotina, 118
spinosa, 118
virginiana, 118
Pseudorchis albida, 286
Pseudotsuga menziesii, 24
Ptychotis saxifraga, 162
Pulicaria dysenterica, 240
vulgaris, 238
Pulmonaria angustifolia, 194
longifolia, 194
mollis, 194
officinalis, 194
Pulsatilla patens, 76
pratensis, 76
vernalis, 76
vulgaris, 76
Pyracantha coccinea, 304
Pyrola chlorantha, 172
media, 172
minor, 172
norvegica, 172
rotundifolia, 172
Pyrus communis, 116
cordata, 116
pyraster, 116
salvifolia, 116

Radiola linoides, 136
Ranunculus aconitifolius, 70
acris, 68
aquatilis, 70
arvensis, 70
auricomus, 70
baudotii, 70
bulbosus, 68
circinatus, 70
cymbalaria, 68

ficaria, 68
flammula, 68
fluitans, 70
glacialis, 70
hederaceus, 70
hyperboreus, 68
illyricus, 68
lanuginosus, 68
lapponicus, 68
lingua, 68
montanus, 68
nemorosus, 68
nivalis, 70
ololeucos, 70
omiophyllus, 70
ophioglossifolius, 68
oreophilus, 305
paludosus, 68
parviflorus, 70
poltatus, 70
penicillatus, 70
platanifolius, 70
polyanthemos, 68
pygmaeus, 70
repens, 68
reptans, 68
sardous, 68
sceleratus, 70
sulphureus, 70
trichophyllus, 70
tripartitus, 70
Raphanus raphanistrum, 88
raphanistrum ssp.
maritimus, 88
sativus, 88
Rapistrum perenne, 88
rugosum, 88
Reseda alba, 100
lutea, 100
luteola, 100
phyteuma, 100
Reynoutria japonica, 42
sachalinensis, 42
Rhamnus catharticus, 144
saxatilis, 144
Rhinanthus minor, 220, 306
Rhodiola rosea, 100
Rhododendron lapponicum,
174
luteum, 174, 304
ponticum, 174, 304
Rhynchosinapis cheiranthos,
88
monensis, 88
wrightii, 88
Rhus typhina, 304
Ribes alpinum, 144
nigrum, 144
petraeum, 144
rubrum, 144
sanguineum, 304
spicatum, 305
uva-crispa, 144
Robinia pseudacacia, 124
Romulea columnae, 276
Rorippa amphibia, 86
austriaca, 86
islandica, 84
palustris, 84
pyrenaica, 84
sylvestris, 84
Rosa agrestis, 305
arvensis, 110
canina, 110, 305
elliptica, 305
foetida, 304
gallica, 110
micrantha, 305
mollis, 305

pimpinellifolia, 110
rubiginosa, 110
rugosa, 304
scabriuscula, 305
sempervirens, 304
sherardii, 305
stylosa, 110
tomentosa, 110
villosa, 305
virginiana, 304
Rubia peregrina, 190
Rubus arcticus, 110
caesius, 110
chamaemorus, 110
fruticosus, 110, 305
idaeus, 110
odoratus, 304
phoenicolasius, 304
saxatilis, 110
spectabilis, 304
Rudbeckia hirta, 240
laciniata, 240
Rumex acetosa, 44
acetosella, 44
alpinus, 44
aquaticus, 44
arifolius, 44
conglomeratus, 44
crispus, 44
frutescens, 44
hydrolapathum, 44
longifolius, 44
maritimus, 42
obtusifolius, 44
palustris, 42
patientia, 44
pseudonatronatus, 44
pulcher, 44
rupestris, 44
sanguineus, 44
scutatus, 44
thyrsiflorus, 44
triangulivalvis, 44
Ruppia cirrhosa, 300
maritima, 300
Ruscus aculeatus, 272

Sagina apetala, 56
caespitosa, 305
intermedia, 56
maritima, 56
nodosa, 56
procumbens, 56
saginoides, 56
subulata, 56
Sagittaria natans, 262
rigida, 262
sagittifolia, 262
Salicornia europaea, 48, 305
Salix alba, 26
alba ssp. coerulea, 26
alba ssp. vitellina, 26
appendiculata, 26
arbuscula, 28
arenaria, 305
atrocinerea, 305
aurita, 26
babylonica, 26
bicolor, 305
borealis, 305
caprea, 26
cinerea, 26
coaetanea, 305
daphnoides, 26
elaeagnos, 26
fragilis, 26
glandulifera, 305
glauca, 28
hastata, 28

herbacea, 28
hibernica, 305
lanata, 28
lapponum, 28
myrsinites, 28
myrtilloides, 28
nigricans, 28
pentandra, 26
phylicifolia, 28
polaris, 28
purpurea, 26
repens, 28
reticulata, 28
rosmarinifolia, 305
starkeana, 28
stipulifera, 305
triandra, 26
viminalis, 26
xerophila, 305
Salsola kali, 48
Salvia glutinosa, 206
nemorosa, 206
pratensis, 206
verbenaca, 206
verticillata, 206
Sambucus ebulus, 226
nigra, 226
racemosa, 226
Samolus valerandi, 180
Sanguisorba minor, 108
officinalis, 108
Sanicula europaea, 158
Saponaria officinalis, 60
Satureja montana, 202
Saussurea alpina, 248
Saxifraga adscendens, 106
aizoides, 106
cernua, 106
cespitosa, 106
cotyledon, 104
foliolosa, 104
× geum, 104
granulata, 104
hartii, 305
hieracifolia, 106
hirculus, 106
hirsuta, 104
hypnoides, 106
nivalis, 106
osloensis, 106 —
oppositifolia, 106
paniculata, 104
× polita, 104
rivularis, 106
rosacea, 106
rotundifolia, 104
spathularis, 104
stellaris, 104
tenuis, 106
tridactylites, 106
umbrosa, 104
× urbium, 104
Scabiosa canescens, 228
columbaria, 228
ochroleuca, 228
Scandix pecten-veneris, 162
Scheuchzeria palustris, 224
Scilla autumnalis, 270
bifolia, 270
verna, 270
Scleranthus annuus, 56
perennis, 56
Scolymus hispanicus, 252
Scorzonera austriaca, 254
hispanica, 254
humilis, 254
laciniata, 254
purpurea, 254
Scrophularia auriculata, 212

canina, 212
nodosa, 212
scorodonia, 212
umbrosa, 212
vernalis, 212
Scutellaria columnae, 200
galericulata, 200
hastifolia, 200
minor, 200
Sedum acre, 102
album, 102
alpestre, 102
andegavense, 102
anglicum, 102
annuum, 102
dasyphyllum, 102
forsteranum, 102
hirsutum, 102
hispanicum, 102
reflexum, 102
rubens, 102
sexangulare, 102
spurium, 102
telephium, 102
villosum, 102
Selinum carvifolia, 164
pyrenaeum, 164
Sempervivum tectorum, 102
Senecio aquaticus, 246
bicolor, 246
cambrensis, 306
congestus, 246
erraticus, 246
erucifolius, 246
fluviatilis, 246
helenitis, 246
integrifolius, 246
jacobaea, 246
nemorensis, 246
paludosus, 246
rivularis, 246
squalidus, 246
sylvaticus, 246
vernalis, 246
viscosus, 246
vulgaris, 246
Serratula tinctoria, 252
Sesamoides canescens, 100
Seseli annuum, 162
hippomarathrum, 162
libanotis, 162
montanum, 162
Sherardia arvensis, 188
Sibbaldia procumbens, 114
Sibthorpia europaea, 220
Sideritis montana, 206
Silaum silaus, 170
Silene acaulis, 58
alba, 60
armeria, 58
conica, 62
dichotoma, 58
dioica, 60
furcata, 58
gallica, 62
gallica var. quinquevulnera,
62
italica, 58
linicola, 58
maritima, 58
noctiflora, 60
nutans, 58
otites, 58
rupestris, 58
tatarica, 58
viscosa, 58
vulgaris, 58
wahlbergella, 58
Silybum marianum, 248

Simethis planifolia, 264
Sinapis alba, 88
arvensis, 88
Sison amomum, 166
Sisymbrium altissimum, 86
austriacum, 86
irio, 86
loeselii, 86
officinale, 86
orientale, 86
strictissimum, 86
supinum, 94
volgense, 86
Sisyrinchium bermudiana, 276
Sium latifolium, 164
Smyrnium olusatrum, 170
perfoliatum, 170
Solanum dulcamara, 210
luteum, 210
nigrum, 210
nitidibaccum, 306
sarrachoides, 210
Soldanella alpina, 178
Soleirolia soleirolii, 38
Solidago canadensis, 234
gigantea, 234
graminifolia, 234
virgaurea, 234
Sonchus arvensis, 256
asper, 256
maritimus, 256
oleraceus, 256
palustris, 256
Sorbaria sorbifolia, 304
Sorbus aria, 116, 305
aucuparia, 116
intermedia, 116
latifolia, 116, 305
torminalis, 116
Sparganium angustifolium, 302
emersum, 302
erectum, 302
glomeratum, 302
gramineum, 302
hyperboreum, 302
minimum, 302
Spergula arvensis, 56
morisonii, 56
pentandra, 56
Spergularia bocconii, 56
echinosperma, 56
marina, 56
media, 56
rubra, 56
rupicola, 56
segetalis, 56
Spiraea alba, 306
douglasii, 306
salicifolia, 304
tomentosa, 306
Spiranthes aestivalis, 286
romanzoffiana, 286
spiralis, 286
Spirodela polyrhiza, 300
Stachys alpina, 206
annua, 206
arvensis, 206
germanica, 206
officinalis, 206
palustris, 206
recta, 206
sylvatica, 206
Staphylea pinnata, 144
Stellaria alsine, 54
calycantha, 54
crassifolia, 54
crassipes, 54
graminea, 54
holostea, 54

humifusa, 54
longifolia, 54
media, 54
neglecta, 54
nemorum, 54
pallida, 54
palustris, 54
Stratiotes aloides, 262
Streptopus amplexifolius, 272
Suaeda maritima, 48
vera, 48
Subularia aquatica, 294
Succisa pratensis, 228
Swertia perennis, 186
Symphoricarpos albus, 305
Symphytum asperum, 192
ibiricum, 192
officinale, 192
orientale, 192
tuberosum, 192
× *uplandicum*, 192
Syringa vulgaris, 304

Tamarix gallica, 144
Tamus communis, 270
Tanacetum corymbosum, 244
parthenium, 244
vulgare, 244
Taraxacum, 258, 306
Sect. *Erythrosperma*, 258
Sect. *Palustria*, 258
Sect. *Spectabilia*, 258
Sect. *Vulgaria*, 258
Taxus baccata, 24
Teesdalia nudicaulis, 96
Telekia speciosa, 240
Tellima grandiflora, 104
Tetragonolobus maritimus, 132
Teucrium botrys, 200
chamaedrys, 200
montanum, 200
scordium, 200
scorodonia, 200
Thalictrum alpinum, 72
aquilegifolium, 72
flavum, 72
lucidum, 72
minus, 72
morisonii, 72
simplex, 72
Thesium alpinum, 38
bavarum, 38
ebracteatum, 38
humifusum, 38
linophyllon, 38
pyrenaicum, 38
rostratum, 38
Thlaspi alliaceum, 98
alpestre, 98
arvense, 98
montanum, 98
perfoliatum, 98
Thuja plicata, 24
Thymelaea passerina, 142
Thymus pulegioides, 208
serpyllum, 208, 306
Tilia cordata, 36
platyphyllos, 36
× *vulgaris*, 36
Tofieldia calyculata, 264
pusilla, 264
Tolmiea menziesii, 104
Tordylium maximum, 168
Torilis arvensis, 160
japonica, 160
nodosa, 160
Trachystemon orientalis, 196
Tragopogon dubius, 254
porrifolius, 254

pratensis, 254
Trapa natans, 262
Tribulus terrestris, 136
Trientalis europaea, 180
Trifolium alpestre, 134
arvense, 134
aureum, 132
badium, 132
bocconei, 134
campestre, 132
dubium, 132
fragiferum, 134
glomeratum, 134
hybridum, 134
incarnatum, 134
medium, 134
michelianum, 134
micranthum, 132
montanum, 134
occidentale, 134
ochroleucon, 134
ornithopodioides, 134
patens, 132
pratense, 134
repens, 134
resupinatum, 134
rubens, 134
scabrum, 134
spadiceum, 132
squamosum, 134
striatum, 134
strictum, 134
subterraneum, 134
suffocatum, 134
Triglochin maritima, 224
palustris, 224
Trigonella foenum-graecum, 130
monspeliaca, 130
Trinia glauca, 162
Trollius europaeus, 68
Tsuga heterophylla, 24
Tuberaria guttata, 152
Tulipa sylvestris, 266
Turgenia latifolia, 160
Tussilago farfara, 244
Typha angustifolia, 302
latifolia, 302
minima, 302
shuttleworthii, 302

Ulex europaeus, 124
gallii, 124
minor, 124
Ulmus glabra, 34
laevis, 34
minor, 34
procera, 34
Umbilicus rupestris, 100
Urtica dioica, 38, 204
urens, 38
Utricularia australis, 306
bremii, 306
intermedia, 294
minor, 294
ochroleuca, 306
vulgaris, 294

Vaccaria pyramidata, 62
Vaccinium microcarpum, 176
myrtillus, 176
oxycoccos, 176
uliginosum, 176
vitis-idaea, 176
Valeriana dioica, 226
montana, 226
officinalis, 226
sambucifolia, 306
tripteris, 226
Valerianella carinata, 306

coronata, 226
dentata, 306
eriocarpa, 306
locusta, 226
rimosa, 306
Vallisneria spiralis, 296
Veratrum album, 264
Verbascum blattaria, 212
densiflorum, 212
lychnitis, 212
nigrum, 212
phlomoides, 212
phoeniceum, 212
pulverulentum, 212
thapsus, 212
virgatum, 212
Verbena officinalis, 196
Veronica acinifolia, 216
agrestis, 218
alpina, 218
anagallis-aquatica, 210
anagalloides, 218
arvensis, 216
austriaca, 218
beccabunga, 220
catenata, 218
chamaedrys, 218
dillenii, 216
filiformis, 218
fruticans, 218
hederifolia, 218
hederifolia ssp. *hederifolia* 218
hederifolia ssp. *lucorum*, 218
longifolia, 218
montana, 218
officinalis, 218
opaca, 218
peregrina, 216
persica, 218
polita, 218
praecox, 216
prostrata, 218
scutellata, 218
serpyllifolia, 218
serpyllifolia ssp. *humifusa*, 218
spicata, 218
triphyllos, 216
urticifolia, 218
verna, 216
Viburnum lantana, 226
opulus, 226
Vicia angustifolia, 305
bithynica, 126
cassubica, 126
cracca, 126
dumetorum, 126
hirsuta, 126
lathyroides, 126
lutea, 126
narbonensis, 126
orobus, 126
pannonica, 126
pisiformis, 126
sativa, 126
segetalis, 305
sepium, 126
sylvatica, 126
tenuifolia, 126
tenuissima, 126
tetrasperma, 126
villosa, 126
Vinca major, 182
minor, 182
Vincetoxicum hirundinaria, 182
Viola alba, 150
arvensis, 146

biflora, 150
canina, 150
collina, 150
elatior, 150
epipsila, 150
hirta, 150
hispida, 146
kitaibeliana, 146
lactea, 150
lutea, 146
mirabilis, 150
odorata, 150

palustris, 150
persicifolia, 150
pumila, 150
reichenbachiana, 150
riviniana, 150
rupestris, 150
selkirkii, 150
suavis, 150
tricolor, 146
uliginosa, 150
Viscum album, 38
Vitis vinifera, 304

Wahlenbergia hederacea, 230
Wolffia arrhiza, 300

Xanthium spinosum, 234
strumarium, 234
strumarium ssp. italicum, 234

Zannichellia palustris, 300
Zostera angustifolia, 300
marina, 300
noltii, 300